Ehrhard Behrends · Nuno Crato ·
José Francisco Rodrigues

Editors

Raising Public Awareness of Mathematics

 Springer

Editors

Ehrhard Behrends
Fachbereich Mathematik und Informatik
Freie Universität Berlin
Berlin, Germany

José Francisco Rodrigues
CMAF
Universidade de Lisboa
Lisboa, Portugal

Nuno Crato
Inst. Superior de Economia e Gestão,
    Depto. Matemática
Universidade Técnica de Lisboa
Lisboa, Portugal

ISBN 978-3-642-25709-4   ISBN 978-3-642-25710-0 (eBook)
DOI 10.1007/978-3-642-25710-0
Springer Heidelberg New York Dordrecht London

Library of Congress Control Number: 2012941839

Cover illustration by courtesy of John M. Sullivan.

Printed on acid-free paper

Springer is part of Springer Science+Business Media (www.springer.com)

Raising Public Awareness of Mathematics

Preface

This book arose from the presentations given at the international workshop held in
Óbidos, 26–29 September 2010, as a result of a joint initiative of the Centro Internacional de Matemática and the Raising Public Awareness (RPA) committee of
the European Mathematical Society (EMS). The objective was to provide a forum
for general reflection with an international mix of experts on building the image
of mathematics, ten years after the World Mathematical Year 2000 (WMY 2000).
Óbidos, a charming town situated one hour by car to the north of Lisbon, Portugal,
was also the site of the re-creation in the year 2000 of the international mathematics exhibition "Beyond the Third Dimension" (http://alem3d.obidos.org/en/) and a
meeting of the EMS WMY2000 Committee.

The opening of the workshop was also a public "mathematical afternoon" organised by the Portuguese Mathematical Society (SPM) in cooperation with the town
of Óbidos. At this event mathematical films and lectures to the general public were
presented. The first lecture was given by H. Leitão, from the University of Lisbon,
on mathematics in the "Age of Discoveries", and the second one by G.-M. Greuel,
the current president of ERCOM (the EMS committee of the European Research
Centres on Mathematics), on the topic "Mathematics between Research, Application and Communication", which text is included in this book.

During the Óbidos public awareness event, the website www.
mathematics-in-europe.eu of the EMS was officially launched and an itinerant mathematical exhibition, *Medir o Tempo, o Mundo, o Mar*, on the use of geometry to
measure the universe and help astronomical navigation, jointly organised by the
SPM and the Museum of Science of the University of Lisbon, took place at a local
art gallery. This book aims to encourage and inspire action to raise the public awareness of the importance of mathematical sciences for contemporary society through a
cultural and historical perspective, and to provide mathematical societies, in Europe
and in the world, with ideas and details of concerted actions with other national or
international organisations and societies with regards to raising the public awareness
of science and technology and other important areas of society that have a strong
mathematical component. The book is divided into four parts:

- National Experiences
- Exhibitions and Mathematical Museums
- Popularisation Activities
- Popularisation: Why and How?

National Experiences   During and after the World Mathematical Year 2000 several European countries started extensive RPA projects in mathematics. In this part of our book activities in the following countries are described: the UK (John D. Barrow and Robin Wilson), France (Jean-Pierre Bourguignon), Germany (an article about the German website www.mathematik.de by Wolfram Koepf and another about the German Mathematical Year 2008 by Günter M. Ziegler and Thomas Vogt), the United States (Reinhard Laubenbacher), Portugal (Renata Ramalho and Nuno Crato) and Spain (Raúl Ibáñez Torres).

Exhibitions and Mathematical Museums   Over the last few years there have been a number of (temporary or permanent) mathematical exhibitions. The experiences of the organisers are given: How can one present mathematics successfully? Ehrhard Behrends (the exhibition "Mathema—Is Mathematics the Language of Nature?" during the German Mathematical Year 2008) , Albrecht Beutelspacher (the "Mathematikum" in Gießen), Manuel Arala Chaves ("Atractor"), Ana Eiró, Suzana Nápoles, Jorge Nuno Silva and José Francisco Rodrigues (exhibitions in collaboration with the Museum of Science in Lisbon), Enrico Giusti ("Il Giardino di Archimede" in Florence) and Andreas Daniel Matt ("IMAGINARY").

Popularisation Activities   The large variety of RPA projects that happened in various countries was really impressive: films, popular websites, RPA using computer games or the history of mathematics. Surprisingly most of these activities were unknown to the majority of participants until this workshop. A number of them are described in more detail here: Ehrhard Behrends describing the international mathematical popular website www.mathematics-in-europe.eu; Franka Brueckler on the problem of how to organise RPA projects with a low budget; Mireille Chaleyat-Maurel on her experiences during the Word Mathematical Year, 2000; Krzysztof Ciesielski on how to explain "strange" geometries to an audience of non-mathematicians; João Fernandes, Carlos Fiolhais and Carlota Simões on various projects at the University of Coimbra; Steen Markvorsen on his experiences of an event that had a large impact on raising the public awareness of mathematics; Yasser Omar on RPA projects in developing countries; and John M. Sullivan on the role of pictures in mathematics, art and RPA activities.

Popularisation: Why and How?   A number of talks were of a more "fundamental" character. This part of the book starts with an article by F. Thomas Bruss, who explains why it is of fundamental importance to improve the image of mathematics. Then Jorge Buescu and José Francisco Rodrigues stress that it is necessary to present "useful" mathematics to convince people that it is important. Barry Cipra asks what advice Martin Gardner would have given us for creating successful popularisation projects. Maria Dedó also starts her contribution with a question: "How important is rigour in communicating maths?" In Gert-Martin Greuel's article the focus is whether it is possible or necessary to impart an understanding of mathematics to the general public. Vagn Lundsgaard Hansen advises us to keep mathematical awareness alive, and in the last article António Machiavelo explains why the question of what mathematics really is and what it exactly deals with can only be satisfactorily understood within an evolutionary perspective.

The editors and organisers of the workshop wish to express their gratitude to the Centro Internacional de Matemática (CIM), the Portuguese Mathematical Society (SPM), the town of Óbidos, the Portuguese Fundação para a Ciência e Tecnologia, the Fundação Calouste Gulbenkian, the Munchen RE and the Portuguese mathematics research centres, CMAF/U Lisbon and CEMAPRE/TU Lisbon for their support.

Berlin, Germany   Ehrhard Behrends
Lisbon, Portugal   Nuno Crato
Lisbon, Portugal   José Francisco Rodrigues

Contents

Contributors[1]

Manuel Arala Chaves Universidade do Porto, Porto, Portugal

John D. Barrow DAMTP, Cambridge University, Cambridge, UK

Ehrhard Behrends Fachbereich Mathematik und Informatik, Freie Universität Berlin, Berlin, Germany

Albrecht Beutelspacher Universität Giessen, Giessen, Germany

Jean-Pierre Bourguignon Institut des Hautes Études Scientifiques, Bures-sur-Yvette, France

Franka Miriam Brueckler Department of Mathematics, Faculty of Science, University of Zagreb, Zagreb, Croatia

F. Thomas Bruss Département de Mathématique, Université Libre de Bruxelles, Bruxelles, Belgium

Jorge Buescu FCUL (Faculdade de Ciências da Universidade de Lisboa), Departamento de Matemática, Lisbon, Portugal

Mireille Chaleyat-Maurel University Paris Descartes, Paris, France

Krzysztof Ciesielski Mathematics Institute, Faculty of Mathematics and Computer Science, Jagiellonian University, Kraków, Poland

Barry Cipra Northfield, MN, USA

Nuno Crato ISEG, Universidade Técnica de Lisboa, Lisbon, Portugal

Maria Dedò Universitá degli Studi di Milano, Milan, Italy

Ana Maria Eiró Museu de Ciência and Faculdade de Ciências da Universidade de Lisboa, Lisbon, Portugal

---

[1]This is a complete list of the lecturers of the Óbidos workshop.

João Fernandes CFC—Centro de Física Computacional, Departamento de Matemática, Observatório Astronómico da Universidade de Coimbra, Coimbra, Portugal

Carlos Fiolhais CFC—Centro de Física Computacional, Departamento de Física da Universidade de Coimbra, Coimbra, Portugal

Enrico Giusti Il Giardino di Archimede, Firenze, Italy

Gert-Martin Greuel University of Kaiserslautern, Kaiserslautern, Germany

Vagn Lundsgaard Hansen Technical University of Denmark, Copenhagen, Denmark

Raúl Ibáñez Torres Universidad del País Vasco, Lejona, Spain

Wolfram Koepf Fachbereich 10 Mathematik und Naturwissenschaften, Institut für Mathematik, Universität Kassel, Kassel, Germany

Reinhard Laubenbacher Department of Mathematics, Virginia Polytechnic Institute and State University, Blacksburg, VA, USA; Virginia Bioinformatics Institute, Virginia Polytechnic Institute and State University, Blacksburg, VA, USA; Society for Industrial and Applied Mathematics, Philadelphia, USA

António Machiavelo Centro de Matemática da Universidade do Porto, Porto, Portugal

Steen Markvorsen Department of Mathematics, Technical University of Denmark, Copenhagen, Denmark

Andreas Daniel Matt Mathematisches Forschungsinstitut Oberwolfach, Oberwolfach, Germany

Suzana Nápoles FCUL (Faculdade de Ciências da Universidade de Lisboa), Departamento de Matemática, Lisbon, Portugal

Yasser Omar SiW–Scientists in the World & CEMAPRE, Department of Mathematics, ISEG, Technical University of Lisbon, Lisbon, Portugal

Zdzisław Pogoda Mathematics Institute, Faculty of Mathematics and Computer Science, Jagiellonian University, Kraków, Poland

Renata Ramalho Sociedade Portuguesa de Matemática, Lisbon, Portugal

José Francisco Rodrigues FCUL/Centro de Matemática e Aplicações Fundamentais da Universidade de Lisboa, Lisbon, Portugal

Jorge Nuno Silva FCUL/Centro Interuniversitário de História das Ciências e da Tecnologia and Associação Ludus, Lisbon, Portugal

Carlota Simões CFC—Centro de Física Computacional, Departamento de Física da Universidade de Coimbra, Coimbra, Portugal

John M. Sullivan Institut für Mathematik, Technische Universität Berlin, Berlin, Germany

Thomas Vogt DMV Mathematics Media Office, Institute of Mathematics, Freie Universität Berlin, Berlin, Germany

Robin Wilson Open University, Milton Keynes, UK; Pembroke College, Oxford, UK

Günter M. Ziegler DMV Mathematics Media Office, Institute of Mathematics, Freie Universität Berlin, Berlin, Germany

Part I
National Experiences

Raising Public Awareness in the UK
—Some Snapshots

John D. Barrow and Robin Wilson

Abstract The United Kingdom has a track record of events to raise the public awareness of mathematics, although mathematics still remains a closed book to the vast majority of UK citizens, and prominent figures are quite happy to admit to their lack of knowledge of mathematics in situations where they would never dream of admitting they knew nothing of Shakespeare or music. In the UK there is a two-pronged approach to mathematics outreach—outreach to the general public, and finding ways to encourage students to pursue mathematics in schools and universities. Below are some of the attempts that have been made in these two directions to improve the situation.

Gresham College, London

Gresham College was founded in London in 1596 for the purpose of providing free public lectures in a range of subjects for anyone who wished to attend—and 400 years later, this remains its purpose. The wealthy financier Sir Thomas Gresham left money to the City of London and the Mercers' Company to provide for professorships in geometry, astronomy, physic, rhetoric, music, divinity and law. Early Geometry professors included Henry Briggs (the co-inventor of logarithms) and Robert Hooke. In the 1890s Karl Pearson's lectures on statistics introduced the terms 'histogram' and 'standard deviation' for the first time. For a brief history of Gresham College, see Robin Wilson's article in the June 2007 issue of the European Mathematical Society's *Newsletter*.

J.D. Barrow (✉)
DAMTP, Cambridge University, Cambridge, UK
e-mail: jdb34@hermes.cam.uk

R. Wilson
Open University, Milton Keynes, UK
e-mail: r.j.wilson@open.ac.uk

R. Wilson
Pembroke College, Oxford, UK

E. Behrends et al. (eds.), *Raising Public Awareness of Mathematics*,
DOI 10.1007/978-3-642-25710-0_1, © Springer-Verlag Berlin Heidelberg 2012

The original professors were appointed for life and were required to live in the College (within the City of London), to remain unmarried, and to lecture in both Latin (for visiting foreign scholars) and English (for the general public). Nowadays, professors are appointed for three or four years and recent appointments have included Sir Christopher Zeeman, Ian Stewart, Sir Roger Penrose and Robin Wilson. The current geometry professor is John Barrow, a previous Gresham professor of astronomy.

Each professor gives six public lectures per year, and recent lectures have ranged from Greek geometry to the Riemann hypothesis, from Hilbert's problems to continued fractions, and from the mechanics of a superball to Einstein's theory of relativity. In addition, there are one-off lectures by such figures as Tim Gowers and Simon Singh, and also annual joint lectures with the London Mathematical Society and the British Society for the History of Mathematics. The funding of the lectures is provided (as 400 years ago) by the City of London and the Mercers' Company, and now amounts to about £700,000 per year for about 150 events.

When the second author first held the Geometry Chair in 2004, the total face-to-face audience for the year (for all the lectures) was about 4000. By 2009, this had risen to around 17,000 and continues to rise, with live attendances greatly boosted by online views through the web site. All lectures are now webcast, and from an initial 1000 downloads per month there are now over one million downloads per year worldwide. In particular, all recent Gresham maths lectures can be viewed online on www.gresham.ac.uk.

The Open University

One particularly effective means for raising the public awareness of mathematics (especially when maths programmes are broadcast on television) has been through the Open University (OU), a major distance-learning institution, which was founded in the UK in the 1960s and initially presented courses in four subjects (mathematics, arts, science and the social sciences) to 16,000 adult UK students. There are now some 260,000 students studying a wide range of topics, including 40,000 students outside the UK, and the number of OU graduates already exceeds two-and-a-half million.

Teaching is by a range of methods, of which the main vehicle has traditionally been the printed correspondence text, prepared by a course team of academics over a period of time. In addition there have been radio and television programmes (over the national BBC network), cassette tapes and videos, home experiment science kits, CDs and DVDs. At one stage, some of our mathematics TV programmes, though designed for 4000 students, were watched by up to 400,000 viewers. There is an increasing use of online teaching methods, including much use of conferencing and iTunes.

Face-to-face teaching is given by over 7000 part-time tutors around the UK who mark the students' written work and give tutorial classes at venues around the country, as well as communicating with students via e-mail and electronic course conferences. For many years, there were also week-long summer schools that students were required to attend (in one summer over 40,000 students attended summer schools in 15 venues over an eight-week period) or residential weekend courses. Virtual summer schools are now increasingly used.

The assessment of most courses (or modules) is by regular assignments throughout the year (tutor-marked by hand or online, or computer-marked), followed by a three-hour written examination at the end—over many years of studying for a BA degree in Humanities with Music, the second author sent about 60 assignments to his tutors and sat 20 hours of written exams.

Open University students now have a wide range of choices of courses to study, ending with many possible degrees and diplomas, including a range of taught and research higher degrees (MSc and PhD). Most MBA students in Eastern Europe study with the OU, and one-third of the UK's master's students in mathematics take the OU's taught MSc courses (about 600 students per year). Current courses in mathematics include a range of foundation-level courses (about 5000 students per presentation), introductory courses in pure and applied mathematics (800), and higher-level courses in complex analysis (350), groups and geometry (350) and fluid mechanics (250). There are also courses in mathematics education for teachers, and MSc courses in such areas as analytic number theory, fractal geometry, applied complex variables, functional analysis and coding theory.

An early example of a course that the second author originated was a 30-point course on *Graphs, Networks and Design*, written by a team of about a dozen mathematicians and technologists (including a BBC producer). Over a three-year period this involved the production of 1600 pages of printed text, sixteen 25-minute television programmes, and several audio-tapes; this course was studied by more than 5000 students over a 14-year period. More recently, he has written a 10-point history of mathematics course to accompany the BBC TV series *The Story of Maths*, produced jointly by the BBC and the OU and presented by Marcus du Sautoy, Simonyi Professor for the Public Understanding of Science at Oxford University; this course regularly attracts about 300 students per year.

Broadcasting Mathematics

Much public awareness of mathematics in the UK has been through BBC radio and television programmes—not only the hundreds of OU programmes mentioned above, but also various series and one-off programmes. In his teenage years, the second author used to watch Jacob Bronowski's television programmes (including a memorable one on perfect numbers) which appeared before Bronowski presented his well-known series on *The Ascent of Man*. More recently, there were the four one-hour documentary programmes, mentioned above, entitled *The Story of Maths*; these were watched by up to half-a-million viewers in the UK, were broadcast by BBC World to two million homes worldwide in January 2009, and won first prize

for the best BBC World documentary series of 2009. Other TV programmes have
included Simon Singh's *Fermat's Last Theorem* and a series on *Codes*, and Marcus
du Sautoy's *Music of the Primes*.

BBC Radio features a weekly series entitled *In Our Time*, chaired by Lord
Melvyn Bragg, in which three experts discuss a topic of interest (the telescope,
Attila the Hun, humour in Greek plays, etc.). Over the years these broadcasts have
included several mathematical topics, including *Archimedes*, *Mathematics and music*, *Prime numbers*, *Newton's 'Principia'*, *Negative numbers*, *Gödel's theorem* and
Henri Poincaré. They regularly attract audiences of about two million listeners, and
all remain available online.

The Royal Society also features all its public and prize-winning lectures on its
TV website; these can be found online at http://royalsociety.org/royalsociety.tv/
?from=footer. Here you will find the Faraday prize lectures of Marcus du Sautoy
and John Barrow, as well as earlier popular maths lectures: these lectures attract
very large audiences and are webcast live. The Royal Society also has an advisory
committee, ACME, on mathematics education.

Every September there is a major public outreach effort at the time of the
annual British Science Festival of the British Science Association, formerly
named the British Association for the Advancement of Science (see http://
www.britishscienceassociation.org/web/BritishScienceFestival/). This week-long
programme of talks and outreach activities is always hosted by a UK university
(it will be in Aberdeen in 2012). It is one of the largest science festivals in Europe,
and the week of the Festival is one of the few occasions when science is front-page
news in the quality national newspapers. There is a separate mathematics section,
and a new president for the section is elected each year. Supported by the mathematics sectional committee, the president is responsible for giving a presidential lecture
and organising a stimulating series of talks during the Festival week. Other sections,
particularly physics, astronomy and engineering, also regularly feature talks that are
of mathematical interest.

The Royal Institution (RI) is another famous institution dedicated to the public promulgation of science and dialogue about its role in society. Although it is
best known for its annual series of televised Christmas Lectures, it is also active
throughout the year running popular lectures on mathematics and mathematics master classes for keen and able school students (see www.rigb.org).

The Millennium Mathematics Project

The *Millennium Mathematics Project* (MMP) is an extensive education and public
outreach initiative begun by the mathematics and education departments of the University of Cambridge in 1999. The first author directs the project from the Centre
for Mathematical Sciences in Cambridge and its principal webpage is to be found at
www.mmp.maths.org.

The MMP aims to improve the understanding and appreciation of mathematics amongst school students (ages 5–19), teachers and members of the general public. It was set up in response to a perceived drop in the standards of teaching and learning of mathematics in the UK and a perceived lack of appreciation of the role that mathematics plays in science, business and everyday life. From its inception it has been active both nationally and internationally. It combines web-based activity with face-to-face activities in schools, professional development for teachers, video-conferencing to schools, placing students in local schools to support teachers, visits to schools with talks, articles, podcasts and videos about mathematics and its applications for the general public. It is divided into sub-projects that have different audience focuses and different styles of delivery.

The MMP comprises about 17 members, some part-time. They have a range of expertise, spanning research in mathematics and its applications, mathematics teaching, public understanding, editing, fund-raising, administration, and technical aspects of video, web and other modes of presentation. All of the MMP's outputs are free to individual users. The project is supported by grants, donations from benefactors, and some of the science and mathematics departments within the University of Cambridge contribute to the project. In 2006 the MMP was awarded the Queen's Anniversary Prize, the highest award for educational achievement in the UK, and its sub-project PLUS won the Webby Award for the best science education site on the web. In the academic year 2009–2010 the MMP's web-based mathematical resources attracted more than 4.3 million visitors worldwide, while over 39,000 pupils and more than 3500 teachers were involved in its face-to-face activities and events.

The project consists of a family of complementary programmes, each of which has a particular focus:

- The *NRICH website* provides thousands of free resources, designed to develop problem-solving skills and subject knowledge. High-school students send in solutions to problems (only those solutions are published—if a problem is unsolved, it remains so), and there is an Ask NRICH discussion board staffed by Cambridge team members and students who give help and hints to students and participate in discussion threads. All of these threads are archived and indexed: they can be seen and searched at http://nrich.maths.org/discus/messages/board-topics.html. Recently, NRICH has been extended to include maths for science and technology (STEM) under the acronym stemNRICH, with dedicated fast-developing webpages at http://nrich.maths.org/stemNRICH: stemNRICH provides free post-16 resources that explore the mathematics underlying biology, physics, chemistry and engineering, to support the school-to-university transition.
- The *Plus website* is an online magazine opening a door to the world of mathematics, and including a careers library of interviews with people who use mathematics in interesting ways in a variety of careers. It focuses on the applications of mathematics and provides articles and interviews with leading mathematicians (see www.plus.maths.org). Plus also runs a 'Plus maths news' for registered participants, giving regular updates on what is happening in the world of mathematics, together with updates on mathematical aspects of the news, as well as a blog. The first author has a regular column in Plus called *Outer Space*.

- *Motivate* is a videoconferencing scheme that links mathematicians to schools in the UK and other countries (time zones permitting). It provides bespoke mathematics enrichment sessions for schools by live video-conference, and online multimedia enrichment resource packs. Recently, the project work by schools who participated in this project was published as a research paper in the *Proceedings of the Royal Society B* (see http://rspb.royalsocietypublishing.org/content/early/2010/10/28/rspb.2010.1807.full).
- The *Hands-On Maths Roadshow*, *Enigma Project*, and *Risk and Probability Show*: staff visit schools all over the UK and abroad to run hands-on mathematics enrichment activities and workshops; the Risk and Enigma Projects also provide shows for the general public. The Enigma Roadshow is about the mathematics of codes and codebreaking and features a real WWII Enigma machine, as well as an electronic simulation on CD for the participants to keep. The Risk and Probability Show is about many aspects of probability, uncertainty and risk: it is closely linked to the programme for the public understanding of risk, which is run by David Spiegelhalter, the Winton Professor for the Public Understanding of Risk in the Department of Pure Mathematics and Mathematical Statistics at Cambridge, who works very closely with the MMP. The website of his programme, *Understanding Uncertainty*, can be found at http://understandinguncertainty.org/.
- A programme of mathematics lectures and events for schools, families and the general public, held in Cambridge. Each year in the UK there is a National Science and Engineering Week to showcase the achievements of science and mathematics: all universities take part in this. At Cambridge, for example, there is a very extensive programme that runs over two weeks in order to accommodate everything on offer. The mathematics departments offer a programme of talks, exhibitions and interactivities for all ages. The Isaac Newton Institute also participates in these activities, and last year it housed a major display of mathematical art.

The MMP has recently developed resources on maths and sport, with a special focus on the London 2012 Olympic and Paralympic Games. This activity, with a new accompanying roadshow, was launched in early 2011. The MMP was chosen as the official Olympic education provider in this subject area by the Local Organising Committee of the London 2012 Games (see www.sport.maths.org).

The MMP also maintains close links with other mathematics outreach activities in the UK. These include the series *Maths in the City* by Marcus Du Sautoy and sponsored by the Engineering and Physical Sciences Research Council, of guided mathematical walks in Oxford and East London, (see http://gow.epsrc.ac.uk/ViewGrant.aspx?GrantRef=EP/H047158/1), which one of the Plus editors (Rachel Thomas) helps to organise, and Rob Eastaway's *Maths Inspiration Project* (http://www.mathsinspiration.com/jsp/index.jsp?lnk=200), which runs at major public venues around the UK. Several other outreach projects exist that aim to enthuse young people about mathematics, see (for example) the FunMaths Roadshow, (www.maths.liv.ac.uk/lms/funmaths) of the Liverpool Mathematical Society.

In 2004, the British Government's Department for Education and Skills (now the DCSF), led by the Secretary of State Charles Clarke, gave financial support for the creation of the National Centre for Excellence in the Teaching of Mathematics (see https://www.ncetm.org.uk/). Its present director is Professor Celia Hoyles, and its role is to encourage mathematics teachers to develop their mathematical knowledge and pedagogical skills. Its webpages and local centres act as sources of information and activity in these areas, with the aim of sharing and promoting best practice. They also provide all sorts of stimulating material to help teachers enliven and enrich the teaching of mathematics, for example *the Early Years, Primary, Secondary* and *FE* electronic *Magazines*.

The Future

We have provided a few snapshots of the maths outreach programmes in the UK in which we participate. They can be followed up in more detail by looking at the web links we have provided in the text. Over the past ten years the profile of mathematics has steadily risen within the UK, but there is more still to be done. This is something that has also been recognised by the national agencies that fund university research in science, engineering and mathematics. Researchers are encouraged to include outreach activities in their research programme and are asked to report on the impact that they have made on society. In the future this evaluation of research impact will play a significant role in the nationwide evaluation of university departments and their research groups. As a result, the profile of public outreach in mathematics is likely to continue to rise and more young mathematicians will be encouraged to participate in this exciting mission. The possibilities of collaborating with European colleagues will add significantly to what can be achieved by sharing best practice and exchanging new ideas.

Raising the Public Awareness of Math: Discussing Recent Initiatives in France

Jean-Pierre Bourguignon

Abstract The article presents some initiatives taken by the Institut des Hautes Études Scientifiques (IHÉS) and a few other French organisations to raise the public awareness of mathematics. They are organised in categories, some rather standard, some more original corresponding to opportunities that have been seized. They all address a public wider than the public of scientists and have involved contributors coming from different backgrounds.

A possible sub-title for this lecture is: *A tour of some initiatives taken by the Institut des Hautes Études Scientifiques (IHÉS) and a few other French organisations to raise the public awareness of mathematics.*

The types of initiatives that will be discussed here are diverse, and they represent only a sample of the actions undertaken to raise the public awareness of mathematics and science in general in France. Here are some typical categories:

1. *Open days* at the Institute
2. Conferences in *unusual* locations
3. A *special lecture series* organised by the Bibliothèque Nationale de France and the Société Mathématique de France
4. Events involving *artists*
5. A *very special adventure* with pictures of scientists, who contributed short texts, and a photographic exhibition that has travelled the world.

This lecture is an opportunity to try and identify:

- Conditions for the success of events whose objective is to raise public awareness,
- Conclusions that can be drawn from these experiences,
- Networks that need to be mobilised.

These initiatives were all addressing a public wider than the public of scientists, and some of them have attracted high school students, and also the general public. This was often made possible thanks to the help of those organisations of teachers

J.-P. Bourguignon (✉)
Institut des Hautes Études Scientifiques, 35, route de Chartres, 91440 Bures-sur-Yvette, France
e-mail: jpb@ihes.fr

E. Behrends et al. (eds.), *Raising Public Awareness of Mathematics*,
DOI 10.1007/978-3-642-25710-0_2, © Springer-Verlag Berlin Heidelberg 2012

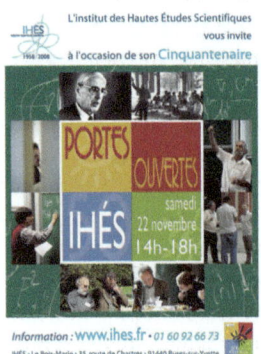

Fig. 1 Open Days at IHÉS

and mathematicians whose purpose is to create networks. As far as reaching out to high school students is concerned, involving their teachers is clearly critical. The purpose of these events was mainly to create opportunities to meet scientists and get a better understanding of what research is about, and also of how it functions.

Open Days at IHÉS

Almost every year IHÉS opens its doors to the public, most of the time on the occasion of the Science Week, see Fig. 1, and proposes a variety of activities all centered on the researchers at the Institute and their appearances in different media.

This is of course not very original. Nevertheless each time, the event has been successful bringing a few hundred people to the Institute. It is in particular a very good basis for further contacts and mixing the generations.

Conferences in Unusual Locations

Over the years IHÉS has organised conferences in *unusual* locations, see Fig. 2. These events were almost all of the same format, namely 20 minute lectures followed by 10 minutes of questions for a sequence of typically eight such lectures. The audiences attending the events were mixed with always a large component of young students.

Another common feature of these events is that they took place in *unusual* places: two of them were held at the Centre Pompidou, the epoch-making building by Renzo Piano and collaborators in downtown Paris. The first event, entitled *Voyage dans l'imaginaire mathématique*, organised in 2000 in the framework of the *World Mathematical Year*, was the first ever mathematical event held there. The 2004 event was called *La face cachée des mathématiques*. In 2008 the event *à la rencontre des*

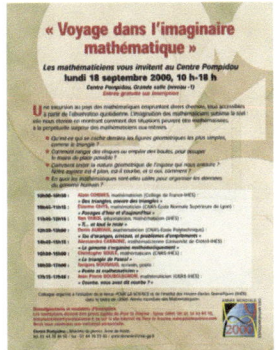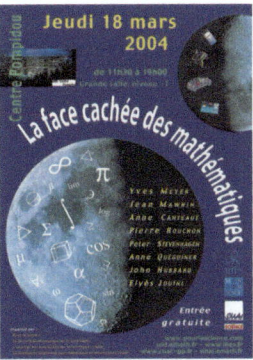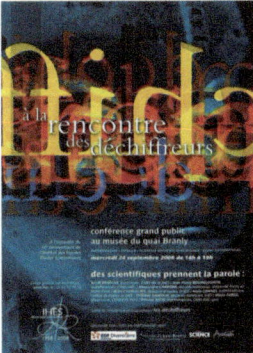

Fig. 2 Conferences in unusual locations

déchiffreurs was held at the Musée des Arts Premiers, a brand new museum in Paris in an impressive building designed by the architect Jean Nouvel.

The science magazine *Pour la Science*, the French edition of Scientific American, was the partner of IHÉS for all these events. The first one was also sponsored by the CNRS who provided the means to record it very professionally. The second one was organised jointly with the Société Mathématique de France (SMF) and the Société de Mathématiques Appliquées et Industrielles (SMAI). For the 2008 event, several foundations helped cover the cost of renting the facilities. It was the major public event in the programme to celebrate the 50th anniversary of IHÉS. Each time these events attracted hundreds of people and generated interesting reactions.

Some of them were held abroad: one took place in Tokyo as part of a week-long series of events organised jointly with the University of Tokyo and Keio University, again during the celebration of the 50th anniversary of IHÉS. In 2010 the Institute took advantage of the special interest raised by the Shanghai World Expo to hold another edition of the *à la rencontre des déchiffreurs* conference at the *Pavillon France* there, partly in French and partly in Chinese, see Figs. 3 and 4, as the Tokyo event had been partly in English and partly in Japanese, with simultaneous translation. In both cases young high school students were involved and could meet prominent mathematicians. In Tokyo, the help of Professor Heisuke Hironaka and of colleagues from the Mathematical Society of Japan was instrumental in the success of the enterprise.

A Special Lecture Series

Since 2005, the *Bibliothèque Nationale de France* (BNF), in association with the SMF, *France-Culture*, the national cultural radio channel, and the mathematical magazine *Tangente*, has been hosting a special lecture series, entitled *Un texte, un mathématicien*, see Fig. 5. In recent years, the general science magazine *La Recherche* has also been a partner. The purpose is to take a historic text produced

Philippe LAGAYETTE   **Jean-Pierre BOURGUIGNON**
法国高等科学研究院董事会主席   法国高等科学研究院院长

恭请您参加法国高等科学研究院于2010年10月12日
在上海世博会的法国展馆举办的"会见解码者"会议。

时间为下午**2点30分**至**7点**，会议之后将于法国展馆的餐厅6SENS举行鸡尾酒会。

由于位置有限，请有意参加者用以下方式注册：
• 会议注册方式：法国高等科学研究院的网站（*www.ihes.fr*）
• 鸡尾酒会注册方式：请于2010年9月27日前发电子邮件至 *touchant-landais@ihes.fr*

注册会议参加者将获得一张当日的世博会入场卷。

SOCIETE GENERALE   **SAINT-GOBAIN**
LCL Banque Privée 提供赞助

Fig. 3 Invitation for *à la rencontre des déchiffreurs* in Shanghai

演讲名单

Jean-Pierre BOURGUIGNON *(CNRS-IHÉS)*, 数学家,
Flexaedrons不冒烟

Josselin GARNIER *(Université Paris 7)*, 数学家,
噪声生成图像

HU Sen *(*中国科学技术大学*)*, 数学家,
寻找宇宙的几何

LIU Kefeng *(*浙江大学*)*, 数学家,
物理对数学的启示

LONG Yiming *(*陈省身数学研究厅，南开大学*)*, 数学家,
数学家眼中的太阳系

George PAPANICOLAOU *(Stanford University)*, 数学家,
金融中的数学问题

LI Ta-Tsien *(Institut Sino-Français de Mathématiques Appliquées, *复旦大学*)*, 数学家,
暂时未定

Cédric VILLANI *(Institut Henri Poincaré, Université de Lyon)*, 数学家,
星系会休息吗？

Fig. 4 Programme for *à la rencontre des déchiffreurs* in Shanghai

by a mathematician and to show the way it changed the course of mathematics and discuss its descendance and its most recent consequences.

The series has mostly taken place in the BNF main auditorium, which can hold several hundred people, although some lectures have been given in Grenoble, Avignon and Lens. For each conference, about half of the auditorium is occupied by high school students, who come with their teachers after some preparatory lectures

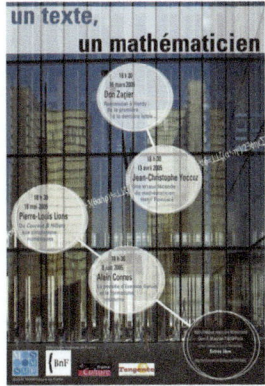  

Fig. 5 *Un texte, un mathématicien*

organised through the extensive network of the association *Animath*, which coordinates the work of several special initiatives in schools lead by teachers. On many occasions, because of overbooking, some people have to watch the lecture on TV sets outside the auditorium. This has become a reference event in terms of reaching out to a wide public. The lectures are now videotaped in a good format and are watched by many people.

Events Connected with Artists

Connections between mathematicians and artists have been developed in different contexts:

- Some artists are asked to take part in a project involving mathematics;
- Some artists have, on their own, an interest in mathematics, and want to be associated with mathematicians;
- Some mathematicians have been in contact or have inspired artists.

In the first category, there is the sculpture *Lump Bumps and Windy Figures* by the American artist Jessica Stockholder, see Fig. 6. The piece was commissioned by primary school pupils from Longjumeau, near Paris, who explored, during the World Mathematical Year 2000, a combinatorial problem introduced by the Norwegian mathematician Skolem: given a number n, find sequences of the $2n$ numbers $1, 2, \ldots, n$ repeated twice, so that the distance between the two occurrences of any number k is exactly k. One can represent such a sequence by putting n two-legged structures, the knights, whose legs are apart by $1, 2, \ldots, n$ units on a board, in such a way that no two knights step on each other's toes. Jessica Stockholder produced a piece, located in the Bois-Marie domain of IHÉS, in two parts: one, merely esthetical, shows the arrangement of 8 objects of sizes 1 to 8 on a chessboard so that one object begins or ends on each line of the board, producing a Skolem sequence in

Fig. 6 *Lump Bumps and WIndy Figures* by Jessica Stockholder, the knights in the foreground

Fig. 7 Catalogue of the Hiroshi Sugimoto Exhibit

each direction; the other fulfils a request of the children, namely to be able to playfully look for Skolem sequences by moving 8 metallic knights on an 8-line board.

In the second category, there is for example the work of the Japanese photographer Hiroshi Sugimoto, who made magnificent pictures of the collection of geometric surfaces from the nineteenth century kept at the Mathematical Department of the University of Tokyo. The pictures were exhibited for the first time in 2004 at the *Fondation Cartier pour l'art contemporain* in another beautiful building by Jean Nouvel in Paris. I was charged with the task of explaining in the catalogue why these surfaces are mathematically significant, see Fig. 7.

In the third category, one finds René Thom, a former Permanent Professor at IHÉS who passed away in 2002. During his life he interacted with a number of artists: Salvador Dali, who dedicated to him his last series of paintings, among which is the *Topological abduction of Europe* inspired by Thom's work on catastrophes; Jean-Luc Godard, the Swiss film director who produced a movie called *A René*, showing Thom in a provocative way; Pascal Dusapin, a French composer of modern music, whose piece *Loop* is again inspired by Thom's work on catastrophes;

Fig. 8 Poster for a lecture on
René Thom

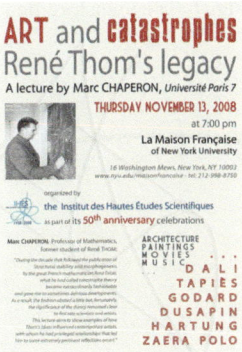

the Spanish architect Zaera-Polo, who designed the ferry terminal in Yokohama according to a *Thomian* perspective; the German painter Hans Hartung, and the Catalan painter Antoni Tàpies can be added to the list. A lecture on these interactions was given by Marc Chaperon (Université Paris-Diderot) at the Maison française of New York University during the US celebration of the 50th anniversary of IHÉS, see Fig. 8.

A Very Special Adventure

The following is a most improbable story:

- A couple of film makers employed by CNRS, Anne Papillault and Jean-François Dars, had to find a place to install their equipment after the CNRS decided to stop paying the rent on their professional studio;
- I knew them because we made two movies together, a rather long one on Jacques Tits on the occasion of his retirement and a shorter one on Henri Cartan in connection with the Bourbaki seminar, and they found refuge at IHÉS where they spent four years;
- Jean-François Dars, also a photographer, took thousands of pictures of researchers at work in many instances of the life at the Institute;
- Alain Connes suggested making a book containing the best pictures, accompanied by short texts written by those whose picture was selected. Early in 2008, *Les déchiffreurs* was produced by the French publisher Belin, and is now in its third printing (see Fig. 9); shortly after, an English version was published by A.K. Peters, Boston; a Japanese version was produced later in 2008 by Springer Japan, and a Chinese version in 2010 by Higher Education Press (see Fig. 10); a Korean edition is being prepared.

Some pictures in the book are the result of the exceptional patience of the photographer, like the one in Fig. 11 showing Étienne Ghys lecturing on the Lorenz attractor and the butterfly effect at the École polytechnique (on the occasion of my 60th birthday). A historic text in the book was written in 2007 by Ngô Bao Châu,

Fig. 9 The French edition of
Les Déchiffreurs

Fig. 10 The Chinese edition

Fig. 11 Étienne Ghys

shown on Fig. 12, who describes his personal experience when he foresaw that he
had put his hands on a very simple but very powerful idea while being a CNRS visitor at IHÉS. The conclusion of his article is prescient because, in 2010, he received
the Fields Medal for the completion of this work. He wrote: "*Je pense maintenant*

Fig. 12 Ngô Bao Châu

Fig. 13 Sir Michael Atiyah

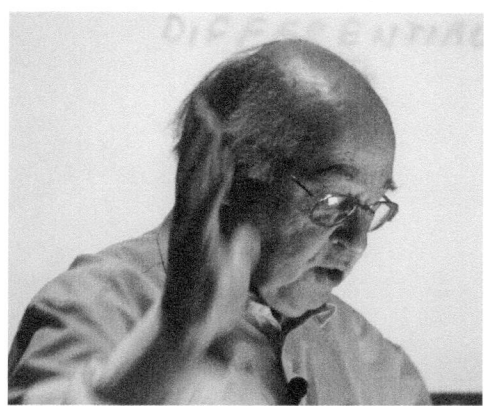

que... cet après-midi-là j'avais vécu l'un des moments les plus décisifs de ma carrière". Some other pictures show historic figures, such as Sir Michael Atiyah, see Fig. 13, or Eugenio Calabi and Shing Tung Yau, one of the very few pictures where they appear together, Fig. 14, an embodiment of *Calabi-Yau* of a sort.

The very best pictures from the book, about 30 of them, were printed in large format on canvas for a photographic exhibition, which has been touring the world. It has been shown in high schools (such as in Rochefort, France, see Fig. 15), in bookshops in downtown Paris, see Fig. 16, but also in universities and institutes in the US (Chicago, New York, the Institute for Advanced Study), in Japan (Tokyo University and Keio University), and in Thailand (Mahidol University in Bangkok). It is presently touring China thanks to the network of the Alliances Françaises in the country, and will be doing a Tour de France later in 2011–2012.

The exhibition was also presented at the International Congress 2010 in Hyderabad, see Fig. 17, with a few additional pictures such as the one of Shiing Shen Chern shown in Fig. 18 with, standing in front, his daughter May Chu and Louis Nirenberg, the first recipient of the Chern Medal awarded on this occasion.

Fig. 14 Eugenio Calabi and Shing Tung Yau

Fig. 15 High school, Rochefort

Fig. 16 A bookshop in downtown Paris

A Few Concluding Words

From these diverse situations, one can see that a variety of approaches were used in order to reach a mix of people but the most important has been:

Fig. 17 Joanna Jammes (IHÉS) in Hyderabad

Fig. 18 Louis Nirenberg (*left*), May Chu (*right*) and a photograph of Shiing Shen Chern (*center*)

- to seize chances and passing opportunities,
- to do with minimal budgets, while looking for appropriate sponsors …

Some ongoing projects will rely on the use of much bigger resources and aim for more international visibility in the media but it is too early to say more.

The task to be achieved is immense, due to the very rapid growth of the discipline during the last century, which continues unabated, and its continued sophistication. Mathematicians need to make an in depth transmission of the heart of their discipline. This of course cannot be carried out through the usual channels of school training, which can barely touch recent developments. It is therefore necessary to find appropriate shortcuts, which may have nothing to do with the technical development of the discipline. This is a task that mathematicians need to understand properly. They also need to estimate all the effort that has to go into such an endeavour, give it the appropriate recognition, and they must do so quickly.

Acknowledgements   Thanks to the City University of Hong Kong for hospitality while these notes were written.

The German Website Mathematik.de

Wolfram Koepf

Abstract In this article we describe the German website Mathematik.de, and the ideas used to try to raise the public awareness of mathematics in Germany.

This website was originally created by Ehrhard Behrends. I joined the team in 2007, and since 2009 I have been the editor responsible for the site.

What Is Our Idea?

The idea behind Mathematik.de other than being the website of a professional association like the Deutsche Mathematiker-Vereinigung (DMV, the German Union of Mathematicians) is to give information and news about mathematics, about mathematics in the media, about mathematicians at work, about the history of mathematics, etc. to the general public. We therefore provide:

- A collection of news and current media stories about mathematics in Germany and elsewhere
- A collection of mathematical resources for students, engineers and otherwise interested people

W. Koepf (✉)
Fachbereich 10 Mathematik und Naturwissenschaften, Institut für Mathematik, Universität Kassel, Heinrich-Plett-Str. 40, 34132 Kassel, Germany
e-mail: koepf@mathematik.uni-kassel.de

E. Behrends et al. (eds.), *Raising Public Awareness of Mathematics*,
DOI 10.1007/978-3-642-25710-0_3, © Springer-Verlag Berlin Heidelberg 2012

23

Fig. 1 Page header: the
polynomial for a sweet

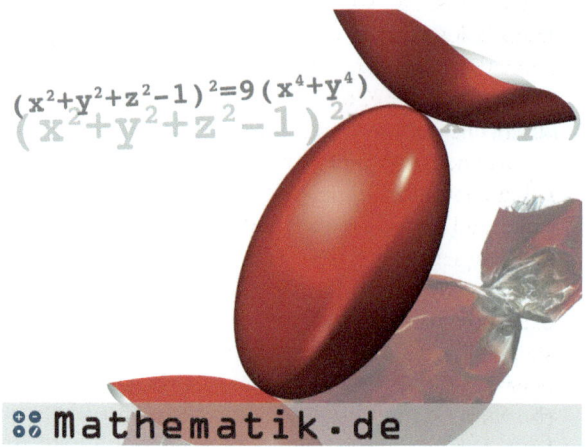

$$(x^2+y^2+z^2-1)^2=9(x^4+y^4)$$

- A forum for people interested in mathematics to get an idea of the working life of a mathematician
- A list of web links to other interesting pages about mathematics

and much more.

Page Structure

Our web page:

- Has a daily graphical header
- Has a left-hand side with daily items
- Has a right-hand side with a menu with many interesting topics, see Fig. 2.

This structure was introduced in 2008 when we had the official Year of Mathematics in Germany and the site was given a facelift.

Our site is served by a database, which is invoked both to present the changing items automatically and to serve the menu.

The Page Headers

Each page header (see Fig. 1) combines a polynomial formula from pretty simple to rather complicated of a three-dimensional surface, a picture of this surface and a photo of a real-world item as an illustration of the model surface. (For more examples see Fig. 3). This type of header was largely influenced by the Imaginary exhibition (www.imaginary2008.de), which visited many German cities in 2008.

Fig. 2 The start page of Mathematik.de on 1 September 2010

News Messages

We add a news entry about every third day, so that each month we come up with about 10 news messages (Fig. 4). These are either news from electronic media in Germany, interesting press releases about mathematics, mathematicians, competitions or anything that we consider to be interesting for the public, for example:

Fig. 3 Page headers: polynomials for a cushion, finger ring and banana

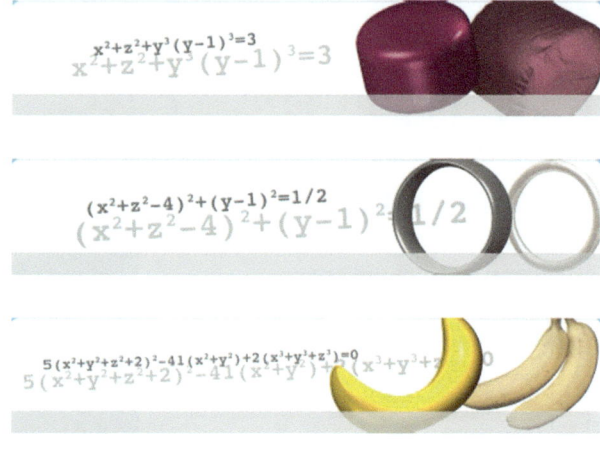

Fig. 4 News items

- We announced Benoît Mandelbrot's recent death giving details about his life and presenting a nice movie of his fractals, see Mathematik.de/ger/diverses/aktuelles/benoit_mandelbrot_gestorben.html.
    Of course this was only possible because:
  - we asked for and received the authorization of the film director Nigel Lesmoir-Gordon [1]
  - we have somebody in our team who can prepare such a Flash animation
  - we were fast enough to get this page online without any substantial delay
- We released another news message about the DMV media award ceremony in 2010, presenting the relevant information about the awardees and publishing several pictures to visualize this event, see Mathematik.de/ger/diverses/aktuelles/medienpreis_2010/dmv_medienpreis_2010.html.
- A news message about Eugen Jost introduced and showed a dozen of his marvellous mathematical pictures, which were used to create nice calendars and for

Fig. 5 Citation of the day

Zitat des Tages

Die Mathematiker, die nur Mathematiker sind, denken also richtig, aber nur unter der Voraussetzung, daß man ihnen alle Dinge durch Definitionen und Prinzipien erklärt; sonst sind sie beschränkt und unerträglich, denn sie denken nur dann richtig, wenn es um sehr klare Prinzipien geht.
- *Blaise Pascal*

Fig. 6 Historic reminder

Heute vor 130 Jahren

Am 31.8.1880 wurde Heinrich Franz Friedrich Tietze in Schleinz (Österr.) geboren.
Er arbeitete in der Topologie und war wesentlich am Aufschwung dieser Disziplin beteiligt. Mit seinem Buch "Gelöste und ungelöste Probleme aus alter und neuer Zeit" machte er viele Laien mit schwierigen mathematischen Problemen bekannt.
Tietze verstarb am 17.2.1964 in München.
Mehr finden Sie hier ▶.

Fig. 7 Book of the day

Rezension des Tages

Alles Mathematik - Von Pythagoras zum CD-Player
- *Martin Aigner, Ehrhard Behrends* - mehr ▶

numerous exhibitions, see Mathematik.de/ger/diverses/aktuelles/mathemacher_november_2010/eugen_jost.html.

- In Mathematik.de/ger/diverses/aktuelles/energie_fuer_den_marathon.html we reported on a mathematical theory for marathon, which was published by Benjamin Rapoport from Harvard University.

Of course not all messages have a film or a slide show but every news message is illustrated by at least one picture to make the website as visually attractive as possible.

Daily Changing Items

Our daily changing items include one of 185 citations of the day (Fig. 5).

There is a large selection of historic reminders (which we obtain from http://www-history.mcs.st-and.ac.uk/history/Mathematicians, Fig. 6).

There are 80 books of the day (Fig. 7).

There are 136 items of Five-minute mathematics (Fig. 8).

The last category was started by Ehrhard Behrends as a weekly column in the German newspaper Die Welt and 100 of these stories were published as a book [2]. Some 36 more stories from different authors were added in 2008. As a result every day we can offer one of this large collection of very interesting short stories about mathematics to our readers.

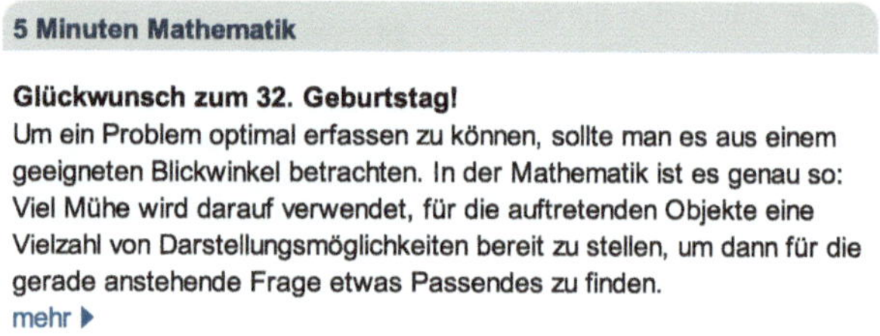

Fig. 8 Five-minute
mathematics

5 Minuten Mathematik

Glückwunsch zum 32. Geburtstag!
Um ein Problem optimal erfassen zu können, sollte man es aus einem
geeigneten Blickwinkel betrachten. In der Mathematik ist es genau so:
Viel Mühe wird darauf verwendet, für die auftretenden Objekte eine
Vielzahl von Darstellungsmöglichkeiten bereit zu stellen, um dann für die
gerade anstehende Frage etwas Passendes zu finden.
mehr ▶

Our selection of book reviews is steadily growing. Currently we have about 250 book reviews accessible on our site, 80 of which are in the selection for the book of the day. The books in the book of the day category were generally published less than five years ago. New reviews are added into the selection for the book of the day, and from time to time we delete the oldest ones.

Main Menu and Welcome

How do we communicate with our readers? Our main menu has the following options:

- Welcome
- First aid
- Exploring mathematics
- University and occupation
- Mathematics in the media
- Information
- School

with suitable subitems. For any of the following typical groups of readers:

- School pupils
- University students
- University lecturers
- School teachers
- Working mathematicians
- Journalists

we have a special welcome page linking to the most relevant submenus for the particular group.

First Aid

The most popular category on our web site is the First Aid. It addresses mainly high-school and university students who have a specific mathematical problem, but also offers mathematical knowledge for other interested people.

Since we don't want to answer daily emails of the type: "How do I solve a quadratic equation?" or "I have a math test tomorrow, and need a lesson on power laws", we offer everybody seeking mathematical help short descriptions, solution procedures and further links to the most frequently asked mathematical questions.

For this purpose we have written 60 detailed articles about topics from algebra, arithmetic, calculus, fractions, equations and systems of equations, geometry, percentage and interest calculation and probability theory, which are accessible on our site. One can easily find these pages from the menu, but also by using suitable keywords in Google.

Currently we are designing and integrating many multimedia-based components and incorporating them in our First Aid articles. Some examples are the following:

Figure 9 shows a JavaScript applet where the user can enter the numerator and denominator of a fraction. Then the applet executes the division calculating the remainder and showing all intermediate results and the complete computation, which gives the periodic decimal approximation of the fraction as shown in Fig. 10. There is a similar applet handling the reverse operation.

An interaction with Riemann sums uses the GeoGebra package [3], which we have incorporated several times after receiving authorization from the author Markus Hohenwarter. In the applet (Fig. 11) the user can change the function, the number of partition points (up to 100), can visualize the upper and lower sums and can compute the numeric value of the integral. This gives the user enough freedom to learn and understand the concept of integrals via an area computation.

In the section Mathematical Calendar Sheet we publish several times a year an article about an anniversary, which is important in the history of mathematics. With this we like to draw attention to mathematical breakthroughs, important publications and other milestones in the organizational and structural development of mathematics.

The recent article "1859: the ζ function, distribution of prime numbers and Riemann hypothesis" about Riemann's article *Ueber die Anzahl der Primzahlen unter einer gegebenen Grösse* [4] gave the historical developments around the ζ function. It contained the graph of $|\zeta(z)|$ from the book [5] (Fig. 12) as well as a *modern variant*, computed with *Mathematica* [6] (Fig. 13), which shows the care taken by the authors Jahnke and Emde.

Mathematics as a Profession

"What does a mathematician do at work?" "What are the working options for a math student?" "Do you want to become a math teacher or remain at the university?" These are typical questions that mathematics students are asked by other people.

In today's life mathematics plays an important although not always obvious role. The career options for mathematicians are excellent, unemployed mathematicians are virtually unknown in Germany and the salary prospects are great. We report these issues frequently. A very interesting story was *Doing the Math to Find the Good Jobs* [7] about a study giving the three best job opportunities in the US as (1) mathematician, (2) actuary and (3) statistician.

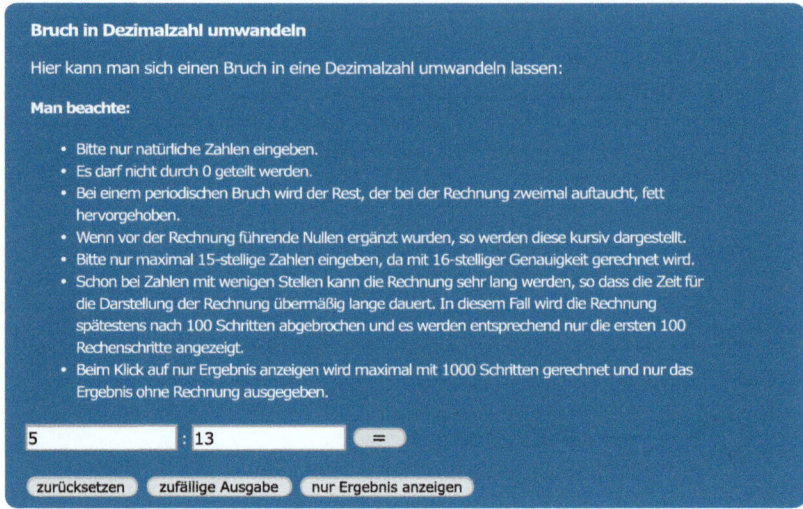

Fig. 9 Applet for converting a fraction to a decimal

Fig. 10 The fraction 5/13 as
a decimal

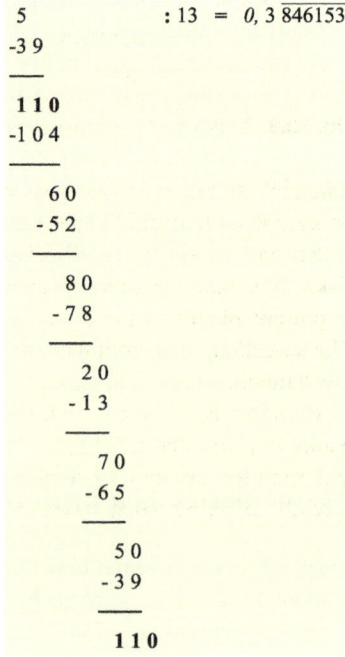

In the exciting category on Mathematics as a profession, which we started recently, we show various career opportunities by interviewing many mathematicians.
Our interviewees work in different professions and give a broad picture and personal
points of view about the professional work of a mathematician.

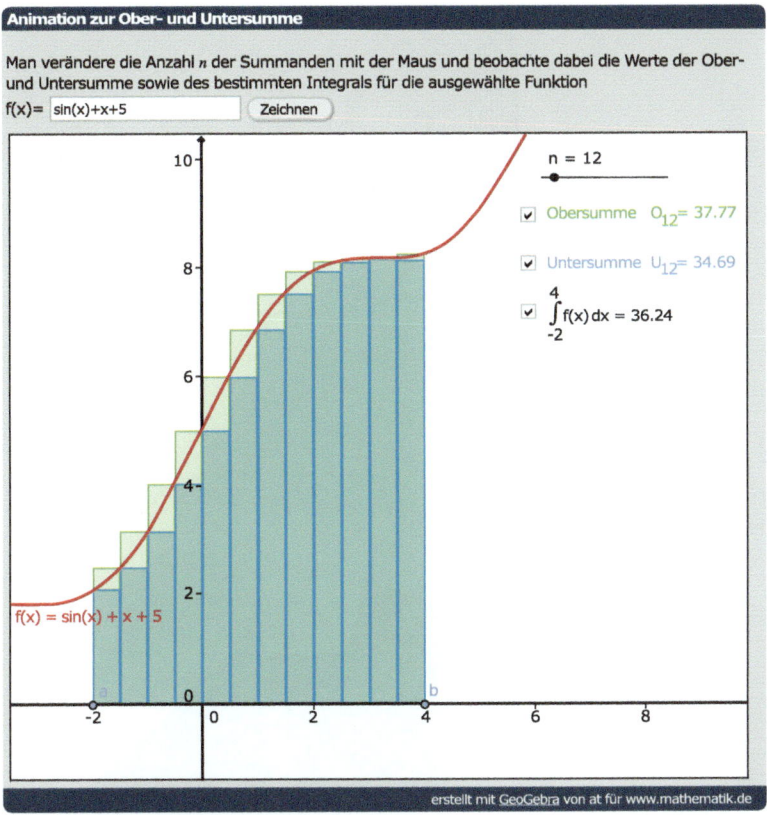

Fig. 11 Riemann sum for $\sin x + x + 5$

Supporting Giftes Students

There are international competitions such as the International Mathematical Olympiad (http://www.imo-official.org), similar national contests, and many institutions cater for them. The section Supporting gifted students features these institutions (Fig. 15). Our site gives these activities a common platform. Institutions that are not yet included in our list can use a web form to submit their details.

Statistics

We are very happy that Mathematik.de is used by many people. Our site has more than 2,000 daily hits showing that the type of information we provide is useful to the public (Fig. 16).

One should realize that most of our readers visit our site via search engines like Google. Note that the search term Mathematik in Google gives Mathematik.de a high ranking.

Fig. 12 $|\zeta(z)|$ function [5]

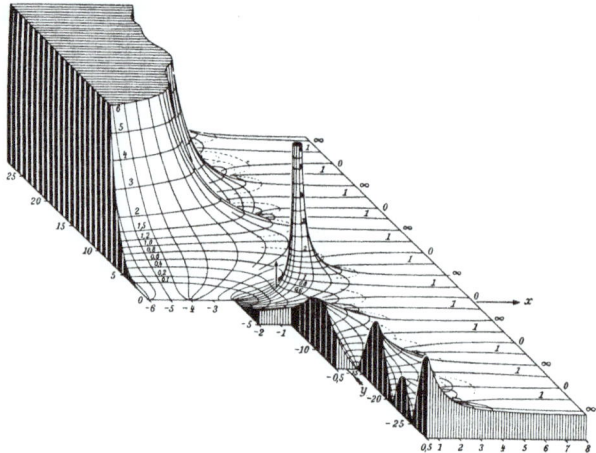

Fig. 13 $|\zeta(z)|$ function
computed by *Mathematica*

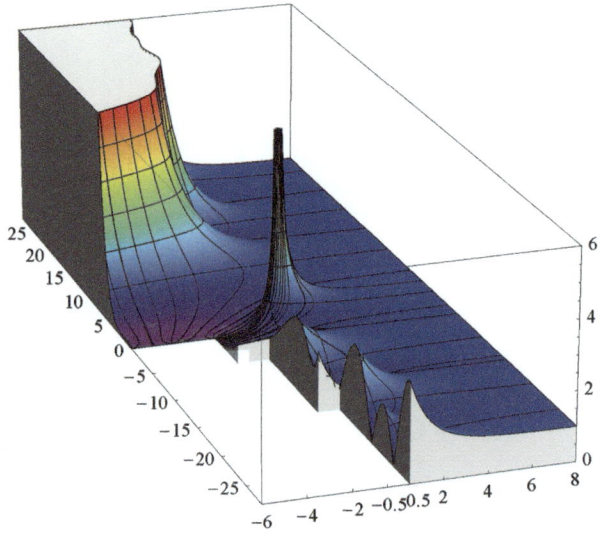

Our website now has been online for ten years and can clearly be called a success story.

Administering the Site

In this section we would like to answer some questions that might be of general interest for those who run a popular website like ours.

- Where do we get our news from? One student in our team is continually searching the Internet for news about mathematics. Important resources are especially the online pages of large magazines and newspapers. Our search finds an interesting

Fig. 14 The polynomial of a street light

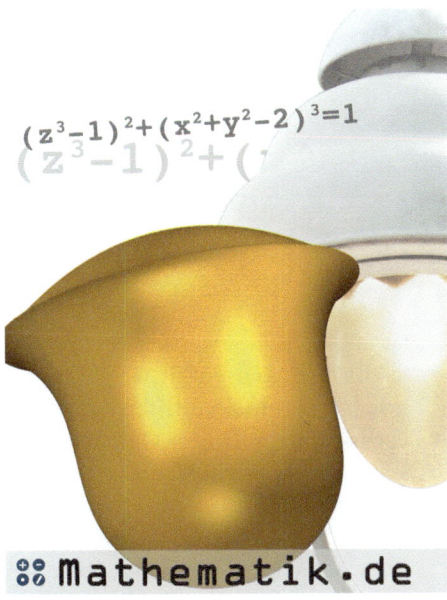

$$(z^3-1)^2+(x^2+y^2-2)^3=1$$

Fig. 15 Locations of institutions supporting gifted students

news article about every third day. Furthermore, the DMV has press releases that we use. We publish only news that seems to be of importance for a wide audience. We link only to sites that seem to be permanent.

- How do we handle copyright questions? For the text of press releases copyright is not an issue, and if we use other sources, we cite them or link to those sites. However, the copyright of pictures is an important legal issue. If we want to use

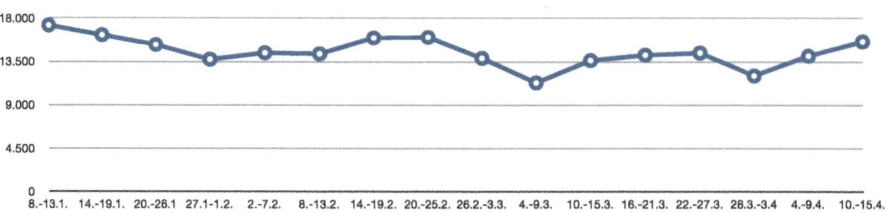

Fig. 16 Number of visitors to Mathematik.de per week

an image from a website or a press release, we ask the owner by email to give us the right to use it. If we get permission, that's fine, and we cite the copyright owner's permission. If this is not successful and in other cases of doubt we do not use such pictures. One other important source for images is Wikipedia [8] since the copyright status of the pictures given there, including many photos of mathematicians, is clearly stated.

- How do we handle book reviews? One student in our team checks new publications, selecting those books about mathematics mainly in German language that would be of interest for the public and ask the relevant publishers to send us reviewer's copies. Some publishers send us their books without asking. The selected books are then distributed to our reviewers. When a review appears, we send the link to the publisher. New reviews are used for the "Book of the Day".
- How do we get interviews with Mathematicians? Some time ago, we sent a letter to the mathematics departments of every German university. We attached questions of the interview, and asked them to be distributed to their alumni. We also sent the questions to business organizations. The complete correspondence for this section uses its own email address.
- How do we guarantee quality? Any of the above processes will create a new or changed article for distribution generated by the responsible student or by our web master. However, before a new article is put online, every new or changed page must get my approval. Although this is a lot of work I find this step necessary for quality control.

Our Team

Currently, our team has nine members, see Mathematik.de/ger/impressum/impressum.html. For each of the following activities one or two student members are responsible:

- Selecting news messages
- Handling the reviewing of new books
- Collecting interviews for "Mathematics as a Profession"
- Producing multimedia resources for the "First Aid".

Fig. 17 Torsten Sprenger
(*left*) and Wolfram Koepf
(*right*)

My assistant Torsten Sprenger (Fig. 17) is responsible for technical questions and for the management of the database, and Margarete Eisele is our technical assistant and web master. If we have a problem that is outside our technical scope, then we ask for external help.

Our sincere thanks go to our sponsor **ERGO** (Ergo.com). Without the financial support of the ERGO Group we could not run our website.

References

1. Gordon Films UK: Gordonfilms.tv
2. Behrends, E.: Fünf Minuten Mathematik: 100 Beiträge der Mathematik-Kolumne der Zeitung DIE WELT, 2nd edn. Vieweg, Braunschweig–Wiesbaden (2008). See Welt.de/wissenschaft/article90306/Fuenf_Minuten_Mathematik.html
3. Hohenwarter, M.: GeoGebra: Free mathematics software for learning and teaching. http://www.geogebra.org
4. Riemann, B.: Ueber die Anzahl der Primzahlen unter einer gegebenen Grösse. Monatsberichte der Königlich Preussischen Akademie der Wissenschaften zu Berlin, 671–680 (1859)
5. Jahnke, E., Emde, F.: Tafeln höherer Funktionen. Teubner, Leipzig
6. Wolfram, S.: The Mathematica Book, Version 4, 4th edn. Cambridge University Press, Cambridge (1999)
7. Needleman, S.E.: Doing the Math to Find the Good Jobs. Wall Street Journal, January 6, 2009
8. Wikipedia Online Encyclopedia: Wikipedia.org

Mathematics for the People

Günter M. Ziegler and Thomas Vogt

Abstract This is a brief report by the Mathematics Media Office of the German Mathematical Society (Deutsche Mathematiker-Vereinigung, DMV), on the 2008 Year of Mathematics in Germany, one of a series of national Science Years in Germany. The Year of Mathematics was seen and successfully used as a great framework for communication among the mathematical sciences, schools, and society at large. It was also a gigantic learning experience, a welcome experimental platform, and perfect opportunity for the professionalization of mathematics communication in Germany. The Year of Mathematics was viewed as extremely successful; it had more impact than most other Science Years. New grounds of communication were broken, new communication formats created. In the following, we start with a short glimpse on the development of science communication in Germany, then we discuss some "do's" and "don'ts" in math communication. We give an account of some of our experiences during the Year of Mathematics and describe some of the communication and event formats that turned out to be especially successful.

Science Communication in Germany

The current, very dynamic development of communicating mathematics in Germany started in 1998, when the International Congress of Mathematicians (ICM) was held in Berlin. This Congress offered a broad science communication program: In parallel to the official program of the Congress, with its Fields Medal awards, a special lecture by Andrew Wiles, and the extensive all-day scientific program, it also included a week of public evening lectures by German mathematicians at the Urania institute, which were designed to be accessible and inspiring not only to a larger Berlin public, but also for journalists. Some of these lectures were later collected

G.M. Ziegler (✉) · T. Vogt
DMV Mathematics Media Office, Institute of Mathematics, Freie Universität Berlin,
Arnimallee 2, 14195 Berlin, Germany
e-mail: ziegler@math.fu-berlin.de

T. Vogt
e-mail: vogtt@math.fu-berlin.de

E. Behrends et al. (eds.), *Raising Public Awareness of Mathematics*,
DOI 10.1007/978-3-642-25710-0_4, © Springer-Verlag Berlin Heidelberg 2012

into a book edited by M. Aigner and E. Behrends *Alles Mathematik: Von Pythagoras zum CD-Player*, which has been published in three editions since then, and is additionally available in English (*Mathematics Everywhere*). The highlight of the week at Urania was a special lecture by the German writer and essayist Hans Magnus Enzensberger. Enzensberger compared the mathematicians of today with medieval knights living their splendid lives in a castle with the drawbridge pulled up. Enzensberger asked the mathematicians to let down the bridge, to leave the castle and to communicate to society at large. The published (English and German) version of this lecture is still available as a book, and is quoted frequently.

Another important milestone was a 1999 memorandum of understanding on science communication in Germany: The large German research organizations such as the Max Planck Society, the Helmholtz Association, and the Leibniz Association together with the German Federal Ministry of Science and Education as well as the Stifterverband für die Deutsche Wissenschaft, the largest German science foundation, agreed upon a joint PUSH initiative. A Science in Dialogue (Wissenschaft im Dialog, WiD) agency was created as a joint platform for science communication to general audiences.

The same partners also decided that from then on each year should be announced as a "Science Year" dedicated to one specific discipline. The first of these Science Years was a Year of Physics in 2000. This year was very successful. It was followed by a Year of Life Sciences, a Year of Geosciences, a Year of Chemistry, and so on. The Year of the Humanities was 2007 and finally 2008 was designated to be the Year of Mathematics. This was our chance to let the drawbridge down, to leave the castle, and talk to the public.

The Year of Mathematics 2008

Lessons Learned: "Do's" and "Don'ts" in Math Communication

The Science Years are large and nationwide events with a multi-million euro budget. For the design, planning, and running of such an event strong professional support is essential. The Mathematics Year had the support of a large communication and advertising agency, Scholz & Friends. Thus the scientists had help in developing communication strategies, defining and addressing target groups, and organizing big events.

A central idea for the Year of Mathematics—and a strong recommendation of the agency—was to have a positive, active message. So we did not complain about the lopsided or negative image that our subject might have in the eyes of the public and in some of the media, but proclaimed that mathematics is an exciting subject, full of interesting challenges, and the basis for many disciplines, technologies, interesting jobs, etc. The motto for the year thus was "*Mathematics. Everything that counts.*" ("Mathematik. Alles, was zählt.").

Another important issue was to define the target groups accurately before the start of the communication project: Who do we want to address exactly? How do we reach this target group effectively? Who do you mean by "the public"? Are they the people who read the tabloids, the people who read the science pages in the papers, the museum-goers, the movie-goers, or others? For the Year of Mathematics, we set a particular emphasis on reaching the schools, and in particular on reaching the teachers as well as the pupils. We did this again with a positive message, claiming *"Du kannst mehr Mathe, als Du denkst!"* (*"You know more math than you think!"*).

We also agreed not to try to *teach* math to society at large; the teaching has to be done in the schools and at the universities. However, it is essential to work on the public perception of what mathematics is about, about what mathematicians do, about mathematics news and challenges.

For this purpose a lot of material was created: posters, brochures, press releases, and websites. One should not underestimate the amount of information material that is published during a Science Year through the demands of the media for texts, pictures, and interviewees. There is also the danger of making mistakes when publishing several texts per day. Since the advertising agency for the Year of Mathematics did not have science journalists or any other scientific competence in their team it was agreed to run a "Math Contents Back Office" at the Mathematics Department of the Technische Universität Berlin, directed by the first author of this report. In this setting trained science journalists (in particular, the second author of this report) generated and checked contents, supported by a team of students, freelancers, and volunteers. We gave advice, checked the mathematics presented in publications, etc. This office turned out to be a very helpful service unit for the media. Also, it did not disappear at the end of the year, but continued its work (building on the years' experience and expertise, on communication contacts and networks), and thus grew into what is now the DMV Mathematics Media Office at Freie Universität Berlin. The office now acts as a service center to answer journalists' inquiries quickly, it prepares press releases, documents media responses, collects mathematics literature, movies and photographs, generates contents for the DMV web site and blog, etc.

WiD and the advertising agency had a big arsenal of well-tested communication formats from former Science Years: this included large public opening and closing events for the year, a web site with a calendar of events, and hiring some VIPs as ambassadors for the year. All of these formats were used during the Mathematics Year; as a VIP role model, for example, the mathematics student who won the casting show *Germany's Next Top Model* was great for addressing teenagers. Other well-tested formats were modified for our purposes. But also many new formats were created for and during the Year of Mathematics. Some of them were so successful that they have been running continuously since 2008. In the following, we give some examples. (There are also web links at the end, where you can find more information.)

Mathematics Exhibitions

Several specialized mathematics exhibitions were surprisingly successful in reaching large audiences. We briefly discuss three examples here. A fourth one, Mathema, is described separately in this volume.

IMAGINARY

The Mathematisches Forschungsinstitut Oberwolfach (MFO) pioneered an exhibition called "IMAGINARY—with the eyes of mathematics" for the Year of Mathematics. It presented images of algebraic surfaces in large color prints and as three-dimensional objects. The exhibition of images included an easy-to-use rendering software package called *surfer*, which was available for download for free, and allowed the visitors to create their own surfaces and images. This was taken up by *Spektrum* (the German edition of *Scientific American*) as well as by www.zeit.de, the online site of the weekly *Die Zeit*, which both ran competitions among their readers for the most creative images. The software is still available and still downloaded at a high rate. There are still competitions and clubs in schools in Germany and abroad, which aim to create new designs of surfaces. The exhibition of the pictures and objects was shown at many places in Germany, and after the Mathematics Year also went to Austria, France, Portugal, Switzerland, the UK, the US and Spain. Core objects have been reproduced and are now part of a permanent exhibition in the framework of a new "Mathematics and Minerals" museum in Oberwolfach and of Deutsches Museum in Munich (Fig. 1). (See also the article of Andreas Daniel Matt on "open IMAGINARY" in this volume.)

The Science Ship

Perhaps the most unusual setting for a math exhibition was the Science Ship, which, run by WiD, is a regular part of each of the Science Years in Germany (see Fig. 2). The ship traveled on the Rhine and Elbe rivers, with stops at larger and smaller cities for one to three days, when school classes and individuals could visit. The ship held an exhibition with many hands-on experiments over some 600 square meters, which was put together from contributions suggested and provided by universities and research institutes. A number of students worked on the boat as tour guides ("pilots"), offering explanations to visitors. The ship had an event area in its back for small events such as lectures and receptions. This event area could be rented during the evening hours when the ship was closed to the general public. It turned out to be a popular location for special events, for universities and science institutes as well as for representatives of the towns and cities in the region. The coverage on "the ship will be in town" by regional media was remarkably extensive.

Fig. 1 Permanent exhibition of IMAGINARY at Deutsches Museum, Munich. (Source: Jürgen Richter-Gebert)

Fig. 2 The Science Ship—a floating math exhibition in the center of Berlin. (Source: Wissenschaft im Dialog)

Jewish Mathematicians in German-Speaking Academic Culture

An exhibition "Terror and exile" on the fate of Jewish mathematicians from Berlin, prepared for the ICM in 1998, was the nucleus for a much larger and comprehensive exhibition "Jewish mathematicians in German-speaking academic culture", which was shown in a dozen German cities during 2008 and the following years. The significance of Jewish mathematicians for German academic life in the nineteenth and twentieth centuries was demonstrated, presenting their lives, their ideas, and their outstanding contributions to the discipline; of course this included records of the Jewish mathematicians who were oppressed, persecuted, expelled, or killed during the Nazi regime. The exhibition contrasted biographical information on display stands with books that were shown as originals, at the heart of the exhibition. Visitors had the opportunity to browse the books and thus learn about, for example, Richard Courant, Max Dehn, Felix Hausdorff, Hermann Minkowski, and Emmy and Max Noether. The exhibition was created by Moritz Epple and his team from Frankfurt University. It was translated into English recently and shown in Israel from November 2011 to February 2012.

Interactive Events and Campaigns

Quizzes

Math quizzes in the style of *Who wants to be a millionaire?* turned out to be successful for audiences of all ages. A quiz is a very flexible format that works for small, medium, and large audiences alike. You can have a few contestants on stage, or have school classes compete against each other. All you need is a good collection of problems. The challenge is to make the questions neither too easy nor too difficult for the particular audience. One way to do this is to pose 12 problems—plus one warm-up problem used to explain the rules and one tie breaker as an extra problem. For each problem four answers, labeled A, B, C, and D, are offered; exactly one of which is correct. For example: "How much does the area of a rectangular garden grow by if you increase both its length and its width by 10 %?" Answers: A: 10 %; B: 20 %; C: 21 %; D: Can't tell, depends on the shape of the garden. Note again: This is not a format to teach mathematics, but to entertain—and to whet the appetite for math problems.

The Math Makers

Another aim for the Year of Mathematics was to activate mathematicians and math enthusiasts throughout the country and to integrate their activities into the year's program. Thus, we developed an activation campaign called the "*Mathemacher*":

This translates as math makers and describes people of all kinds and backgrounds who are actively teaching, promoting or popularizing mathematics outside or beyond their professional duties. Some offered additional and special courses at a university or school, others invented special math programs for children, pupils, or students outside school or university, at a business, on weekends or in the evenings. Math makers could present themselves and their activities after registration on the Year of Mathematics web site. Every week one of them was selected and awarded the title "math maker of the week" in order to reward them and display their outstanding ideas and activities. About 880 math makers registered in 2008. About one third of them are still active today—and an additional one hundred could be recruited anew.

Competitions and Prizes

Math Calendars

By digital mathematics advent calendars we have reached tens of thousands of math enthusiasts of all ages. These are modeled after traditional European advent calendars, which traditionally have a picture or a piece of chocolate behind each of 24 little doors, which are to be opened on the 24 days before Christmas: For each day of advent, December 1 to 24, there is a mathematical problem behind a virtual door, embedded in a little story. The participants register online and each day open a door of the calendar. The calendar now has three different levels: The most difficult, which has been running since 2004, is for advanced high-school students and grown-ups; its problems relate to research projects of the DFG Research Center Matheon in Berlin. The two lower-level calendars, for younger pupils, are provided by the DMV Media and Network offices at Freie Universität Berlin. The calendars are becoming more popular each year: more than 100,000 people registered for one of the three math calendars of the Berlin institutions in December 2011. About 100 winners in various age and prize categories are selected among those who have solved all or nearly all the problems correctly. The winners, the best school class, and the most committed school receive award certificates and attractive prizes at a public ceremony in Berlin (Fig. 3).

DMV Abiturpreis

Following an idea of the German Physical Society, in 2008 the German Mathematical Society started to award a "Mathematics Abitur Prize" for the best mathematics high-school graduates: Every high school is invited to nominate one student each year for the Mathematics Abitur Prize. The prize consists of an award certificate, a book (*Π & Co. Kaleidoskop der Mathematik*—a kaleidoscope of mathematics,

Fig. 3 One of the school classes that won an award after solving correctly all exercises of the advent calendar in December 2011. (Source: Kay Herschelmann)

edited by E. Behrends, P. Gritzmann and G.M. Ziegler specially published for this prize and sponsored by Springer), and one year of DMV free membership, which includes four issues of the *Notices of the DMV* (*DMV Mitteilungen*, a four-color illustrated mathematics magazine). The number of Abitur prizes awarded increased from 1,320 in 2008 to about 2,600 in 2011.

Cartoon Prize

For the Mathematics Year, DMV awarded a Mathematics Cartoon Prize, funded by De Gruyter Publishers, and promoted by the (Berlin-based, international) cartoon web site www.toonpool.com. This contest was an unexpected national and international success with 260 contributions by 158 cartoonists from 40 different countries. Humor is definitely a category that is underused in science communication. In addition to the memorable award presentation ceremony there were two exhibitions presenting the 30 best cartoons, in Bonn and Berlin. Dealing with mathematics in a satirical way was certainly another adventure that was not expected of mathematicians.

Traditional Competitions

Existing national (and international) math competitions were promoted during the Year of Mathematics and there was a strong increase in the numbers of participants during the Year of Mathematics, a trend that continued in subsequent years. The largest math contest for primary and secondary schools in Germany is *Känguru*, the German branch of the international Kangaroo of Mathematics competition. The number of participants in Germany rose from 549,000 in 2007 to 768,000 in the Math Year 2008 and then further to 860,000 with 9,000 schools participating in 2011. Competitions for young mathematics talents were promoted strongly during the Year of Mathematics and became more widely known to the public via press releases and media coverage. Examples are the German National Mathematical Olympiad (MO) and the Federal Mathematics Competition *Bundeswettbewerb Mathematik*.

Conclusion

The Year of Mathematics has produced lasting effects in Germany, which include many of the activities that we have described in this paper. In particular, it has produced the Mathematics Media Office, which has permanent funding, at Freie Universität Berlin. Now, in year four after the Year of Mathematics we have built a broad network, which is experienced and effective in reaching the public via all kinds of media. The Media Office each year runs M^3 (Math Month May), a common framework for about a dozen local mathematics events all over Germany, as well as the DMV Media Awards every second year. It has joined forces with the DMV Network Office Schools-Universities, which is sponsored by Deutsche Telekom Foundation and is located on the same floor at Freie Universität Berlin, to manage and publicize the Abitur Prize and the mathematics advent calendars.

Much effort (and also financial support) is needed for such a long term plan to address the media and the public. But the effort will pay off. Hearing about mathematics and about what mathematicians do is a first step towards doing mathematics. Positive experiences with mathematics will draw young people into mathematical studies and careers some day. And decision makers, reading about mathematics in the daily papers, will gradually become aware of mathematics as a part of our culture and the basis of daily life—and will learn to appreciate its importance for society and economy. We'll get there.

Links

www.bmbf.de/de/12048.php (web site of Germany's Federal Ministry of Education for the Year of Mathematics) www.du-kannst-mathe.de (archived web site of the Year of Mathematics for teenagers)

www.mathematik.de (web site of the German Mathematical Society for the public)

www.w-i-d.de (web site of Wissenschaft im Dialog)

www.dmv.mathematik.de (website of Deutsche Mathematiker-Vereinigung)

www.imaginary-exhibition.com (web site for the exhibition and software on three-dimensional objects)

www.juedische-mathematiker.de (web site of the exhibition "Jewish mathematicians in German-speaking academic culture")

www.mathekalender.de (web site of the mathematics advent calendars)

www.mathe-kaenguru.de (web site of the German Kangaroo of Mathematics competition).

Mathematics in the Public Mind: The USA

Reinhard Laubenbacher

Abstract This article discusses different aspects of the effort to raise awareness of mathematics in the United States of America. After sketching the broader national dialogue about the contribution of mathematics and science to the nation's prosperity and security, the article describes different constituencies that need to be reached and gives examples of ongoing activities and programs to engage them. Finally, the article provides several recommendations.

Introduction

Why is it important to raise the public's awareness of mathematics? Is it important that the mathematics research community are involved in this effort? As mathematical scientists, what is it we want society to do to support mathematics? What are effective ways in which we can highlight the role of mathematics in society and advocate for its support? This chapter will explore some answers to these four questions in the context of the United States of America. The answer to the first question depends, of course, on the particular segment of the public that is being targeted: children, the general public, teachers, policymakers, or university officials. One important goal for the mathematical sciences community needs to be, of course, to recruit the next generation of research mathematicians. But another, maybe equally important, goal is to convince the general public that research in mathematics is important and deserves financial and other resources. These two goals involve different target audiences and require different approaches.

R. Laubenbacher (✉)
Department of Mathematics, Virginia Polytechnic Institute and State University, Blacksburg, VA 24061-0123, USA
e-mail: reinhard@vbi.vt.edu

R. Laubenbacher
Virginia Bioinformatics Institute, Virginia Polytechnic Institute and State University, Blacksburg, VA 24061-0477, USA

R. Laubenbacher
Society for Industrial and Applied Mathematics, Philadelphia, USA

E. Behrends et al. (eds.), *Raising Public Awareness of Mathematics*,
DOI 10.1007/978-3-642-25710-0_5, © Springer-Verlag Berlin Heidelberg 2012

The answer to the second question about the involvement of research mathematicians is, in my opinion, an unequivocal "yes." It is widely accepted among the public, and certainly within the sciences, that mathematics is important as a scientific tool. The life sciences in particular have greatly expanded their use of mathematical and statistical methods for the analysis of evergrowing mountains of data. It is, however, much less widely understood, even within the sciences, that there is a need for mathematical research, that new mathematics is needed for the scientific and engineering problems we face today and tomorrow. Furthermore, scientists not generally identified as mathematicians, such as engineers, physicists, or computer scientists, are making a significant contribution to the mathematical research literature. The strongest possible case needs to be made, in particular to policymakers, that, yes, new mathematics is needed more than ever to meet the challenges of the future, and that key advances will come from what is traditionally considered to be basic, non-applied mathematics, one of the domains of the mathematician. Only the mathematics research community can make this case credibly and effectively.

After we have made a successful case that research mathematics is important for society, we come to the third question of what support the mathematical sciences community needs to play its part in society. Do we need more research funding, more funding for Ph.D. students, more faculty positions, higher salaries, more required mathematics classes for high school students, a more engaging mathematics curriculum for elementary school students, more mathematics programs on television? The differing circumstances in different countries will result in a different final mixture of priorities. And the circumstances in individual countries will also dictate the answer to the fourth question on how to best achieve this support. This chapter focuses on possible answers to these questions in the USA.

Society in Crisis: The Gathering Storm

The discussion about the role of mathematics in American society has been framed in recent years largely in terms of a broader discussion about science and economic competitiveness. The catalyst for this discussion was the 2005 report *Rising Above the Gathering Storm: Energizing and Employing America for a Brighter Economic Future* [1], published by the Committee on Prospering in the Global Economy of the 21st Century: An Agenda for American Science and Technology, National Academy of Sciences, National Academy of Engineering, Institute of Medicine. The charge to the committee was: *What are the top ten actions, in priority order, that federal policymakers could take to enhance the science and technology enterprise so that the United States can successfully compete, prosper, and be secure in the global community of the 21st century? What strategy, with several concrete steps, could be used to implement each of those actions?* The report lays out arguments to show that science and technology research is critical for the economic well being of the country and that the United States is losing ground to other nations in a world that has become flat, in reference to Tom Friedman's influential book *The World is Flat*

[2]. It provides statistics showing that, in comparison to other countries, the US is not doing enough to educate its children in mathematics and science. It makes recommendations for action in the areas of teacher training, higher education, research funding, and government policy, in particular changes in visa and immigration policies to make it easier for highly trained researchers from other countries to live and work in the United States. Specifically, its recommendations are to:

1. Increase America's talent pool by vastly improving K-12 science and mathematics education.
2. Sustain and strengthen the nation's traditional commitment to long-term basic research that has the potential to be transformational to maintain the flow of new ideas that fuel the economy, provide security, and enhance the quality of life.
3. Make the United States the most attractive setting in which to study and perform research so that we can develop, recruit, and retain the best and brightest students, scientists, and engineers from within the United States and throughout the world.
4. Ensure that the United States is the premier place in the world to innovate; invest in downstream activities such as manufacturing and marketing; and create high-paying jobs based on innovation by such actions as modernizing the patent system, realigning tax policies to encourage innovation, and ensuring affordable broadband access.

These high-level recommendations are supplemented with a long list of very specific actions. The actions to realize Recommendation 1 focus primarily on the training, recruitment, and retention of highly trained K-12 teachers. Recommendation 2 calls for increased government spending on long-term basic research, incentives for innovation, and new funding mechanisms and opportunities for innovative research. A call for an increase in the number of Americans with advanced college degrees and a reformation of visa and immigration policies will address the need for the large qualified workforce called for in Recommendation 3. For Recommendation 4, changes in the tax and patent structures are suggested, together with infrastructure improvements. The Executive Summary of the Report concludes: "We owe our current prosperity, security, and good health to the investments of past generations, and we are obliged to renew those commitments in education, research, and innovation policies to ensure that the American people continue to benefit from the remarkable opportunities provided by the rapid development of the global economy and its not inconsiderable underpinning in science and technology."

The Response: America COMPETES

The *Rising Above the Gathering Storm* Report received extensive media coverage and generated a discussion among policymakers, business leaders, and academics. When President George W. Bush announced his American Competitiveness Initiative during his 2006 State of the Union address he incorporated several of its key ideas. Many of the report's recommendations were included in the *America Creating Opportunities to Meaningfully Promote Excellence in Technology, Education, and Science Act*, known as the *America COMPETES* Act, which was signed into law in August 2007, with a large bipartisan majority. Many state governments also acted on some of the report's recommendations, in particular as they pertained to the teaching of science, technology, engineering, and mathematics (STEM). Several initiatives sprung up in response to the report, such as the National Math and Science Initiative (NMSI) (http://www.nationalmathandscience.org/), a public-private partnership that provides "ideas, inspiration, and resources to help America close the competitive gap."

For mathematics, one important aspect of the COMPETES act was its focus on increasing funding for the National Science Foundation (NSF), since the NSF represents the single most important funding source for mathematical research in the United States. Through its funding authorization the bill laid the groundwork for NSF funding increases consistent with an eventual doubling of the NSF budget, similar to the earlier doubling of the budget of the National Institutes of Health (NIH), the main funder of biomedical research in the US. In the three years since then the budget of the NSF's Division of Mathematical Sciences has indeed received increases that are roughly consistent with such a doubling [3]. The largest effect of the budget increases was an average higher percentage of proposals funded. But the increases have also allowed additional investments in workforce programs at different levels.

The initial authorization of the America COMPETES Act was for three years, and the reauthorization in the fall of 2010 met with a broad sentiment among members of the US Congress that the reduction of budget deficits was a high national priority. As this chapter is being written the fate of the reauthorization in Congress is unclear. Despite its uncertain future, the America COMPETES Act represents an important case study of an effective and successful effort to raise awareness among our policymakers of how important science and mathematics are for the economic health of the country and the continued ability to compete in the global marketplace.

Initiatives to Raise Public Awareness of Mathematics

As mentioned in the introduction, there are several different constituencies that must be reached in an effort to raise awareness of mathematics, and the most efficacious approaches to each may differ. This section contains a description of some efforts here in the US that target different constituencies, chosen from a plethora of such activities. This section is by no means exhaustive, but is intended as a collection of case studies. In particular, it does not include a discussion of the many efforts by mathematicians engaged in improving formal K-12 education. While extremely important, they are beyond the scope of this chapter.

A good program with which to start this list of examples is the 2010 *USA. Science and Engineering Festival* (http://www.usasciencefestival.org). The two-day event brought many hundreds of hands-on activities, exhibits, and performances to Washington, DC, and attracted over 500,000 people. A follow-up event in 2012 will feature 1,500 hands-on activities and 70 performances. Many events around the country are coordinated with the Washington activities. While the scope of the Festival is very broad, there was a significant number of mathematics activities, and every professional mathematics society had a presence.

Policymakers   The effort that led to the COMPETES legislation succeeded in part because it made clear that science and mathematics are the key enabling technologies for economic strength, national security, and global competitiveness. The main professional societies that represent the mathematical sciences community in the United States have been using this same approach to engage government policymakers. The American Mathematical Society (AMS), the Mathematical Association of America (MAA), the Society for Industrial and Applied Mathematics (SIAM), and the American Statistical Association (ASA) all maintain a Washington presence and carry out a variety of activities directed at lawmakers and their staff, ranging from individual meetings to larger presentations and exhibits by mathematical scientists. The societies, either individually or in collaboration with other mathematics, science, and engineering societies, regularly prepare position papers, letters, and briefings to many different constituencies.

The General Public   As much as or more so than in other countries, television in the US is the medium that reaches the largest number of people. Maybe the best-known representative of mathematics on television is the weekly network television drama *Numbers*, aired by the CBS Network from 2005 until 2010. The main characters are two FBI special agents, one of whom is a mathematical genius who provides the key insights to solving crimes using mathematical tools, helped by other mathematical consultants. The series was the most popular show airing on Friday evenings for its first four seasons. The creators of the program have received several awards, including the Carl Sagan Award for Public Understanding of Science in 2006 and the National Science Board's Public Service Award in 2007. In January 2011, they received the Communications Award of the Joint Policy Board for Mathematics, an umbrella organization of the above-mentioned professional mathematics societies.

Another hub of public awareness activities are the eight mathematical sciences research institutes (http://www.mathinstitutes.org/), funded by the NSF's Division of Mathematical Sciences, all of which engage in a range of activities to raise the public awareness of mathematics, and we mention here a small sample of events. The Institute for Mathematics and its Applications at the University of Minnesota, for instance, has established a public lecture series. The March 2011 lecture is entitled "Secrecy, privacy, and deception: the mathematics of cryptography." The Mathematical Sciences Research Institute in Berkeley, CA, also holds a series of public events. A February, 2011 event featured a panel discussion and performance of "Kingdom of Number" at the Monday Night PlayGround at the Berkeley Repertory Theatre. The Mathematical Biosciences Institute at The Ohio State University in February 2011 hosted a public lecture with the title "Heart attacks can give you mathematics." More information can be obtained from the various institutes' websites. The NSF mathematical sciences research institutes would be a natural stakeholder and resource for a future broader public awareness initiative.

An important initiative of the Joint Policy Board for Mathematics, mentioned above, is the annual Mathematics Awareness Month, held every April across the United States (http://www.mathaware.org/index.html). The 2011 topic is "Unraveling Complex Systems." A creative poster on each topic is distributed to university mathematics departments and related organizations. These, in turn, organize a variety of events around the theme, some of which involve the general public and K-12 students.

K-12 Students   There is a broad array of mathematics awareness programs at the local, regional, and national level that target children in various age ranges, often coupled with formal educational programs. At the middle- and high-school level an exciting program is Math Circles, a nationwide program organized by the National Association of Math Circles (http://www.mathcircles.org/). The idea of the US Math Circles program is patterned after the European Math Circles that originated in Hungary more than a century ago and spread over Eastern Europe and Asia. The first Math Circle in the US was established in Berkeley, CA, in 1998, as an educational enrichment program that brings mathematical scientists into direct contact with pre-college students. The setting is informal, after school or on weekends, and students work on mathematical problems and listen to research mathematicians talk about their field. The program has grown to approximately 100 Math Circles around the country.

Competitions and challenges have always had a special appeal to children and youth. The Society for Industrial and Applied Mathematics, in cooperation with the Moody Foundation, runs the annual *Moody's Mega Math Challenge* (http://m3challenge.siam.org/), a mathematics contest that has groups of high-school students compete in solving real-world problems using mathematics, such as determining the effectiveness of ethanol as a biofuel. Teams of three to five students have 14 hours to solve open-ended problems. The goal of the competition is to motivate students to study and pursue careers in applied mathematics, economics, finance, and related fields. Winners receive scholarships totaling US $100,000 toward the pursuit of higher education. The 2010 Challenge involved 531 teams comprising 2409 students in total.

There are fewer programs targeted at younger students, despite the fact that attitudes toward mathematics and science are developed early on in a child's school career. An example of a regional outreach program targeting nine-to-twelve-year-old children is *Kids' Tech University* (http://kidstechuniversity.org). The program is modeled after the extremely successful *Children's University* programs started in Germany in 2003, which now exist all over Europe (http://eucu.net/cu/projects?id=135). The goal of the program is to engage scientists and mathematicians directly with children and their parents. The program creates a university experience, complete with hands-on laboratories and virtual activities that complement the on-campus parts of the semester-long program. Children and parents come to a university campus on four Saturdays over the course of a semester for a full-day learning experience. Approximately 450 children and their parents from around Virginia and neighboring states participate every year. The program currently exists at

Virginia Tech, sponsored by the Virginia Bioinformatics Institute and the Virginia 4H Extension Program, with a satellite program at Virginia State University. Mathematics activities have focused on a variety of topics. For instance, the inaugural mathematics program in 2009 featured Keith Devlin from Stanford University, who discussed the question "Why are there animals with spotted bodies and striped tails but no animals with striped bodies and spotted tails?"

Conclusions and Recommendations

In the end, the imperative for the mathematical sciences community to raise awareness of mathematics among the general public is clear: we want to recruit the next generations of mathematicians; we want our university administrations to value mathematics as an important contributor to the institutional research mission and support it accordingly; and we want the general public and its elected representatives to support mathematics and provide adequate funding for the agencies that support our research; we want our funding agencies to view mathematics as the central enabling technology for much of scientific progress and fund it accordingly; we want our K-12 educational system to train students adequately in mathematics so that they are wellprepared when they enter college.

In the United States, there used to be a broad political consensus that mathematics and science are key to the prosperity and security of the country. As this chapter is being written this consensus has eroded and funding for mathematics and science research and education is being threatened seriously in the effort to cut budgets and reduce the role of government. There is no doubt that in some part a lack of appreciation of the role of mathematics and science in our lives among the general public and, by extension, their elected representatives, is to blame. Also, in an effort to look for "return on investment" and an immediate positive impact of research on society increasing emphasis is placed on applications of mathematics, and basic research is under growing pressure to prove itself relevant at many levels, ranging from university departments to funding agencies and government appropriations committees. The mathematical sciences community needs to act strongly at several different levels:

- We need to influence the teaching of K-12 mathematics to make sure that students are well prepared for the challenges of the modern workplace and to inspire our children to embark on mathematical careers. We need to provide resources and guidance for teachers so they can show their students the importance and beauty of mathematics.
- We need to launch outreach programs that enrich the mathematical experience of children and teenagers and that show them the fundamental importance of mathematics and the breadth of career options available for mathematicians. Programs need to provide tangible evidence that mathematics research can make a difference.

- We need to educate the general public and our governments about the pivotal role of mathematics research. Even the notion of research in mathematics is alien to most people, including many scientists.
- We need to connect our undergraduate and graduate training programs more closely to the other disciplines, with more interdisciplinary training that prepares our students for a variety of careers. As an example, in 2007 the American Mathematical Society gave its Award for Exemplary Achievement in a Mathematics Department to the University of California, Los Angeles (UCLA) [4]. Among other achievements, the award cites the broad range of choices that the over 800 undergraduate mathematics majors have among interdisciplinary specializations within the major.
- We need to engage other scientific disciplines more proactively. There is no better way to demonstrate the centrality of mathematics in the scientific enterprise than to show that mathematics is deeply networked, with many interdisciplinary collaborations and training activities. Many of us still remember the attempt by the administration at the University of Rochester in 1996 to abolish the mathematics graduate program [5]. Apparently, few other university research and training programs thought that they would be impacted negatively by this decision. This led the university president Thomas Jackson to state that: "At Rochester, we don't care very much about mathematics. It has low priority."
- We need to engage our policy makers and their staff in federal and state governments and provide them with information about the role mathematics research plays in the economic well-being and security of our countries.

The burden of this outreach falls on all of us, whether we are educators, researchers, or administrators. Mathematicians with credentials in education might be best suited to engage in K-12 or undergraduate education, whereas our top researchers have the credibility to authoritatively talk about the impact of mathematics research. While this takes us away from the things we love, whether it is teaching students or proving theorems, the continued health and prosperity of the profession demands that we all become engaged. And we need to leverage other stakeholders in a strong mathematics culture, such as industry and hightech companies, as well as governments. The 2008 *Year of Mathematics* program in Germany provides an excellent case study of such a partnership, which has led to the creation of many outreach programs and networks of individuals and institutions, which continue to grow and be active. Corporations like Intel, Google, Microsoft, and General Electric are spending many millions each year on outreach efforts to increase the awareness of mathematics and science, with very effective advertising efforts and outreach programs. They are natural partners for the mathematical sciences community.

As this chapter documents, many individuals, groups, and professional societies are busy carrying out efforts at all these levels. Since the mathematics community is quite small compared to, say, chemistry or engineering, it is important that we partner with the other sciences to amplify our message. But at the same time mathematics is quite different from the sciences and engineering, and we must control the narrative of mathematics awareness efforts. We must redouble our efforts in this time of unprecedented opportunity for and danger to our profession.

References

1. National Academy of Sciences: Rising Above the Gathering Storm: Energizing and Employing America for a Brighter Economic Future. The National Academies Press, Washington, DC (2007)
2. Friedman, T.: The World Is Flat. Farrar, Straus, and Giroux, New York (2005)
3. National Science Foundation, Directorate of Mathematical and Physical Sciences. Report (2010)
4. American Mathematical Society, 2007 Award for an Exemplary Program or Achievement in a Mathematics Department. Notices of the AMS, May 2007
5. Jackson, A.: Downsizing at Rochester: Mathematics Ph.D. Program Cut. Notices of the AMS, March 1996

Balancing Math Popularization with Public Debate: A Mathematical Society's Continued Efforts to Raise the Public Awareness of Mathematics and for Youth Mathematical Education

Renata Ramalho and Nuno Crato

Abstract After briefly describing the Portuguese Mathematical Society (SPM) recent history, we discuss the science popularization and communication activities developed by the Society from 2005 to 2010. We describe the contacts maintained with the media, the organization of events for the public and for scientists, and the publishing and publicizing of magazines and books. We argue that some of the approaches adopted can result in better public understanding of science role and in larger public support for math education and research.

Introduction

For the last six years, the Portuguese Mathematical Society (Sociedade Portuguesa de Matemática, SPM) has been in the eyes of the public in a way that is not common among scientific societies in Portugal and in Europe. During that period, the society has often been on the cover of newspapers and magazines, it has played a major role intervening in math education themes, and it has been at the forefront of mathematics popularization, by organizing talks, exhibitions, conferences, and other events.

These were not accidental accomplishments. A new board for the society was elected in July 2004, and it set as one of its main goals the establishment of a communications office within SPM, as a means to boost science popularization activities and to increase the society's public role. Two factors were crucial for the success of these goals: (i) the constant support from mathematicians on the board and other

The authors report in this paper a recent experience in which both have been deeply involved.
N.C. was a member of the SPM Board from 2000 to 2010 and SPM president from 2004 to 2010.
R.R. was public relations chief officer at SPM from 2005 to 2010.

R. Ramalho (✉)
Sociedade Portuguesa de Matemática, Lisbon, Portugal
e-mail: renata.ramalho@gmail.com

N. Crato
ISEG, Universidade Técnica de Lisboa, Lisbon, Portugal
e-mail: ncrato@iseg.utl.pt

E. Behrends et al. (eds.), *Raising Public Awareness of Mathematics*,
DOI 10.1007/978-3-642-25710-0_6, © Springer-Verlag Berlin Heidelberg 2012

mathematicians by generating material that could be popularized and (ii) the hiring of a journalist.

Over time, the communications department grew, as well as its activities. Additional staff were hired, and further mathematicians took an interest in the society's popularizing actions. Some of the steps in this strategy may be of interest for other countries.

A Bit of History

During the 1930s and the beginning of the 1940s, mathematics in Portugal knew a particularly fruitful period. Research centers were created, scientific and popularization magazines and books were published, and international contacts were boosted as a new generation of mathematicians arose. As a result, the Portuguese Mathematical Society was founded to organize mathematics activities in the country.

This dynamic scientific environment, though, contrasted with the political ambiance of the time. A dictatorship ruled Portugal from 1933 until 1974 and forbade all association activities, including scientific conferences that intended to bring together Portuguese and foreign mathematicians. By the end of the 1940s, the government had expelled most of these young scientists from the universities and prevented others from finding jobs in Portugal. The whole generation was forced to emigrate and the movement they had started came to a halt. Only after the Carnation Revolution, in 1974, was SPM able to register its by-laws and resume its work.[1]

Since that time, many people have been on the board of SPM, and three generations of mathematicians have dedicated themselves to accomplish the goals that were set for the society by its founders: to promote research, education, and the popularization of mathematics. During the World Mathematical Year, in 2000, talks were organized and books published. These activities set the ground for the work that would be done a few years later, which is described in this paper.

SPM's Three "Pillars of Wisdom"

Since its creation in 1940, the Portuguese Mathematical Society has had three pillars on which its activities have been based.

Being a scientific society, SPM's first pillar is *research*. As with other societies, some of its activities include the organization of conferences, inviting key international mathematicians to give talks at research centers, and reaching out to graduate students with specific events. In a partnership with the International Center for Mathematics (CIM), the society organizes "mathematical journeys" that bring

---

[1]From SPM's web site. Further information is available at http://www.spm.pt/spm/historia/ (in Portuguese).

together researchers from all over the country interested in a specific area, thus fostering cooperation and allowing them to discuss research topics.

The second pillar is *mathematical education* at all levels. Though for some time the society was focused on undergraduate and graduate mathematical education only, recently it has turned its attention towards primary and secondary education, becoming a key player in the debate over education policy. SPM does not avoid sensitive topics, takes a very clear public stance on the need for high standards in curricula and evaluation, criticizes the continued watering down of textbooks, and voices the need for demanding teacher training. Instead of simply complaining about math education flaws, SPM has developed and defended clear guidelines for math curricula, math evaluation, and math teaching. These ideas have been based on applied scientific research, namely on cognitive psychology, and not on the sloppy thinking and nonsense ideas of some educationalists.

Among the initiatives organized by SPM, we highlight: a biennial summer school for math teachers; careful reviewing and certification of textbooks; publishing math education essays and support books; and what is now recognized as the largest math teacher training center in the country.

SPM's third pillar is *raising the public awareness of mathematics*. This pillar needs the first two in order to generate material for exhibitions, public debate, public relations work, popular talks, and many other activities undertaken by the society. It has proven to be an area at which SPM has been very successful. The Society has been able to assist mathematicians all over Portugal in approaching the general public.

Communicating Mathematics

At the beginning of 2005, less than one year after the election of the new board, it was decided to create a communications office. The society recruited a journalist, who specialized in science communication, to be in charge of the new office and its activities, in direct contact with the society's board.

From the very beginning, it became clear that the communication activities would be divided into three main areas: connecting with members, dealing with the media, and communicating with the general public. Although these areas would often overlap, neither should be neglected and a specific strategy should be set for each of them.

Connecting with members was something that had been done for a long time through three publications edited by the society: *Portugaliae Mathematica*, an international research journal, *Boletim da Sociedade Portuguesa de Matemática*, a member's bulletin, and *Gazeta de Matemática*, a math popularization magazine. Later, SPM supported a fourth publication: the *Jornal de Matemática Elementar*, a long-running newsletter that was edited for a long time by a member of the society.

Two of these publications went through extensive changes in order to broaden their appeal. The publishing of *Portugaliae Mathematica* was transferred to the European Mathematical Society to consolidate the journal's internationalization, and

Fig. 1 Cover of *Gazeta de
Matemática*. This cover has
recently been changed
following a redesign of the
magazine

to continue the modernization process started in the 1980s. The executive editorial board, though, remained in Portugal, so that the magazine did not lose its Portuguese identity. The editing of *Gazeta de Matemática* was transferred to the communications office of SPM, which has guaranteed an enhanced professionalization of the magazine (Fig. 1). Articles are still mostly written by teachers and research mathematicians and the editorial board is composed almost exclusively of society members, but the magazine was redesigned in order to attract more readers such as high-school teachers, undergraduate students, and others with an interest in mathematics. The magazine is now sold at selected news-stands and bookstores. The other publications are still to go through a similar process.

These four paper publications were insufficient in an age where information should be rapidly distributed. The lack of an electronic newsletter with a constant feed of information about events and other news of interest to the mathematical community became apparent. When SPM set up a monthly electronic newsletter, it received a flow of news from all over the country and had to establish of a limit of ten topics per newsletter, which was in itself a barometer of its success.

An electronic resource that came to attention was the society's web site. While SPM already had a site with information about it, its redesign made it more accessible and more user friendly. The new platform has been developed as a way of connecting with members and as a resource for the general public and the press. Its content was reorganized, and the new look has increased traffic.

This brings us to the second challenge that was faced by the communications office: *working with the media*. Getting mathematics into newspapers and on national television and radio was achieved through contacts with specific journalists and through targeted press releases. The subjects of choice for journalists have often been the polemic ones, mostly on mathematics education (Fig. 2). However, many more topics also made the front pages; we highlight among these research news, popularization events, exhibitions, and the mathematics Olympiad (Fig. 3).

The strategy for approaching the media was based on the idea that getting press coverage would be of interest to the society and its members. Therefore, the regular answer to questions asked by the press was "Yes, we can." Even when the topic was not of interest to the Portuguese Mathematical Society—for instance, the mysticism around the number 7 or what happens in pyramid schemes—SPM made a significant

Fig. 2 Article in *Diário Económico* on Miguel Abreu, elected president of SPM in July 2010 "The best school is the one closer to the parents"

Fig. 3 Poster for the 29th Portuguese Mathematics Olympics

effort to find a credible source for the journalist. The concept behind this attitude is based on three assumptions: firstly, if the Society is productive as a news source, then journalists will come back whenever they have new questions about mathematics, and eventually these demands would meet SPM's interests; secondly, by providing a credible source for even the most absurd requests weakens journalists' resistance to mathematical subjects; thirdly, a credible source would do its best to remove all mysticism from the subject, and take the place of other possible sources who might mistakenly reinforce misconceptions around an issue.

Another principle adopted by the press office is to try and deliver information "just in time." Whenever contacted by members of the press, one of the questions always asked had to be the journalist's deadline, so that the news information was produced on time. Though time and deadlines tend to be very different for scientists and journalists, making an effort to meet journalists' not always reasonable deadlines gives the possibility of getting news coverage and of establishing good relationships with journalists. In most cases, providing an answer after the deadline only gives a sense of failure to all parties, especially if the news source is not used by the media.

The writing of press releases for selected events is a major way of getting into the news. Some rules, however, must be taken into account in order to achieve the goal of supporting the communications outreach, because a poor press release could also easily undermine it. It is important to avoid having too many releases as this usually creates a resistance among journalists. When writing one, the style should allow for easy transformation into a news article. It is crucial to show the journalistic angles that can be used to cover the story. If the information is about an event, the press release should also have an "agenda" at the end of the text, including the main information: name, date, time, and location. Contacts should be listed separately at the end of the text, preferably with the mobile phone numbers of all indicated sources.

SPM has also collaborated with the press through media partnerships for radio and TV shows, newspapers, and the Internet. One of the first partnerships established was with *2010*, a science magazine for RTP, the public television channel. The project was called "Mathematics Memoirs", and consisted of interviews with senior and retired mathematicians. The interviews were long and yielded two versions: a short one for the TV show, and a longer one, edited by SPM to be distributed to schools and historians specializing in mathematics.

Another fruitful partnership was with *Antena 1*, the national public radio station, which requested that SPM nominate mathematicians to give interviews about specific research topics. This was an excellent opportunity to make the general public aware of the "Mathematical Afternoons", since the interviewees could discuss the topic that would be presented at these programs.

An example of a partnership with a newspaper was *Kalkular*, a mathematics magazine for children that was distributed along with *Público*, one of the main reference newspapers in Portugal. On the Internet, there was a media partnership with *Mundus* magazine, for which SPM provided monthly articles written by different mathematicians, covering research, education, games, the Math Olympiads, and other topics of their choice.

Though many of the examples above are ultimately aimed at the general public—which, one should keep in mind, includes scientists too—there was also the need to organize events which would allow for direct contact between mathematicians and a generalist audience. *Communicating face to face with the public* was therefore identified as a priority, and a series of events were organized with this in mind.

Talks for the general public had been promoted for many years in different places. As the result of this experience, the so-called *Mathematical Afternoons* series was created with a different format (Fig. 4). First developed in Lisbon in 2001 with *Ciência Viva*, the governmental science popularization agency, from 2005 on it was expanded throughout the country, reaching 11 different towns on the continent and the Portuguese islands. Sessions would have a specific theme, involve mathematicians and often other professionals working in related fields, and be followed by questions from the audience. Policemen, musicians, writers, historians, religious leaders, magicians, architects, bankers, green activists, and even TV stars have been invited to speak about the influence of mathematics on their work, while a mathematician would explain how the subject can be explored through mathematics. Talks

Fig. 4 Poster for the
Mathematical Afternoon
sessions at Guarda, 2010

Fig. 5 Children's Day
Mathematical Fair at Museu
de Ciência

were organized in science centers, schools, universities, shopping malls, bookstores, libraries, museums, and music venues and had as partners all kinds of institutions, whether academic or not.

Exhibitions were another successful initiative that traveled throughout the country popularizing mathematics. Themes included the 25 years of the math Olympiads to "Measuring the Earth, the sky and the sea", sundials, Escher, topics from the history of mathematics, and symmetry. They were adapted for public venues of all kinds, were lent to schools and activities were organized around them. The opening of these sessions was often attended by opinion makers and the press, besides the general public.

Sometimes exhibits were integrated into bigger fairs organized by the Society or at which it took part. In 2007, with the launching of the mathematics club, a fair was organized for Children's Day, 1 June, which attracted over 1,000 people that day alone (Fig. 5). On SPM's 70th birthday, another such event had activities for all ages that involved origami, games, math movies, magic, history, and so on. Both events were organized in partnership with the University of Lisbon's Science Museum, another regular partner of SPM's popularization activities.

One should note that these partnerships were vital to SPM's popularization activities, since the society itself does not have the personnel, the money, or the venues to organize such large events on its own. Academic institutions such as universities, schools, and museums, government ministries, agencies and municipalities, private

Fig. 6 Cover of SPM's
edition of Felix Klein's book

sector institutions such as banks, and foundations and editors were always sought
to co-organize events, edit books, support competitions, and participate. For SPM
it has been a big achievement to be able to engage all these different institutions in
math popularization events.

Books

It would seem that books, being a traditional means for communicating, are somehow outdated and should not be considered as sufficiently important to deserve
special mention. It would also seem that trade books, which are printed and sold
throughout the world, should not be promoted by a small society in a small country.
However, we consider books as a very important means of communicating mathematics and math-related subjects such as education.

As the Italian writer and scholar Umberto Eco famously observed, after World
War II some political currents won because they wrote more books ... in fact, books
are means of developing at length matters and arguments. Charles Darwin at a certain point considered his *On the Origin of Species* as simply "a long argument."

Books published in the national language are still important for developing at
length theories, arguments, and topics. This is also important for mathematics and
for math education.

For the World Mathematical Year, the society developed a publication plan that
included three book series. The first one, dedicated to the national history of mathematics, published historic textbooks and texts on the history of Portuguese mathematics that were long (some of them almost two centuries) out of print. No commercial or international publisher could have done it. The second series (started in
2000) is a series of booklets that included Klein's books for school teachers and
other useful texts (Fig. 6). The third series, started in 2004 in collaboration with a
Portuguese trade publisher, reached a large audience. One of the books, for instance,
went through eleven prints. Others were less successful but included themes, notably
on math education, that no one else would publish.

Fig. 7 Cover of *Público* mentioning the Portuguese Mathematical Society in the headline

Debating Education

During the last few years, many requests were received from the media around the time of the national exams. As demands at these dates started arriving on a regular basis, SPM decided to write exam answer proposals and reports evaluating the quality and adequacy of the exams. This became quite a task, involving teachers, professors, and the press office. All this material had to be delivered promptly to the media, which had been counting on them for reporting the exam news. In 2008, SPM assessments were very harsh, though completely correct, and originated a public debate. Some of the issues raised were discussed at the national Parliament and generated a number of reports on national TV and the front pages of newspapers (Fig. 7).

SPM did not shy away from taking polemic stands on education. The society is very proud of the fact that its public stance has earned it public recognition for clear, well-grounded and sensible positions in defense of an adequate math education and demanding math standards for elementary and secondary schooling.

Results

The results obtained through mathematics popularization activities may sometimes not be easy to identify and include a growth in the interest of children and adults in the discipline, an increase in their knowledge, the ability to positively influence public education and research policy. None of these are easily measurable, though one can feel their impact through media coverage, demands from the public, the number of students that enroll in math at the universities and individual or collective recognition of mathematicians.

The public awareness of mathematics could be evaluated through the interest of the media in the different subjects surrounding it, as well as how an interest in mathematics has increased for different stakeholders. The writers of a teenage soap opera asked SPM to help them introduce the math Olympiads into the program through a character that would participate in the competition. A TV commercial to foster milk consumption had a child saying that the bottle contained "good grades in mathematics." The number of news clips relating to the Portuguese Mathematical Society rose dramatically, and the number of clips received each year kept increasing.

The fund-raising capability of the society also substantially increased. Its activities led to one of the major Portuguese commercial banks asking SPM to become a partner for its financial literacy initiatives. This national bank became a major sponsor of the Olympics as well as other activities such as scientific meetings. There were other sponsorship deals, and the teacher training center became the biggest for mathematics in the country, which also provided substantial funds for the society. The number of books sold also increased, and the financial health of the society reached its highest point.

Another clear measure of the success attained was a 70 % increase in enrollment for the Portuguese Mathematical Olympiads. Though this initiative started in 1983, it had never known such publicity. During that period the renewed communications initiative was supported by an increase in funding and better results in international competitions. Altogether, this provided for the competition to be expanded in 2011 to reach students at the lower grades: instead of having the youngest category for seventh graders, it now reaches students from the third grade.

Most of all, a significant increase in SPM's capability to appear publicly and to be recognized as a major player in mathematics, namely in mathematical education, was felt. The goal of having journalists call the society whenever they had needs relating to mathematics was more than achieved, and SPM's opinions were taken into account by opinion makers and stakeholders in policymaking and public debate. Popularizing mathematics successfully is possible. In our experience, it depends mostly on the availability of capable mathematicians and the support of professional public relations staff.

The Butterfly Effect and the Popularization of Mathematics: Spain

Raúl Ibáñez Torres

Abstract World Mathematical Year 2000 was an inflection point in the popularization of mathematics in Spain, as in many other countries around the world. In this article we shall show some interesting and important activities of popularization of mathematics organized by the Royal Spanish Mathematical Society after WMY2000: web pages, lectures, literature contests, expositions, programs like math in the school libraries, . . .

Despite the social and cultural importance of mathematics, and also in education, their image and that of mathematicians is negative. This gap between society and mathematics was undoubtedly crucial in the decision of UNESCO, at the request of the International Mathematical Union (IMU), to declare the year 2000, World Mathematical Year (WMY 2000). This was an inflection point in the popularization of mathematics in Spain, as in many other countries around the world. Before WMY2000 there were very few actions for the popularization of mathematics, almost all of them came from the regional Spanish societies for teachers of mathematics (which belong to the Spanish Federation of Mathematics Teachers, FESPM, www.fespm.es) and they were organized in schools, and rarely for the general public. There was no raising of the public awareness of mathematics in and by the "university world", and the few mathematicians carrying out this work were frowned upon. There was almost no mathematics in the mass media, in the publishing world, the Internet, expositions, . . . with isolated exceptions (some of them really good).

But in the WMY2000, the mathematical community, in an extraordinary way, turned to work to improve the social image of mathematics, to disseminate mathematical culture to society so that society became aware that mathematics is a fundamental part of our society, our daily life, our culture, and that the economic, scientific, and technological development of a country would be impossible without them. All around the world, mathematicians—although for most of them this was a new challenge—put a lot of effort into this task by organizing a series of lectures,

R. Ibáñez Torres (✉)
Universidad del País Vasco, Lejona, Spain
e-mail: raul.ibanez@ehu.es

E. Behrends et al. (eds.), *Raising Public Awareness of Mathematics*,
DOI 10.1007/978-3-642-25710-0_7, © Springer-Verlag Berlin Heidelberg 2012

expositions, or round tables, and collaborating with the mass media or by writing books.

The Royal Spanish Mathematical Society (RSME) [1] decided to continue working in this vein after WMY2000. In the middle of 2003, the RSME asked the author of this article to create the Commission for the Popularization of Mathematics, with the aim of designing and developing a program for raising the public awareness of mathematics. The commission was created to include everyone involved in primary schools, secondary schools and universities, to collaborate with other societies and in general with the whole mathematical community. The targets set initially by the commission were: (i) to change the negative attitudes of people toward mathematics; (ii) to develop the mathematical culture of our society; (iii) to remove the conflict between science and the humanities; (iv) to communicate mathematics, its beauty and its power, to a wider public; (v) to encourage people to be more active mathematically; (vi) to increase the appreciation of the mathematics of our surroundings, by learning to watch reality with mathematical eyes; (vii) to stimulate mathematical activity; (viii) to show society mathematical investigations; (ix) to demonstrate to politicians, the business community, ... that a developed country needs developed mathematics.

Now we will describe some of the actions organized by the Commission for the Popularization of Mathematics of the RSME, under the chair of the author of this article, and also of the new chair, Pedro Alegría (UPV-EHU).

Divulgamat, Virtual Center for the Popularization of Mathematics

The first action of the Commission was the creation of a webpage for the popularization of mathematics, divulgamat (www.divulgamat.net), which attempted to capture the essence of the program. The Internet is an ideal means for disseminating mathematical culture in society. It is a global tool that gives to a large number of different people in different places, all around the world, access to any online activity, and offers a wide range of different resources. And also it is a modern medium that young people can identify with.

Divulgamat is a webpage in Spanish (Figs. 1 and 2) [2] with original work by Spanish and Latin-American authors. It is an open webpage, sponsored by the RSME but for the whole Spanish-speaking mathematical community. A large number of mathematicians, educators, writers, journalists, artists, ... collaborate with divulgamat, a total of more than 150 people. Inside divulgamat we can find culture, information, entertainment, inspiration, teaching resources ... a whole world of mathematics.

In 2008–2009 divulgamat underwent a major change in its structure and design, modernizing its aesthetics, but also the internal organization in order to facilitate both navigation and the work behind the scenes.

The home page of divulgamat displays beautiful images, as well as the most outstanding news for the portal. There are also five tabs to organize the page: the

Fig. 1 Home page of divulgamat

main menu, information, news, literary text, and image of the month. Sections of
the main menu are:

– *Mathematical Challenges*: Two mathematical problems or challenges (one elementary and the other more advanced) and solutions to the previous two problems.
– *History of Mathematics*: Biographies of mathematicians, mathematical topics,
fundamental work in mathematics, different cultures, articles in *La Gaceta de
la RSME*, history through images, cartoons, ...
– *Virtual Expositions*: Exhibitions related to mathematics including art, photography, books, history, and more.
– *Culture and Mathematics*: Mathematics in science fiction, music, magic, film,
theater, art, origami, the media, and in advertisements.

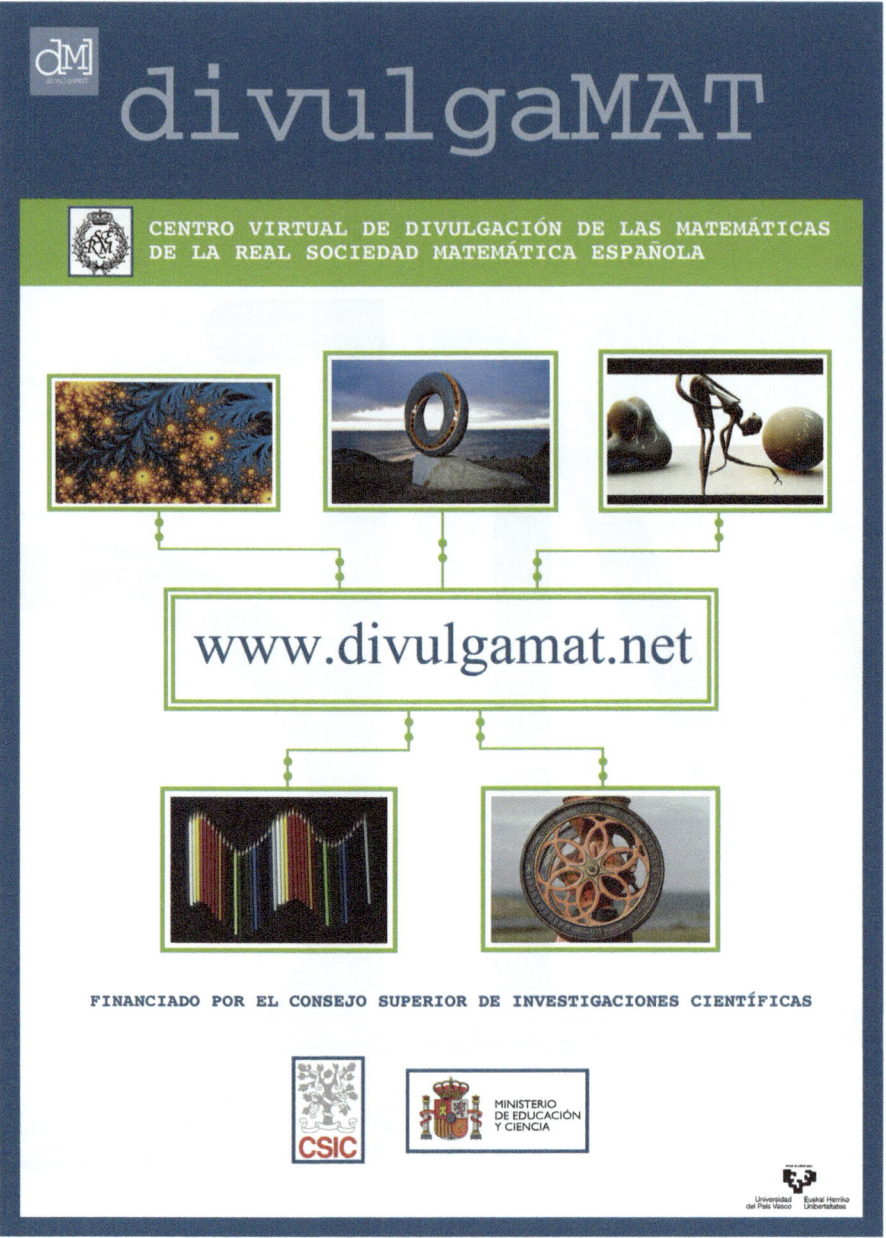

Fig. 2 Poster for divulgamat

– *Math Stories*: "Once upon a time … a problem", stories with problems … a mixture of mathematical problems, literary stories, and humor. Also a section for graphical humor.

- *Publications about the popularization of mathematics*: Databases with information and reviews of books (about 700 currently), videos, articles from newspapers and magazines, . . .
- *Texts on line*: Unpublished mathematical texts, seminars, workshops, lessons, inaugural talks in universities, in the Royal Academy of Sciences . . .
- *Didactic resources*: Guides and experiences for the classroom to be used as resources for learning and enjoying mathematics.
- *Mathematical applications*: The objective is to show the many applications that have math in our daily lives: industry, economics, medicine, architecture, art, . . .
- *Tributes*: Miguel de Guzmán (2004), El Quijote (2005), ICM (2006), the International Mathematical Olympiad (2008), and RSME (2010).
- *Math surprises*: Quotes, jokes, anecdotes, illusions, paradoxes, puzzles, cartoons, . . .

The information tab includes: new publications, events, news, suggestions, RSME–ANAYA literary contests, RSME expositions, commented links, among others. Next there is a section for news, which includes divulgamat news, news of the RSME relating to mathematical culture, or other news from society that would be of interest for visitors of divulgamat. The final two sections are literary texts and images of the month, in which every fifteen days one literary snippet and one image, relating to mathematics, are included.

Since its inception in mid-2004, divulgamat has become a reference portal for the popularization of mathematics in Spanish, and it reached, in its first two years of existence, more than 9,000 visits per day. And it has continued working . . .

Conference on the Popularization of Science: Mathematics

After the first year of the Commission for the Popularization of Mathematics of RSME, we decided to reflect about the problem of the dissemination and popularization of mathematics. This reflection would allow us to analyze the problem, and could focus on the needs and possible actions to take in the future. However, it seemed important to study this problem with other scientists, educators, journalists, writers, editors, artists, politicians, directors of the Science Museum, etc. to give us different perspectives on the problem we were interested in. For this reason we organized "The Workshop on the Popularization of Science: Mathematics", held in Miramon Kutxaespacio de la Ciencia (Donostia-San Sebastian) on 18–19 November 2004.

The workshop was split into three working groups: Press and Media (coordinated by A. Pérez), Education and Research (coordinated by M. Lezaun), and Museums, Publishing companies and Magazines (coordinated by R.M. Ros). The reflections and conclusions of the workshop were collected in the book *Divulgar las Matemáticas* [3].

Collaboration with the Media   In this workshop some future actions were considered, in collaboration between mathematicians and the media. It was decided that it was important to establish a press office for the Spanish mathematical community (an initiative that has now picked up the "i-math project"), it was important to establish specific collaborations with the media (interviews, scientific advice, contacts, articles, . . .), that if there was any news related to mathematics, we would have to make the effort to inform the media and provide information, and finally, to establish more permanent partnerships. An example is the collaboration that was developed by the author of this article, R. Ibáñez: on the program *Graffiti* of Radio Euskadi, since July 2005, every Wednesday from 18:40 to 19:00 we talk about mathematics (to a large audience) and simple problems are posed for listeners. There has been a good reception from the public.

Lectures

Very important tools, which are known as being useful in raising the public awareness of mathematics, are the series of conferences, seminars, courses and scientific conferences for other mathematicians and scientists, teachers, students, or for the public in general. One example is the series of conferences "Mathematics in everyday life" held in a public library (Bidebarrieta, Biblioteca Municipal de Bilbao) (Fig. 3). It is an open series of talks for the general public, organized since 2004, and with between 130 and 150 people per conference (even more than 200 in some talks). Some examples of conferences are:

2006: "Challenger would not have launched if NASA had known statistics", J.A. Cuesta, "Life is a lottery, gambling and financial management", J.L. Fernández, "Look at art with mathematical eyes", F. Martin;
2007: "From the secret messages to security in our daily lives", P. Morillo, "Origami and Mathematics", J.I. Royo, "The secret of Google", P. Fernández;
2008: "Mathematics for magic tricks", P. Alegría, "My Pentagon of Beauty", R. Pérez Gómez, "A mathematical look at elections", A. Quirós; and so on.

Other examples are the series of conferences for university students "A walk on geometry", organized in the Universidad del País Vasco since 1997, and a series of conversations between mathematicians and writers, held in a public library (Bidebarrieta), as for example between Michele Audin, a French mathematician belonging to Oulipo, and Antonio Altarriba, writer and expert on Oulipo.

RSME–ANAYA Literature Contests: School Stories and Short Stories

These contests have been organized by the RSME in collaboration with the publishing houses ANAYA, Nivola, elrompecabezas, and Projecto Sur, with the philosophy

Fig. 3 Publicity the series of conferences "Mathematics in everyday life"

of showing students, teachers and the general public that mathematics is part of human culture. There is no separation between the humanities and science in culture.

School Stories The aim of this competition, for young people between 12 and 18, is the dissemination of mathematical culture to them (and by extension to society), thus promoting their interest in this science, its history, and its protagonists. The competition consists in the creation of a fictional story based on a mathematical result, a character related to this science, or a situation where mathematics emerges … but through the critical eye and the imagination of the author. This activity is an interesting teaching tool that has managed to reach young people, and also to involve teachers. In 2006 there were 328 stories, 591 in 2007, 490 in 2008, 477 in 2009, and 850 in 2010.

Short Stories Participants submit a short story, on any theme related to mathematics. In 2006 there were 31 stories, 91 in 2007, 59 in 2008, 46 in 2009, and 67 in

2010. The authors are from all parts of Spain, but also Argentina, Chile, Colombia, Cuba, USA, France, Guatemala, Mexico and Uruguay. The jury, in both competitions, consists of mathematicians, writers, journalists, ...

The quality of the work, in both competitions, encouraged the RSME and ANAYA publishing house to publish the finalists and winners of the literary contests in two collections under the titles "Mathematical fiction" (school stories) and "Math Stories" (short stories), see Fig. 4 [4, 5].

Expositions

The use of art exhibitions in the popularization of mathematics shows this science as a part of culture, but also contributes to the normalization of the relationship of people with math and encourages them to approach mathematics. Exhibitions are also a tool for the popularization and education of mathematics.

The Frontier Between Art and Mathematics (2003)

This is an international exposition of fractal art organized for the WMY2000 that has visited various cities in the world, and it was transferred to RSME in 2003. In the same way that painters expose their personality and sensitivity in their works through their techniques, fractal artists express themselves through formulas and algorithms (fractals), gradually changing them to achieve the desired objective, and this type of work lies on the border between art and mathematics. The exposition has visited science museums, exhibition halls, universities, ... in Spain. It is fundamentally an art exhibition that invites people to enjoy the beauty of the work, which also allows us to introduce visitors to the mathematics of fractals and to many of their applications (computing, medical, telecommunications, engineering, economics, acoustic, ...).

Anda con ojo, an Exhibition of Math Photography by Pilar Moreno (2005)

Pilar Moreno has made photography a tool for the dissemination and teaching of mathematics. For twenty years she has used photography as a teaching resource in the classroom. As well as showing everyday images full of beauty and tenderness, her photographs are always related to geometry. Geometry is everywhere, but it is the photography that selects and isolates the elements so that they can be seen by the public (Fig. 5). The exhibition—organized and funded by the RSME—has visited science museums, exhibition halls, universities, ... in Spain, with a surprising success wherever it goes.

(a)

Fig. 4 Title page of the books **a** *Un Teorema en la Biblioteca* from the collection "Math Stories" and **b** *Fragmentos de una Realidad Imaginada* from the collection "Mathematical fiction" (ANAYA-RSME)

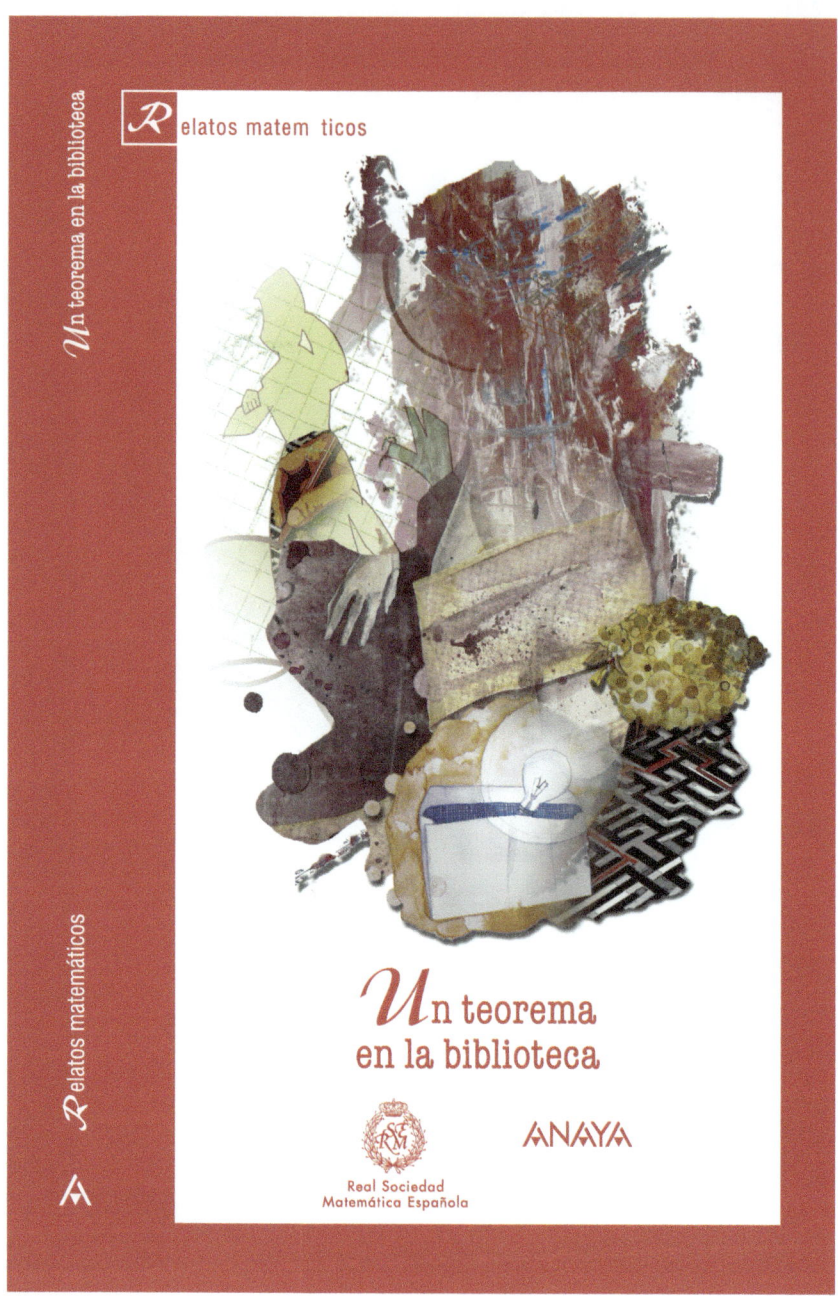

(b)

Fig. 4 (Continued)

Fig. 5 Photograph by Pilar Moreno

The Human Face of Mathematics

This is an exposition of cartoons of mathematicians (by the artists E. Morente and G. Basabe of Viñaspre) with a short biography of each mathematician (Fig. 6) [6]. The aim of this exhibition is to bring the history of mathematics, and in particular the major figures in that history, closer to youth, students and their parents and teachers, and society in general. This exhibition, funded by the FECyT in the Spanish Year of Science 2007, has several formats: the original exposition was for museums and galleries, several flexible copies for schools (secondary schools and universities) and a virtual exposition in the webpage divulgamat. The team is a highly experienced group of mathematicians: R. Ibáñez, S. Fernández, P.M. González, V. Meavilla, Fco J. Peralta, A. Pérez, and A. Salvador. The featured mathematicians are: Pythagoras, Euclid, Archimedes, Apollonius, Al-Khwarizmi, Fibonacci, Cardano and Tartaglia, Fermat, Descartes, Newton, Lagrange, Cauchy, Galois, Abel, Leibniz, Euler, Gauss, Riemann, Hilbert, and Poincaré. But also some women: Hypatia, Madame du Châtelet, Sophie Germain, Sonia Kovaleskaia, and Emmy Noether. Even some Spanish mathematicians: Julio Rey Pastor, Luis Santaló, Puig Adam, Miguel de Guzmán, and Ventura Reyes Prosper. Nivola has published a book called "The Human Face of Mathematics".

Fig. 6 *Top*: "The Human Face of Mathematics" in Casa das Ciencias (A Coruña). *Bottom*: A cartoon of Gauss (by E. Morente)

Collaboration: International Congress of Mathematicians, Madrid 2006

Another important date for raising the public awareness of mathematics was 2006, the year of the International Congress of Mathematicians (ICM) [7] in Madrid (Spain). This event, the most important in the world of mathematics, has been held every four years since 1897 and it was celebrated in Spain for the first time. The organizers of ICM2006 decided to make this event not only an event for the world of mathematics, but for the whole of society. For this reason we made a special effort in organizing cultural activities related to mathematics.

One such activity was the triple exhibition, Fig. 7, held at the Centro Cultural Conde Duque in Madrid (August 17 to October 29, 2006): "Experiencing mathematics?", "Fractal Art: Beauty and Mathematics", and "Demoscene: Mathematics in motion" (the latter two were also held at the congress). The exposition was visited by over 40,000 people. More than 100 schools made guided visits. The presence of the exposition in the media was important and included coverage in newspapers, radio, and television. The commissioners of the exhibition were R. Ibáñez and A. Pérez.

The exposition "Experiencing mathematics?" has its origins in the actions of the IMU in the year 2000 to celebrate the World Mathematical Year and falls within the framework of the cultural and scientific missions of UNESCO. This exhibition aims to show that mathematics is amazing, interesting, and useful. Mathematics has a major in everyday life, and is very important in our culture, development, and progress. This exposition is for the general public ... but was particularly created for young audiences and the people around them: parents and teachers. It is an interactive exhibition ... an exhibition to see, watch, play, experiment, and think. Lemma: "please touch!"

The exposition was accompanied by an exhibition of documentary videos, including the TVE series *More for Less* and *Math Universe* (screenplay and presentation by Antonio Pérez Sanz). Also, we prepared a catalog (see divulgamat), with texts developing some of the themes of the exhibition, and an activity book for high school students. This exhibition was organized with the support of the Ministry of Education.

The second exhibition in Centro Cultural Conde Duque was a fractal art exhibition, entitled "Fractal Art: Beauty and Mathematics" (Fig. 8). The paintings were by the twenty-five finalists for the international contest during the celebration of the ICM 2006 in Madrid. The president of the jury was B. Mandelbrot. The exhibition was also accompanied by a very visual catalog (see divulgamat). The third part of this triple exposition was "Demoscene: Mathematics in motion", which included a selection of animated computer graphics. Demoscene is a powerful source of mathematical algorithms creating graphics and visual effects, such as the digital special effects used in movies and video games. To complement this exhibition, demoscene artists explained their art to the public. These last two presentations were funded by FECyT.

Fig. 7 Poster for the triple exposition held at the Centro Cultural Conde Duque in Madrid in 2006, during the ICM

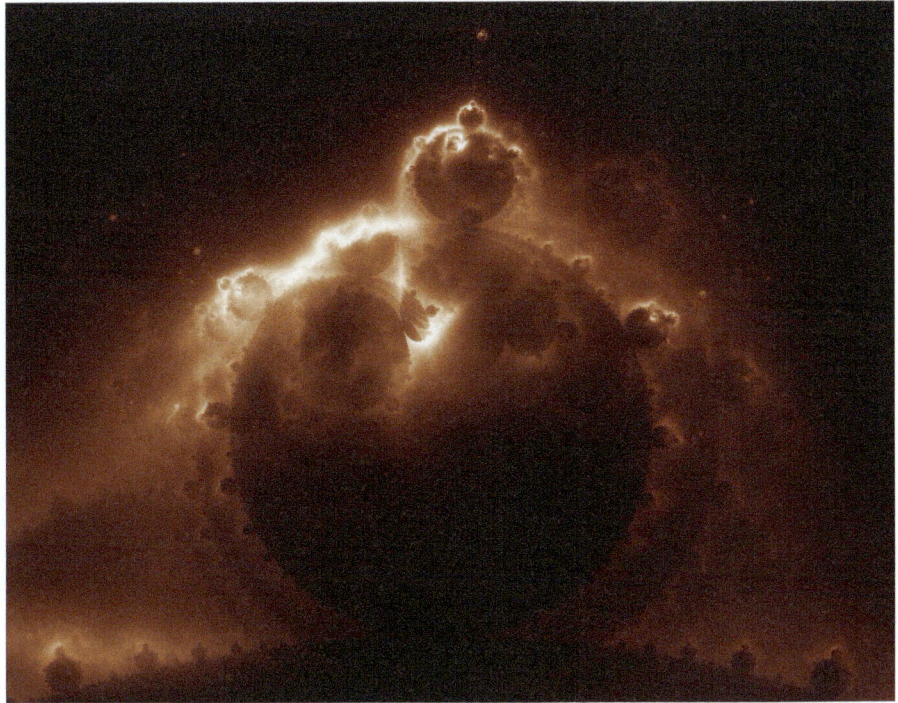

Fig. 8 Painting from the exhibition "Fractal Art: Beauty and Mathematics", *Warmglow* by Kerry Mitchell

The "Fractal Art: Beauty and Mathematics" exposition has continued to tour Spain, being shown in different exhibition halls and universities, even in the Bilbao underground.

Other expositions were "The Life of Numbers" in the Spanish National Library, "History of Mathematical Knowledge" in the Library of the Universidad Complutense de Madrid, and the well-known Japanese sculptor Keizo Ushio placed a geometrical sculpture facing the public in front of the congress site (Fig. 9).

Centennial of the Royal Spanish Mathematical Society (1911–2011)

Another important date was the year 2011 when the Royal Spanish Mathematical Society celebrated its centenary [8]. It is as a modern and active society, which works to support Spanish mathematicians in their efforts to improve research, education at all levels, the range of applications, public esteem, and recognition from institutions. For this reason it is active in research, education, and the popularization of mathematics.

Fig. 9 Sculpture by Keizo Ushio at the International Congress of Mathematicians held in Madrid in 2006

An important activity is the exposition "RSME-Imaginary" (Fig. 10) [9]. This is a joint exhibition with the Mathematisches Forschungsinstitut Oberwolfach. It consists of a permanent section in the Cosmo Caixa Museums of Barcelona and Madrid, which contains an interactive workshop and a part of the collection, and the original traveling exhibition, which has visited Salamanca, Valladolid, Palma de Mallorca, Bilbao, Pamplona, Gijón, Zaragoza, Madrid, Seville, Valencia, Malaga, La Laguna, and Barcelona. The exhibition is accompanied by guided visits, didactic guides, surfer contests, and other activities. More information can be found at www.rsme-imaginary.es.

Another activity was the Colloquia of the Centennial comprising ten talks in faculties, institutes of mathematics, and public libraries throughout 2011 including John D Barrow on "Maths is everywhere" (Zaragoza), Alfio Quarteroni on "Mathematical modeling in medicine, sports and the environment" (A Coruña) and Michèle Audin on "The two ideas of Sofia Kovalevskaya" (Bilbao).

Program "BBK-matics—Mathematics in the School Libraries"

This is an exciting project in which we are currently working with the Basque Government and the BBK Foundation (https://portal.bbk.es/bbkportal/memoriabbk2010). Our objective is to show mathematics in action in school libraries. As already discussed, math is part of our culture, but also important in

Fig. 10 RSME-Imaginary in Bilbao (May 2011)

Fig. 11 Math storytelling with E Morente (illustrator) and R Ibáñez (mathematician and story-teller)

schools, in the education of young people. The aim of this program is to include mathematics as an important element of the extracurricular activities of schools and to use literature related to mathematics in the education of our youth. The program is aimed primarily at students in primary and secondary schools.

The program "BBK-matics—Mathematics in the School Libraries" is held in 85 schools in Vizcaya. Each year this program delivers many books for children relating to mathematics to each participating school. This program also has math storytelling (Fig. 11) and conversations with writers; loads of puzzles, accompanied with didactic guides; workshops on puzzles, mathematical magic or origami (for teachers, librarians and students); and lectures for young people. There are micro-exhibitions , which go from school to school: (i) "Numbers in our cities" (photographer and designer: Axi Olano); (ii) "Daily Geometry daily" (photographer: Pilar Moreno), (iii) "Human face of mathematics", etc.

References

1. RSME, web page of the Royal Spanish Mathematical Society, www.rsme.es
2. DivulgaMAT, Virtual Center for the Popularization of Mathematics, www.divulgamat.net
3. Ibáñez, R. (editor-coordinator), Divulgar las matemáticas, Nivola (2005)
4. Alegría, P., Ibáñez, R., Pérez, A. (coordinators of the project), "Math Stories" collection, ANAYA-RSME (2007–2011): Sobre letras y números, Un teorema en la biblioteca, Todo por demostrar, La conjetura de Borges
5. Alegría, P., Ibáñez, R., Pérez, A. (coordinators of the project), "Mathematical fiction" collection, ANAYA-RSME, (2007–2011): Entre lo real y lo imaginario, Fragmentos de una realidad imaginada, Ensoñaciones desde mi pupitre, El despertar de una ecuación
6. Ibáñez, R., Pérez, A. (coordinators of the project), Human face of mathematics, Nivola (2008)
7. Curbera, G.: Mathematicians of the world, unite!, A.K. Peters (2009)
8. Web page of the Centennial of the Royal Spanish Mathematical Society, http://www.rsme.es/centenario/
9. Web page of the exhibition Imaginary, http://www.imaginary-exhibition.com

Part II
Exhibitions and Mathematical Museums

Mathema—Is Mathematics the Language of Nature?

Ehrhard Behrends

Abstract One of the projects in the "Year of Mathematics 2008" was a large exhibition on mathematics in the Deutsches Technikmuseum in Berlin (DTMB): "Mathema—Is Mathematics the Language of Nature". In this article we describe the intention of the organizers and the structure of this exhibition.

The Prehistory of "Mathema"

It has been known for many years that mathematical exhibitions are very appropriate for raising public awareness. In Germany the exhibitions of A. Beutelspacher are the most well-known. He started with a traveling exhibition created together with his students from Gießen university, which was widely shown. His contribution to the "public" events during the ICM 98 in Berlin was a big success. The gigantic soap bubbles and other interesting hands-on exhibits attracted not only the public but also the media. Beutelspacher's efforts culminated in the realization of the "Mathematikum" in Gießen, a science museum dedicated solely to mathematics. (For more details see Beutelspacher's article in this book.)

But mathematical exhibits are not only presented in Gießen, there is, e.g., a mathematical section in the huge science museum "Deutsches Museum" in Munich. And many colleagues supplement special events for school children or the general public by presenting posters and hands-ons of various complexity to demonstrate mathematical facts and the beauty hidden behind mathematical theories: brachistochrones, mirrors, tessellations, etc.

The idea of a large mathematical exhibition in the Berlin Technical Museum arose in the middle of the last decade. Hadwig Dorsch, custodian at this museum with some experience in computer science exhibitions, contacted the author of this article with this proposal. She had visited some of my projects "Mathematik für alle Sinne" that had been organized at the "Long Night of the Sciences" where

E. Behrends (✉)
Freie Universität Berlin, Berlin, Germany
e-mail: behrends@math.fu-berlin.de

E. Behrends et al. (eds.), *Raising Public Awareness of Mathematics*,
DOI 10.1007/978-3-642-25710-0_8, © Springer-Verlag Berlin Heidelberg 2012

Fig. 1 The logo of
"Mathema"

mathematical posters and various mathematical exhibits had been shown. She had
also read the articles in my weekly column "Fünf Minuten Mathematik".[1]

Our Intention

As soon as we had decided to try to start this project some initial major questions
had to be answered: Who would be the "typical" visitors (school children, parents
with their children, a general audience interested in mathematics, . . .)? Would the
emphasis be on "recreational" mathematics? Or were we going to convince people
that mathematics is really useful?

After some round-table discussions with experts from the museum, professional
mathematicians, students, teachers and school children it was clear to us that the
most attractive challenge would be to combine two aspects, namely:

1. Mathematics is useful
2. Without mathematics it is impossible to comprehend our current understanding
   of the world

It was not difficult to find an appropriate motto, namely Galileo's "The book of
nature is written in the language of mathematics."[2]

We were advised to complement this motto to indicate the connection with mathematics, and in this way we arrived at "Mathema—Ist Mathematik die Sprache der
Natur?" ("Mathema—Is Mathematics the Language of Nature?"). Here "Mathema"
was created as a fantasy word with no German equivalent, in fact it is the name of
a logistic enterprise, who gave us permission to use it. The logo of the exhibition is
shown in Fig. 1.

But how could this be achieved? Our intention was to offer interesting information to various groups of visitors: some spectacular items, many hands-on exhibits

---

[1]This column was published between 2003 and 2005 in the newspapers "Die Welt" and "Berliner
Morgenpost." In the meantime these articles appeared as a book ("Fünf Minuten Mathematik,"
Vieweg 2005), and Japanese, English and French translations are now also available.

[2]Of course we were aware that this quote is only a short version of Galileo's original formulation
that appeared 1610 in his *Sidereus Nuntius: "Philosophy is written in this grand book—I mean
the universe—which stands continually open to our gaze, but it cannot be understood unless one
first learns to comprehend the language in which it is written. It is written in the language of
mathematics, and its characters are triangles, circles, and other geometric figures, without which
it is humanly impossible to understand a single word of it; without these, one is wandering about
in a dark labyrinth."*

Fig. 2 The plan of "Mathema"

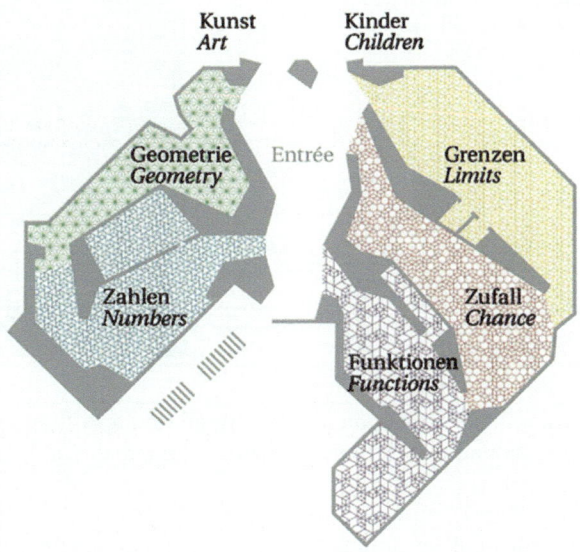

to illustrate mathematical facts, real historical exhibits (old mathematical games, encryption machines, ...) and also a survey of "mathematics through the centuries". The most important decision, however, was to use every day experience as the system for the presentation of the material. Each of us "knows" what "Zahl" (number), "Raum" (space), "Bewegung" (motion) and "Zufall" (chance) are, and these aspects of mathematics were chosen as the names of the main sections of "Mathema." Since we also had in mind to touch the "philosophical" aspects we added the section "Grenzenlose Mathematik?" (Mathematics without borders?) so that we arrived at five sections that were called "islands". In addition we wanted to have an entrance area where the visitors could themselves tune into the exhibition.

The Structure of "Mathema"

We had at our disposal an area of roughly 1000 square meters and additionally some hundred square meters for "Mathematics and Art" and mathematics for very young children. The latter was called "Mathemachen", a play of words oscillating between "Mathe machen" meaning "to make mathematics" and "small Mathema." It attracted many groups of children and it was overbooked for the whole duration of the exhibition.

We decided to partition the area as shown in Fig. 2.

Below we describe the most important ideas for the five islands and the foyer that were finally created.

The entrance area   This area prepares the visitors for the rest of the exhibition. We had in mind visual information (a light show, mathematical art prepared by a

Fig. 3 Two Vietnamese girls counting in their language from one to ten

professional artist) and audio material (quotes of famous people concerning the role of mathematics: Einstein, Frederick the Great, Galileo, Wigner, etc.).

The island "number"   As in all of our islands a "Zeitband" ("timeline") would illustrate the development of this part of mathematics over the centuries. Not surprisingly, the island "number" was the record holder: no other time line started earlier. We wanted to present the following aspects of "number":

- Mathematics is universal (I): A film where people from different countries count from one to ten in their language (see Fig. 3). This should "prove" that elementary mathematics is indispensible in every culture.
- Cryptography: Elementary and not so elementary mathematics is used to encrypt messages from the Caesar code to the Enigma machine to the RSA algorithm.
- Mathematics is for everyone: In the 16th century the German "Rechenmeister" Adam Ries wrote books that enabled everyone to master elementary mathematical calculations. Before this time calculating had been the domain of a few specialists, but by this "Rechnen auf den Linien" even school children could manage addition and multiplication. A special room would remind the visitor that basic mathematical knowledge for every member of society is a rather recent invention. See Fig. 4.
- Mathematics and music. A "red thread" of "Mathema" is to communicate that mathematics can be found in many places in daily life, also in unexpected situations. The visitor should be able to learn that the music that surrounds us has a mathematical background. The Pythagorean scale and the well-tempered scale sound much better than a random scale: what is the reason? What is the difficulty with the Pythagorean comma, and why on earth does $\sqrt[12]{2}$ have an important role?
- The island "numbers" would be incomplete without the golden ratio. In fact it can be found throughout in nature and in architecture, and visitors were invited to find this ratio at their own body by using a "golden circle".
- And where were the "museum exhibits"? Some of them were in their proper place in the timeline, but we also wanted to present more important objects: a real "Enigma" encryption machine, early "computers" like the one designed by Leibniz (a replica), abaci from various countries, etc.

Fig. 4 Demonstration of "Calculating on the Lines". (This man, incidentally, is a descendant of Adam Ries.)

The island "space"   The fact that the shortest distance between two points is a line is known even to animals. Elementary geometry was used by ancient cultures several thousand years ago to "measure the world" and applications ranged from agriculture to navigation to astronomy. And geometry saw around 500 BC the first mathematical proofs. Consequently our "timeline" started very early in history. Here are the most important exhibits:

- How can we "measure the world"? A film explained Eratosthenes' method of calculating the circumference of the earth, we showed nautical instruments from antiquity to the present, the GPS method was explained, etc.
- Mathematicians are able to deal with more than three dimensions. We presented a model of a hypercube (see Fig. 5) together with a film to explain how mathematicians try to achieve this.
- Mathematics is everywhere. At this island we also presented the connection between mathematics and art. In particular in the work of the Dutch artist Maurits Cornelis Escher we find numerous examples. For each of the 17 crystallographic groups there are paintings in his oeuvre with precisely this symmetry, and hands-on exhibits that visualize some of them were available for the visitors.
- How can we depict the surface of the earth on a plane sheet of paper? We indicated how this problem was solved in different ways by cartographers.

The island "motion"   Everybody "knows" what motion is, but a satisfactory mathematical description is not simple. Some of the more subtle facts were presented on the timeline, e.g., the history of infinitesimal calculus and the priority controversy between Leibniz and Newton. For the exhibits we had the following ideas:

- The concept of a function is fundamental, but it is far from easy to find an attractive way to explain this concept in a mathematical exhibition. We adopted an idea from the "Mathematikum" in Gießen: A visitor sees a certain function on a screen and has to try to walk in such a way that the path-time diagram of the walk corresponds to the function as precisely as possible.

Fig. 5 The hypercube

- The brachistochrone is more or less a must for a mathematical exhibition. This curve was important in the history of calculus, and investigations gave rise to Euler's theory of variation. The visitors could learn that this curve allows a faster motion than the straight line, but our hands-on exhibit could also be used to demonstrate another property: the brachistochrone is a tautochrone, that is, the time to travel from any point to the lowest point is the same for all starting positions.[3]
- And again "mathematics everywhere": here the visitor can investigate phenomena related to Fourier analysis.
- The "flight over Vienna" was very spectacular. Standing in front of a huge screen, with strapped on wings, the visitor has the impression of flying over the center of Vienna. This was made possible with the help of a fast tracking system. Many visitors, in particular the young ones, were fascinated. (It has to be admitted, however, that there is no relation with more sophisticated mathematics.)

The island "chance"   This was my favorite island. In other places I also liked to explain how mathematicians deal with chance. Since stochastics is a part of mathematics only since a couple of centuries the timeline was rather short in this case. As exhibits we had prepared the following:

- A chaos pendulum, it was placed at the border between "motion" and "chance". In fact the motion is deterministic, but due to the lack of complete information it appears to be completely random.
- Galton's board. This should convince everyone that randomness disappears somehow when many stochastic events are superimposed.
- A large roulette table with a real roulette bowl. Two people could play "virtual" roulette, and it was explained why a winning strategy is impossible. There is only the option of distributing the probabilities between the gains and losses: a small probability for a high gain and a high probability for a small loss? Or the other way round?

---

[3]It should be noted that even some professional mathematicians who visited "Mathema" did not know this before.

Fig. 6 The birthday paradox

- Lotto. For most people it is hard to imagine the small probabilities associated with winning the jackpot. (In fact I have to explain this over and over again for the TV if a jackpot is considerably high.) We decided to produce a film to illustrate the tiny chance of becoming rich this way.
- Random compositions. There exists a "dice composition" that is due to Mozart. It works as follows. Throw two dice, and according to the sum of the top faces choose one bar from the supply "bar 1": this is a set of 11 bars of music labeled 2, ..., 12. Do the same for bar 2, there is another supply. Continue in this way until there is a little piece of music of 16 bars. In the original version this was to be played on the piano, here we have prepared an electronic version where the choice (with real dice) is transformed first to a Midi file and then made audible (for electronic piano, harpsichord or guitar, as you like it).
- Simulated annealing. Choose a number of cities in Germany (or England or France) using a touchscreen. Then a random algorithm will quickly find a shortest path that goes to each city exactly once using a simulated annealing algorithm. (The complete truth is, of course, that the real optimum is found only to a high probability.)
- Paradoxes. Many facts from stochastics are counter-intuitive. We presented the birthday paradox in the following way. Consider the 23 people from two soccer teams and the referee (see Fig. 6). Do two of these people have their birthday on the same day of the year? Theory says that this should occur with more than 50 % probability. And, in fact, when we investigated the soccer games in the World Cup 2006 there was a "double birthday" in 16 out of 31 games. This was nearly too perfect ...

The island "mathematics without borders?"   This island was a little bit different in that it has no direct connection with daily life. But we wanted to demonstrate that there are also "philosophical" aspects of mathematics. Here are a selection of our exhibits:

- In the timeline we mentioned that mathematics as a science started when people first became aware that "proofs" are necessary. And a proof is much more than the demonstration of a fact by using a special example.[4]
- The "flood of rice." In a well-known problem a chessboard has to be covered with rice as follows: put one grain of rice on the first square, two on the next, then four, then eight, etc. And the question is: How much rice will be needed to reach the last square? We used this problem to indicate that our brain is unable to grasp exponential growth. In fact, the amount of rice would suffice to cover Germany with a layer of more than 50 centimeter.
- The most beautiful formula. This is, as every mathematician knows, Euler's $0 = 1 + e^{i\pi}$. This equation stands for a rather mysterious connection between different branches of mathematics. We consider it as so important here that we have a model of it of about two meters high.
- The limitations of mathematics. For tis we prepared listening stations where the visitor could hear the famous Königsberg speech of Hilbert's ("We must know, and we will know!") and an essay on Gödel's theorem.
- Mathematical records. Mathematical records are hard to imagine. Consider, for example, the fact that the largest known prime number has 10 million digits. What does this mean? It means that when you print all the digits you obtain three books. We printed these three books with the record prime number from 2008, and even for some mathematicians it came as a surprise how large this number really is. We also did it with π where the situation is much more spectacular. You could print several thousand books (a whole library!) using only the digits of π known today. We chose to print only the first two volumes of this π-book (see Fig. 7) ...
- Mathematics is universal (II). When presenting the concept of "number" we showed that everyone in the world uses some—at least elementary—mathematics. Here we presented the universal version of this idea: Since it is hard to doubt that mathematics is not the same in all parts of the universe the idea arose in the last century to use mathematical concepts to communicate with intelligent beings far away in other galaxies. In "Mathema" the visitor can listen to the *Arecibo message* (see Fig. 8). The message is a series of pulses, the number of which is a product of two primes, and one hopes that clever civilizations will be able to decrypt it.

The Realization

As often in life an Italian proverb applies: "Tra il dire e il fare cé di mezzo il mare," and it took a lot of time to transform the first ideas into a real exhibition. First of

---

[4]And we stressed that the usual exhibit in science museums "demonstrating" the Pythagorean theorem—water that fills a container of volume c^2 also fills two containers of volumes a^2 and b^2, respectively—in fact doesn't prove anything. This criticism was complemented by a rigorous proof.

Fig. 7 The prime number record holder and the π-book

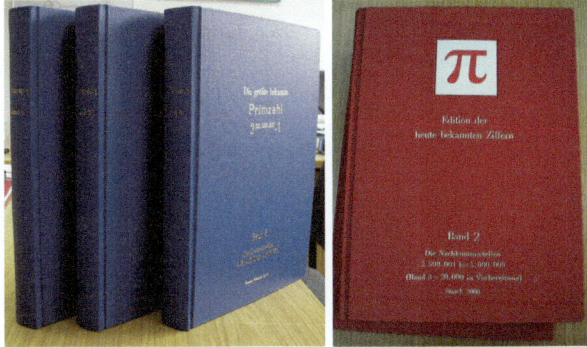

Fig. 8 The Arecibo message

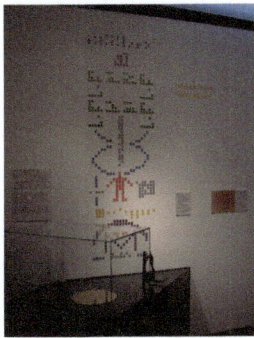

all it was necessary to raise the money. We had in mind a budget of several hundred thousand euros, and there are only a few organizations in Germany where one could hope to acquire funds in this range. A considerable sum was provided by the technical museum, but there remained a large gap. Fortunately it was decided in 2007 that 2008 would be the "Year of Mathematics" in Germany. This was extremely helpful, and rather soon the financial gap was closed, essentially by a large amount from the Telekom Foundation, a foundation that has generously supported mathematical projects for some years.

The next step was to find agencies to prepare the graphics and to take care of the concrete realization. We chose to work with the agency Rimini (graphics) and the agency Schiel (museum architecture), and we were very satisfied with the collaboration.

Let me skip here the many technical problems that had to be overcome: where could we find the museum exhibits (an enigma machine, an abacus, etc.), who could calculate the precise form of a brachistochrone for the carpenter, who could provide the English translations of the texts, who could write a program that produces audible music, etc.

I myself was involved in various ways: providing ideas, writing computer programs, producing little films with the help of my students, negotiating with our sponsors, etc.

Fig. 9 The opening ceremony. *From left*: H. Dorsch, the author, A. Schavan (minister of education), D. Böndel (director of DTMB), K. Kinkel (president of the Telekom foundation), and G. Ziegler (president of the DMV, the main organizer of the "Year of Mathematics 2008"). On the wall behind the group: part of the most beautiful formula $0 = 1 + e^{i\pi}$

Finally everything came to a successful conclusion, many people worked very hard to have everything prepared for November 5, 2008, the day of the official opening ceremony (see Fig. 9).

My Favorite Exhibits

Here I will describe some of my favorite exhibits in more detail.

Mathematics and music: scales Suppose you want to design a musical scale using a stochastic procedure: choose 10 random points in the interval from 1 to 2 and the resulting subintervals correspond to the semitones between the root and the octave. This will produce terrible sounds, we demonstrated this by playing "Happy birthday to you" using such a random scale.

As is wellknown the Pythagoreans had the idea of using proportions when designing a scale. Two tones played simultaneously are pleasant to our ears if and only if the ratio of the lengths is a rational number with a small nominator and denominator. Their scale, the Pythagorean scale, is based on the ratio 3:5, that is, it is constructed by using only perfect fifths. And "Happy birthday to you" in this scale sounds much better.

There is, however, the problem with the Pythagorean comma that sounds really ugly (it's audible). But the invention of well-tempered scales overcame this difficulty, and the song using a well-tempered scale sounds similarly nice. Only a few visitors will be able to hear a difference when compared with the Pythagorean scale.

The fourth dimension How can one visualize the fourth dimension? We produced a little film featuring two students. First they describe how a one-dimensional creature could understand the one-dimensional boundary of a square. Let the creature move along a side of finite length and identify the end points: leaving this side from the right means entering an adjacent side from the left. Then this creature will learn that this space has no boundaries and that the length is finite.

Next we switch to two-dimensional beings moving over the surface of a cube. It is obvious—even if the brain thinks only two-dimensionally—that there are no borders and that the area is finite.

And the last step is to explain the hypercube. After suitable identifications one "sees" that the three-dimensional volume of the boundary of the four-dimensional cube is finite and that nevertheless one can move without ever reaching a border.

Fourier analysis   Our exhibit has several parts. There was, e.g., a screen where the visitor could see the waves associated with various sounds: an orchestra, a guitar concert, a bird, their own voice, a sine wave, a square wave (the latter with various frequencies).

The most important part was an experiment. A square wave of frequency ω is the sum of the sine wave of the same frequency, plus one third of a sine wave of frequency 3ω, plus one fifth of a sine wave of frequency 5ω, etc.

And now the question: what is the highest frequency such that a sine wave can be distinguished from a square wave? The correct answer is one third of the highest frequency ω_0 that you can hear. Different people have different ω_0, it varies considerably. But in any case the experiment was in remarkable agreement with the result predicted by Fourier analysis.

The balance of luck   We invented a new visualization to show to what extent a gambler can influence the long-term outcomes of his games when playing a fair game: the balance of luck, see Fig. 10.

A gambler has at his or her disposal weights that sum up to 100 (this corresponds to 100 %). These weights can be distributed arbitrarily on a balance, the only condition is that the balance is in equilibrium: weights on the right side correspond to possible winnings and those on the left side stand for loses. And then there are strategies such that the winning and losing chances correspond precisely to the chosen distribution. If, e.g., there is a weight of 25 at position 500 to the right then the probability of winning 500 euro is 25 %.

Various strategies are possible, but winning strategies—those where the expectation is strictly positive—cannot be found.[5]

The flood of rice   It is well known that we have no intuition for really large numbers. Such giants sometimes occur in mathematics when the complexity of a problem increases exponentially fast with the length of the input. A famous example is factoring a huge integer; the security of some encryption methods is based on the fact that (at present) factoring is a really hard problem. Also people encounter this phenomenon when joining a pyramid system or when hoping to become incredibly rich by writing chain letters.

---

[5]The first time that the author became involved in the discussion of such questions was in 2005 when during involved in a popular TV show someone claimed to have a perfect winning strategy. In fact, when demonstrating his method, the person won a tiny amount of money several times, and the speculation was that a big loss would occur only when no TV camera was around.

Fig. 10 The balance of luck

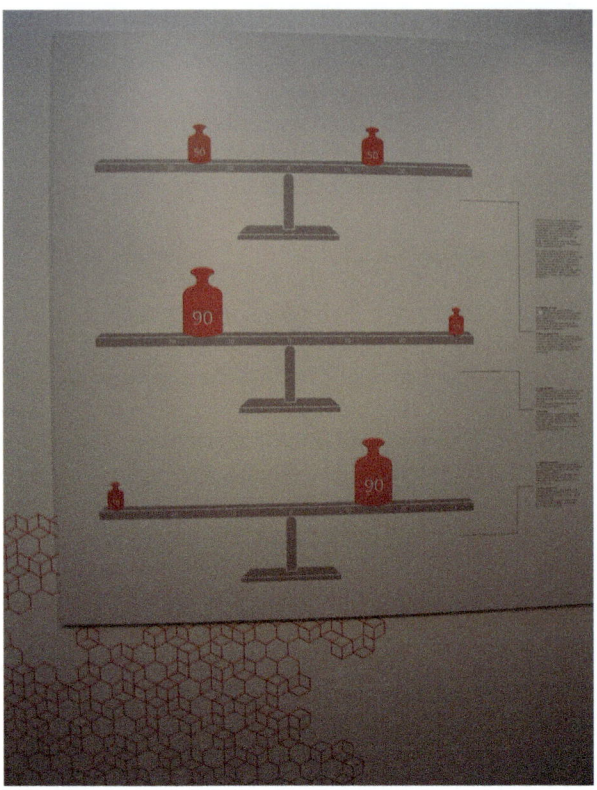

We prepared two visualization of the problem of the grain of rice. First, we took a large chess board (two by two meters) and placed one grain on the first square, then two, four, eight, etc. By the third row the amount of rice was very impressive, and it was clear that it would not be possible to continue with this procedure much longer.

But we also made a film using photomontages:[6] it started rather innocently with an ordinary chess board, but soon it had to move to a much larger chessboard in a park. A few squares later this board also was too small, and it was rather impressive when Germany is covered by a layer of half a meter of rice when the procedure ends at square 64.

The visitors liked this film very much. Several times I listened to discussions at the table next to mine in the cafeteria where a father or a mother explained to the children the mathematical background of this remarkable phenomenon.

---

[6] You can see this film on *youtube*: http://www.youtube.com/watch?v=KnQZ3Mg6upg. Free downloads are available at http://page.mi.fu-berlin.de/behrends/mie/reispotenzen_klein.wmv (low resolution, 3.7 MB) and http://page.mi.fu-berlin.de/behrends/mie/reispotenzen_gross.wmv (high resolution, 24 MB).

"Mathema": The Most Successful Temporary Exhibition in the DTMB

We, the organizers, were very satisfied with the success of "Mathema". It was shown for nine months in the technical museum, and it attracted about 150,000 visitors. Thus it was the most successful temporary exhibition ever shown in the DTMB. Many visitors asked why it could not be shown for a longer period, but this was not possible because of an exhibition on Max Planck starting in the autumn of 2009.

But recently, at the end of 2010, it was decided that the DTMB will create in the near future a permanent mathematical exhibition based on "Mathema."

Lessons Which Can Be Learned from the Mathematikum

Albrecht Beutelspacher

Abstract Mathematikum is the Germany's first mathematical science center. The article describes how it came into existence, how it is now, and what its characteristics are. Furthermore, some rather detailed and concrete advise for establishing and running such a museum are given. Finally, a few typical hands-on exhibits of the Mathematikum are described.

The History of the Mathematikum: How It Came into Existence

The Mathematikum in Giessen (near Frankfurt) is a mathematical science center, which was founded in 2002. It aims to make mathematics accessible to as many people as possible, in particular to young people.

It is based on my work with trainee teachers at the University of Giessen. The history of the Mathematikum is as follows.

- In 1993 I announced a special kind of proseminar for first and second year students. Instead of reading mathematical texts and trying to understand and present them, each student of this seminar had two tasks: They should construct a real model and they should present the "mathematics behind it". After a few problems the students made wonderful models and gave good explanations.
- The models were so brilliant that I thought that it was a pity that only the students and I would see them. So I convinced some of the students to work with me on an exhibition based on their models. This exhibition under the title "Hands-on Mathematics" ("Mathematik zum Anfassen") took place in a small lecture room at the university's campus and was a great success, in particular since we invited school classes to visit it.
- Somehow information about this exhibition spread very fast. Immediately after our exhibition I was asked by colleagues whether we could show the exhibition at their institution. So it became a traveling exhibition. This meant in particular that instead of static models (as, for instance, Platonic solids) we developed and

A. Beutelspacher (✉)
Universität Giessen, Giessen, Germany
e-mail: beutelspacher@math.uni-giessen.de

E. Behrends et al. (eds.), *Raising Public Awareness of Mathematics*,
DOI 10.1007/978-3-642-25710-0_9, © Springer-Verlag Berlin Heidelberg 2012

constructed more and more interactive experiments. A highlight was the International Congress of Mathematics 1998 in Berlin where we could show our exhibition. The traveling exhibition is still a great success. So far, it was been shown in about 300 places and had about 2 million visitors. In the first few years, the traveling exhibition was mostly shown in Germany, in recent years it has traveled through Europe and had also some visits overseas.

- The success of the exhibition was so overwhelming at the beginning that it was obvious to ask: "Could this also work as a permanent exhibition, that is, as a mathematics museum?" Therefore, in 1996, we founded a society whose aim was "to develop and bring to life a mathematics museum at Giessen."
- It took some years to establish this museum. We used those years to develop and test new exhibits, to learn from other science centers, and of course, to search for a building and for money.
- In 2001 the city of Giessen provided a building for the Mathematikum, which is ideally situated next to the train station. Also the Ministry of Higher Education, Research and the Arts of the state of Hesse gave a substantial contribution for reconstructing the building. Another essential source of funds was the European Commission, and, last but not least, about one quarter of the funds needed for the building and the first set of exhibits was given to the Mathematikum by sponsors and many private donors. On November 19, 2002, the president of the republic, Johannes Rau, opened the Mathematikum. In his speech he said: "Mathematics can be fun. This I experienced here."

What Is It Like Now?

From the very beginning the Mathematikum was a success. In the first year more than 130000 people visited the Mathematikum. Since the second year the Mathematikum has been visited by about 150000 visitors each year. About 45 % of the visitors come in groups, mainly as school classes and 55 % of the visitors are private visitors, mainly families. Both types of visitors come from a wide area. A typical visitor has a one hour trip to reach the Mathematikum. In the summer holidays, people from about 50 nations visit the Mathematikum.

The first years of the Mathematikum can be characterized as a development phase. At the opening, the Mathematikum had about 50 exhibits on about 600 square meters. Now there are more than 150 mostly interactive exhibits on about 1200 square meters. The main method of developing the exhibition was the program "each month one new exhibit." Apart from the challenge of developing, designing and manufacturing the new exhibits, this program has enjoyed a monthly report in the newspapers and people living nearby know that "whenever I visit the Mathematikum, I will see something new."

In 2009 we opened an additional section, the Mini-Mathematikum, a special section, equipped with exhibits specially designed for 4 to 8 year old youngsters. These exhibits have the same mathematical and didactical quality as the exhibits in the rest of the Mathematikum.

The Mathematikum has a very clear architecture. The exhibition is on three floors and the Mini-Mathematikum is on the fourth floor. Each of the three floors has the same structure: from a central corridor visitors can go into any of four rooms, two on the right, and two on the left. A visitor does not have to open any doors in the whole exhibition. The ground floor is black, the walls are white. Visitors may start anywhere they like. There is no recommended tour.

The "tables" on which the exhibits are placed have been specially designed. We tried to avoid a "laboratory look" on the one hand and a "living room look" on the other. The wooden tables are inspired by the work of the American artist Donald Judd.

The Mathematikum has a clear color concept. We use "clear" colors (red, blue, yellow, and green). Only those parts of the exhibits the visitor handles are colored. This is usually the "heart" of an exhibit.

Visitors like the Mathematikum. They obviously also like the way mathematics is presented. They are entertained by performing the experiments and trying to understand their experience. The Mathematikum is a house full of communication. In fact some people claim that "it is loud." But when you listen to what people are talking (and, sometimes, shouting) about, you can see that it is about the exhibits, though, sometimes in a very superficial way, but nevertheless they talk about the mathematical exhibits.

The exhibits are not ordered according to mathematical subjects (such as geometry, algebra, analysis and so on). This does not seem to bother the visitors.

There are no explanations at the exhibits. Each exhibit has a label with the title of the exhibit and a few simple sentences, which describe what to do. No explanations and in particular no solutions. Only very few visitors complain about this. The fact that "explanations" are missing is deliberate. It reflects the learning model of the Mathematikum, which claims that the visitors are little researchers: they are confronted with problems and want to solve them. Of course this is sometimes tedious, but there is an enormous reward, if a visitor has succeeded in, for instance, solving a puzzle.

How Expensive Is It to Run Such a Museum?

This first thing you have to recognize is that running a museum is running a business. Therefore there is a simple rule: only spend the money which you have already earned.

The costs can be enormous. There are costs connected with the building: energy, heating, and several insurance policies. Often there are costs for renting the building, or, as with the Mathematikum, you have to deal with a loan.

There are costs concerning the people working in the museum. Typically more than 50 % of the costs are for staff. At the beginning we tried to hire as few people as possible. Many tasks were performed on a voluntary basis or by students from the University—or were simply not performed.

There are also costs for maintaining and enlarging the collection. Do not under-estimate costs for maintenance. We overhaul exhibits on a very regular basis, so that the visitor always feels that "everything is working" and "all is new."

Does It Make Money?

The permanent exhibition is the bread-and-butter business. Basically you earn money with the permanent exhibition. All other activities, such as temporary ex-hibitions, shows, talks, and so on aim at getting the attention of people or, even better, attention in the media. You won't earn money with such activities directly.

In the Mathematikum we offer a lot of additional activities, such as regular chil-dren's lectures, talks, and a kind of talk show ("Beutelspacher's sofa"). Once a year we celebrate a "night of mathematics"; we organize a science festival ("street of experiments"). Also twice a year, we have temporary exhibitions under different themes, such as illusions, puzzles, language, and randomness. Finally, the Mathe-matikum is a place for concerts and art exhibitions.

It is quite easy to get press coverage (including radio and TV) for these events, which is by far the cheapest publicity around.

Important Points for Starting a Mathematical Museum

Of course there are many different types of "mathematical museums" and certainly there are very different local conditions and requirements. For the development of the Mathematikum the following points have been extremely important.

Think Big, Start Small

From the very beginning, try to envisage the ideal size of your museum. This does not necessarily mean "as big as possible." On the contrary: If you would like to have an economically stable institution, you must find the "right" order of magnitude.

Equally important is to be familiar with each detail, in particular, concerning the exhibits. In our case it proved particularly useful that we had quite a few years where we could develop and test experiments. We started with small exhibitions and gradually enriched them.

(Also from today's point of view it was extremely good that at the beginning we got no money. So we were forced to think very thoroughly about the exhibits.)

Build on Scientists, Focus on Visitors

Without doubt, mathematicians want to show mathematical phenomena. And quite naturally each scientist has specific ideas, which they are fond of. It is important to have a large collection of possible ideas and to get continuous support from scientists. But keep in mind that a typical visitor will not be a professional mathematician. On the contrary you want to reach all people, in particular those who are not fond of mathematics. Therefore the second phase of designing exhibits is to radically take the visitor's point of view. At this point museum people, exhibit designers and so on must play a crucial role.

Look at Others, Develop Your Own Profile

Try to get inspiration everywhere. Visit other science centers, natural history museums, and art museums. Look at their content, their presentation, their organization (do you, as a visitor, feel comfortably, and why?). You might also get inspiration from fun parks, McDonald's, shopping malls, and so on.

Try to learn from them. Consider the best of them. Try to avoid their errors.

You should not copy these institutions, but in comparing ideas, you will develop your own profile. In my opinion, it is of utmost importance to have a very clear, recognizable profile. This certainly includes the name of your institution. To find a good name is extremely difficult, at least it was the most difficult task in our case.

Keep It Simple

This should be your leading principle. It applies to all aspects of your institution.

It's certainly true for the exhibits. It is important that they are focused on the main effect. "Simple" means also that there are design principles for your exhibits. And it means that you should keep exceptions to the absolute minimum. I always recall Asger Hoeg, director of Experimentarium Copenhagen, who said: "Use only one type of screw!" This seems to be exaggerated, but it is true: you will be punished if you do not follow this rule.

Keep your building simple. Visitors should always know where they are and where to find the exit (or the toilet). This allows a visitor's brain to focus on the exhibits and not memorizing the shortest way back.

Keep your organization as simple as possible. This includes opening hours (as homogeneous as possible) and entrance fee (give the visitors only a very small choice).

Get Your Priorities Right

When your institution opens, you cannot usually make everything perfect. So you have to compromise. It is important not to fail in making compromises.

When the Mathematikum was opened, we only had very limited means. So, we took two decisions. (a) We opened only half of the space, and (b) we spent money only on reconstructing the building and on exhibits. Everything else had a much lower priority. For instance, we got used desks and chairs for the administration for free from the university. Furthermore the goods we sold in the cafeteria were very limited. Basically we offered coffee, soft drinks, ice cream and some sweet stuff we bought from the nearby bakery.

Have an Organization that Makes Quick Decisions Possible

Especially at the beginning there will be lots of (mostly small) items about which you have to decide immediately. Have someone who is willing, capable, and allowed to make these decisions.

Some Exhibits

When we were planning the Mathematikum, one of the greatest science center directors, Remo Besio from Technorama in Winterthur (Switzerland) gave me very good advice. He said: "In my institution, we have a table with puzzles. And this is the best experiment we have."

Today I can endorse this claim. The puzzle tables in the Mathematikum are very attractive, highly communicative, and full of mathematical experiences. Therefore each mathematical center should have a table with puzzles. They are relatively cheap to construct, attract people, and hold their attention for a long time.

On the other hand each mathematical center should have something "special." These are the exhibits visitors remember most. In the Mathematikum the most popular exhibits are the giant soap bubble and the giant ball machine. Of course, there are individual favorites, such as, for instance, "Who sticks out the farthest?" and "the queue of dice."

Who Sticks out the Farthest?

On a table there are a number of wooden squares all of the same size and the same thickness (see Fig. 1). Your task is to stack the pieces so that one of the pieces is completely beyond the edge of the table. In doing this all pieces should remain horizontal.

Fig. 1 Stack the pieces!

This is a particularly nice exhibit. Firstly there is an "obvious" mathematical solution: Stack all the pieces vertically to form a "tower" so that the lowest piece has an edge in common with the table. Now, push the top piece out as much as you can. Then push the second piece out as much as possible. Continuing this process, the top piece is clear of the table by a distance of $1/2 + 1/4 + 1/6 + 1/8 + \cdots$, and with only four pieces you can achieve the goal. (Clearly, this is an application of the "harmonic series".)

Secondly there are many other, sometimes ingenious and good solutions. They lack the property that top piece sticks out the farthest.

Finally, as far as I know, the world record is unknown (how far can you make them stick out?) for, say, four quadratic pieces.

The Queue of Dice

This exhibit has a few rules, but at the end there is a real surprise-based on mathematical principles (see Fig. 2).

0. As a preparatory step throw about 40 dice and then arrange them in a "queue", so that the dice are in order.
1. Look at the first dice. It shows a number, for instance a "2." This means that you move 2 dice forward, arriving at the third dice. This dice shows a number, for instance a "4." You will move forward 4 and arrive at a new dice. Again, read the number and go forward the corresponding number of steps.

    Repeat this procedure to the end of the queue. Most likely you will not finish at the last dice. For instance, it could happen that you are at a dice showing a "4," but there are only two more dice left. Simply remove the remaining dice.

    In this way you ensure that the game ends at the last dice. So far there is no surprise.
2. For the second round, you throw the first dice once more. Now it shows, for instance, a "3." Follow the same rule as in the first round, which means: on arrival

Fig. 2 The queue of dice

at a dice read its number and move forward the corresponding number of steps. Continue until the end of the queue—and now you will very likely arrive at the last dice.

You can repeat the experiment and you will notice that it is very "robust". Even if you miscount or you don't start at the beginning of the queue—you will end at the last dice.

This experiment shows two characteristic features of a random process: Firstly, the more, the better. The probability of succeeding increases with the number of dice. Secondly there are always exceptions: Imagine a queue, where all dice show a "6." This queue won't work, regardless of length.

Atractor

Manuel Arala Chaves

Abstract This paper describes the methodology used by Atractor Association in its efforts to communicate ideas and relevant results of Mathematics to non-specialists. This description is illustrated with numerous examples of interactive exhibits, both virtual and non-virtual, created in the belief that interactivity is an important factor in the involvement of the target audience. Some examples of non-virtual exhibits, such as a large Ames Room, the three conical billiards or the slit hyperboloid, by their size or difficulty of construction are costly and require a large space; however others are quite simple and can easily be reproduced, even in small schools or groups. The list of examples of virtual exhibits, all produced by Atractor, ends with a reference to the DVD 'Symmetry—the dynamical way' and to the program GeCla, which allows the generation of friezes, patterns and rosettes and the assisted classification thereof, also including the possibility of competition via the Internet.

The Association Atractor was formally founded in April 1999 but the founding group had in previous years already organised several activities, which in current terminology would classify as RPAM (Raising the Public Awareness of Mathematics).

From the beginning of these activities until the present, there was a constant concern to promote interactivity, by either using physical objects that could be manipulated, or by creating programs with a lot of user intervention. That interactivity was always considered an important way to get people more involved.

Another principle, which the organisers have always tried to follow, is that the choice of topics to be addressed should not be conditioned or limited by any school curriculum. Although obviously a large proportion of those interested in Atractor's activities has always been students and teachers of mathematics at various levels, that never restricted the choices of subjects. The form of presentation takes into account the potential audience, but no subject was excluded because it was not in the school curriculum at some level. On the contrary, the broadening of horizons

M. Arala Chaves (✉)
Universidade do Porto, Porto, Portugal
e-mail: machaves@fc.up.pt

E. Behrends et al. (eds.), *Raising Public Awareness of Mathematics*,
DOI 10.1007/978-3-642-25710-0_10, © Springer-Verlag Berlin Heidelberg 2012

that occurs when one is faced with wider areas of mathematics or its use than those taught in school was always considered a positive factor.

The independence or closeness of the topics in relation to school curricula is only one of the points that, sooner or later, anyone who wishes to develop activities for the popularisation of mathematics has to face, especially if they occur in a school. Another important point concerns the degree of accuracy and precision of the formulations and approaches, but that point is treated in a complete and comprehensive way in another chapter of this book.

Another aspect that is generally much emphasised is the social relevance of mathematics, highlighting its applications in various fields of science and technology, ranging from very advanced areas to those used in everyday life, but this aspect although undeniably important, must not overshadow the importance of also highlighting purely mathematical aspects for their beauty and the educational value of the methods used. And in our choices, we have sought to prioritise exhibits with actual mathematical content, rather than those which are entertaining, but irrelevant in terms of the underlying mathematics. We will return to this aspect, after we have given some concrete examples of exhibits from our exhibitions, which were intended to have relevant mathematical ideas.

However, it is worth quoting here from our 2006 article about Atractor published in the EMS Newsletter [1], concerning the spirit in which the "Matemática Viva" exhibition (created in 2000, the International Year of Mathematics) was conceived:

> The organisation of *Matemática Viva* was a very hard task considering the short period, the absence of structures and the lack of previous experience of Atractor for a task of such a dimension. . . . The principle of running interactive mathematical exhibitions was itself not consensual. And even for those in favour there were different possible approaches. I quote two written statements from that time:
> *It is not evident, even for some mathematicians, that mathematics is adequate for "interactive" presentations and on the other hand some feel that such presentations may not have a scientific quality and may pervert the ideas (or the problems) they want to present. The plan should not follow the mathematics curricula for any academic degree, although school population will certainly form a large part of the visitors. But . . . the exhibition should not be reduced to a mere tool for guided school visits.*
> *The exhibits, although they may have an informative component, should be planned mainly to awake the visitor's active curiosity . . . . Leading the visitor to voluntarily make some effort, small though it may be, to grasp an idea is certainly better than to give him/her a higher volume of encyclopaedic knowledge or mathematical results. . . . The question of scientific rigour . . . is controversial. . . . Although it is obviously necessary to avoid use of technical specialised vocabulary . . . it is crucial to avoid actually distorting the ideas to transmit under the temptation of making popularisation accessible at any cost.*
> The general philosophy when designing the *Matemática Viva* Exhibition was to follow these general principles.

Examples of Physical Exhibits

The *Matemática Viva* exhibition was temporarily closed in August 2010 due to works started in the building where it was located, see [1]. We reproduce here photographs of three exhibits, which, by their dimensions (and features), are somewhat

Fig. 1 Slit hyperboloid

Fig. 2 Ames room

emblematic of it: a *slit hyperboloid* outside (see Fig. 1), with two rotating rods, representing two generatrixes for the hyperboloid, an *Ames room* (see Fig. 2) and three *conical billiards*, with elliptical, parabolic and hyperbolic shapes (see Fig. 3).

Photographs of more exhibits exist in the same publication, and for a complete list in Portuguese (of all 75 exhibits), see [2].

We will detail below some examples of exhibits in *Matemática Viva*, other than those already described in [1].

1. *Game of dice*—Fig. 4 shows a game, which allows the player to experimentally confirm an interesting and at first sight unexpected property involving probabilities. The number of dots on the faces of the four dice used are different from

Fig. 3 Conical billiards

Fig. 4 Game of dice

normal dice. The dice are all different from each other, and the probability of a player winning against one of his neighbours is different from ½, which would be the value obtained with normal dice. The way the dice are placed around the table is such that each player has a probability of winning against his or her right-hand neighbour, which is (sharply) greater than ½. The possibility of creating a situation like this clearly highlights the non-transitivity of this "comparison". More precisely, player A is more likely to win against B, B to win against C, C to win against D, but … A is not more likely to win against D, on the contrary A is likely to lose against D. A diagram with the description of the dice and tables, which allows you to easily find the odds, can be found in [3]. Of course it is

Fig. 5 Möbius strip with a map

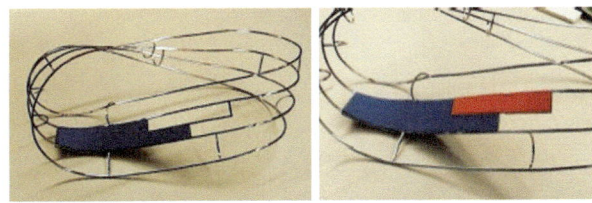

prudent to establish beforehand that each pair of players will throw the dice a reasonable number of times before declaring the winner, to increase the likelihood of the "expected" behaviour. Before the creation of Matemática Viva, an exhibit of this type was shown in a two-day exhibition, which I attended, and we advised throwing the dice five times. At one point, a child came to tell me that she had played and won, which was not expected. I suggested that she should repeat the game, and watched to check if they were playing correctly. Everything was fine and ... the same child won again in my presence, contrary to natural expectation. I mention this incident because I believe it deserves two comments. The first is that what happened that day can in no way be seen as a failure. Rather, it should be, and was, exploited to highlight the limits of what "being more likely" means. The other comment refers to the interest that this very young girl (probably primary school) showed in relation to this experiment: the visit had taken place on a day set aside for school groups and the next day (Saturday) the exhibition was open to the general public; on the second day, the same girl approached me. She had wanted to show the exhibit and explain how it worked to her family, including her parents, and at the end she came to tell me happily that "today it gave the expected result" ...

2. *Möbius strip with a map*, which the visitor is asked to colour with the minimum possible number (seven) of colours "according to the proper rules"; this exhibit allows the visitor to find out that generalisations of the famous four-colour problem to other surfaces may involve different numbers of colours (see Fig. 5).

3. *The problem of the ant*. "Which point on the surface of a parallelepiped is farthest from a particular corner?" This question is illustrated by a simple wooden parallelepiped, with a string that allows the visitor to conjecture the position of the desired point.

4. *Pi to a million decimal places:* A wall chart allows the visitor to easily find any finite sequence of digits (e.g. a date) with the help of a magnifying glass and a program, and raises the problem of whether the digits of Pi have a normal distribution (see Fig. 6). Other adjacent exhibits are concerned with the meaning of Pi.

5. *Sperner Game:* This exhibit is one of several in a corner where issues directly or indirectly related to topology are addressed. The most interesting feature of this game lies not in winning, but in trying to discover if the game may or may not tie. See below a description of an applet for this game.

All these examples illustrate the attitude of Atractor in relation to the question raised at the beginning of this chapter on whether to include or not "recreational"

Fig. 6 Pi to a million
decimal places

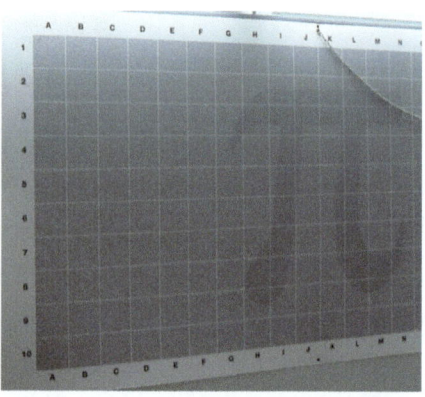

Fig. 7 Cylindrical
anamorphosis

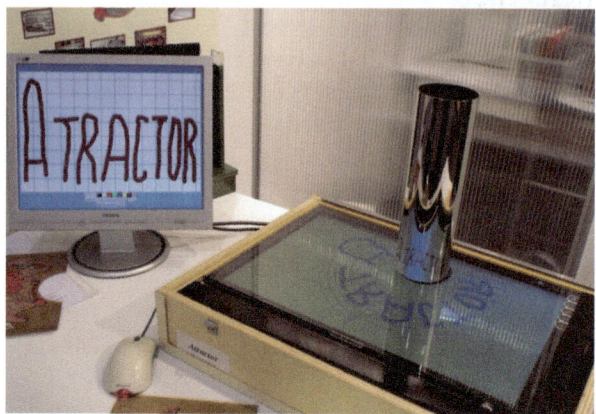

and "entertainment" activities for the popularisation of mathematics. The policy
chosen has always been to only include activities of that kind if they bring some
added understanding of a relevant mathematical idea.

Another example illustrating the same policy was created after *Matemática Viva*:
an interactive exhibit illustrating *cylindrical anamorphosis*. The originality in relation to the usual practise of a reflecting cylinder and anamorphosis images on a
physical medium near the base of the cylinder is the following: the cylinder is placed
on a (horizontal) screen connected to a computer, which has a program that allows
image editing on a vertical screen, just by using the mouse (Fig. 7). As the image is
drawn on the vertical screen, a deformed image, built by the program, will emerge
on the horizontal screen so that its reflection in the cylinder is precisely the correct
image, identical in form to the one edited on the (vertical) screen. It is therefore
clear that this exhibit mixes the usual components of a physical exhibit and a virtual
one.

Besides this example, the creative efforts of the past five years have been spent
mainly producing purely virtual content. Before analysing in some detail the most
ambitious and large-scale work so far undertaken by Atractor, a DVD *Symmetry—*

Fig. 8 Sperner's game

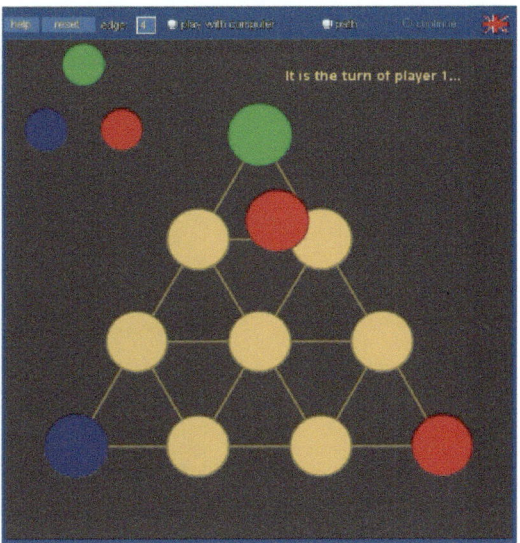

the dynamical way, we will mention some other examples of virtual exhibits that can be seen on Atractor's site.

Examples from Atractor's Site

Sperner's game Consider a triangle with vertices of three different colours. Each side of the triangle is divided into the same number of equal segments and the triangle itself into a corresponding subdivision of smaller triangles (see Fig. 8). Each player takes turns in colouring the vertices of the small triangles; the interior vertices can be coloured with any colour, the ones on the sides of the big triangle must use one of the two colours on the ends of that side. The loser is the first player to form a small triangle with the three colours [4]. Apart from the game itself, what is more interesting from a mathematician's point of view is whether this game can result in a tie or not, in other words: is it possible to colour all the vertices according to the rules without creating a small triangle with three colours? An applet provides not only an interactive board to play this game, but also provides the background for a text that leads the visitor to very gradually take the initiative and to seek answers to partial questions so as to be convinced that the game is never tied. The set of texts, many screenshots and some animated gifs provide the germ of a real proof of this fact in a way that is accessible to people who may never have seen proofs of this kind (Fig. 9).

The result that this conclusion reflects can be used to prove some strong results in elementary topology. One, called the *hairy ball theorem*, which "says" that "*one*

Fig. 9 Completed game

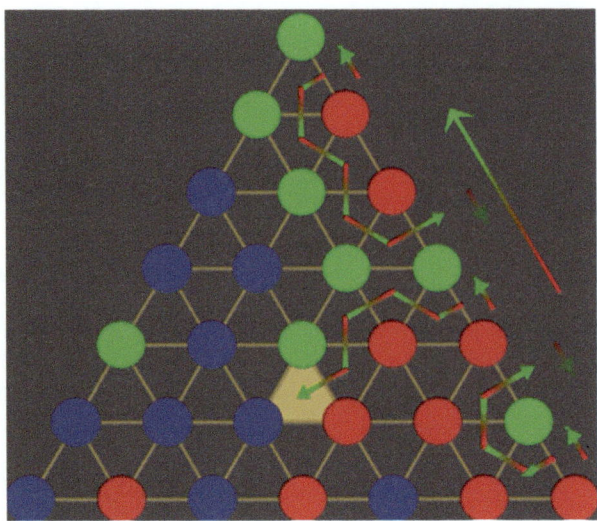

Fig. 10 Stereoscopic shell
model

cannot comb a hairy sphere without making eddies appear",[1] already has two applets (see [5] and [6]) ready to illustrate the proof using Sperner's result, but there is
no text yet.

Mathematical shells It is amazing how you can find mathematical models of
surfaces describing the approximate shape of a large variety of shells so closely.
Stereoscopic images of these surfaces, formulas to produce them and applets allowing you to see the effect of varying the parameters for the shape of the models can
be found in [7] (Fig. 10).

Grass, rabbits and foxes An interactive applet [8] produced with StarLogo using a kind of cellular automaton allows observation of the evolution of a dynamic

---

[1]Using more technical language: every continuous vector field tangent to a sphere vanishes at least
at one point.

Fig. 11 Grass, rabbits and foxes

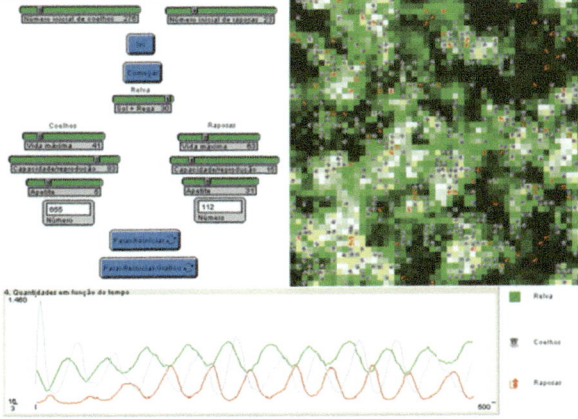

model representing the interaction of a system involving grass, rabbits who eat the grass and foxes who eat the rabbits … Several parameters can be controlled by the user: the initial number of rabbits and foxes, the duration of sunlight and how long the grass is watered, the maximum life span of the rabbits and foxes, their rates of reproduction and appetites. The sometimes unexpected evolution of this system as a result of changes in some of the parameters, is shown either by direct observation of the field of grass and of its "inhabitants" or by the graphs that are drawn (Fig. 11). This observation captures the attention of visitors for long periods and is quite instructive, not only from the mathematical point of view.

15-puzzle An applet illustrating this well-known puzzle and generalisations of it to several surfaces (cylinder, torus, Möbius strip, Klein bottle, projective plane and cube) allows visitors to, among other things, gain some familiarity and intuition about surfaces, in particular non-orientable ones. And users can even choose one of their own pictures for use in the game … [9] (Fig. 12).

With no accompanying text yet and sometimes without detailed help, there is a set of applets, programmed directly in Java (and JavaView), some of which are very powerful:

Hexlet An applet [10], which illustrates the problem known by this name. Generically three spheres are fixed, each of which is tangent to the other two, new spheres

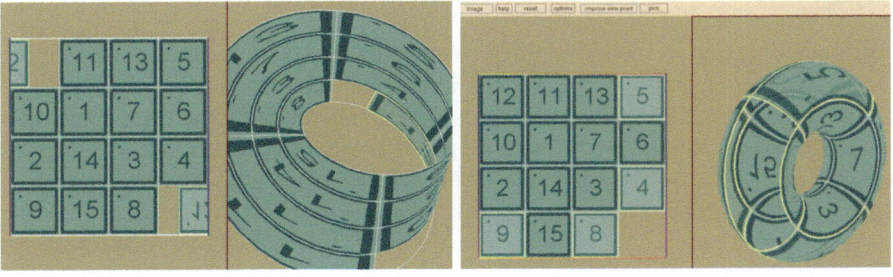

Fig. 12 15-puzzle applet

Fig. 13 Hexlet

Fig. 14 Hexlet

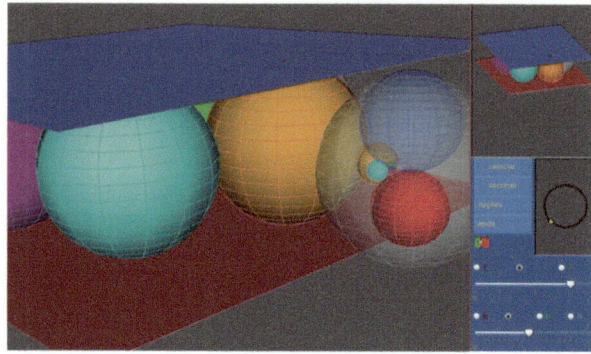

Fig. 15 Hexlet

are added tangent to the initial three spheres to create a ring of spheres, each new sphere being tangent to the previous sphere as well as to the three initial spheres.

It so happens that this ring consists of six spheres and is closed in the sense that the last is tangent to the first (Figs. 13, 14 and 15). This result is somewhat unexpected, because the number of spheres in the ring and the fact that it closes does not depend on the radii of the three initial spheres nor on the position of the first sphere chosen on the ring. In some exceptional isolated situations, one (or two) of the spheres can degenerate into a plane. The proof of the stated result is a particularly simple and elegant example of a very common process in mathematics: reducing the

Fig. 16 Harmonic oscillator and resonating bridge

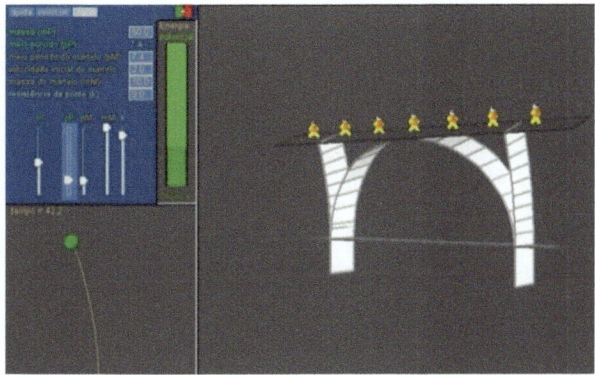

proof of a general result to the proof of a particular case in which the conclusion is trivial. In this case the reduction is made by an inversion in space with the centre at one of the points where the initial spheres are tangent. This inversion preserves tangency and maps all the other spheres into spheres. The two spheres tangent at the centre of inversion are mapped onto two parallel planes and all the others (the fixed one and the ring), being tangent to both planes, have the same diameter (which is equal to the distance between the planes). Now in this case, the conclusions to be drawn are trivial. This whole process is clearly illustrated in the applet.

Pendulums   Twelve applets [11] deal interactively with several classic examples of dynamics, allowing the user to vary the parameters and observe the effects on the evolution of the system. There are animations which show simultaneously and in real time the movement of the pendulums, their representation in the phase plane or, where relevant, in the space of configurations, the variation of the different types of energy involved, the vectors of the forces, accelerations and velocities and the graphs of the relevant quantities. These examples deal with the following cases: simple rigid pendulums (one and two, with and without friction) and string pendulums, a double pendulum, a spherical pendulum, coupled pendulums and a cycloidal pendulum (one and two), an excited pendulum and associated resonance phenomena (Fig. 16), horizontal and vertical oscillators. Since it is not so frequently considered, the case of a simple string pendulum is perhaps worth mentioning, because the applet highlights the difference in behaviour in relation to a simple rigid pendulum for large oscillations and also a specific example of resonance phenomena, which evokes an event which occurred with a bridge in Porto in the last century.

Complex dynamics   An application in Flash [12] allows the user to analyse the dynamics of the real function $x \to x^2 + c$, when c varies. By gradually varying transparency with a slider, the Mandelbrot set (Fig. 17) is displayed thus linking (for each real value of c) the dynamics (possible presence of periodic or pre-periodic orbits and respective periods) detected through the graph in the real case, with on one hand the component of the interior of the Mandelbrot set, which the corresponding value of c belongs to, and, on the other, the distance of c from its centre. Several applets

Fig. 17 The Mandelbrot set

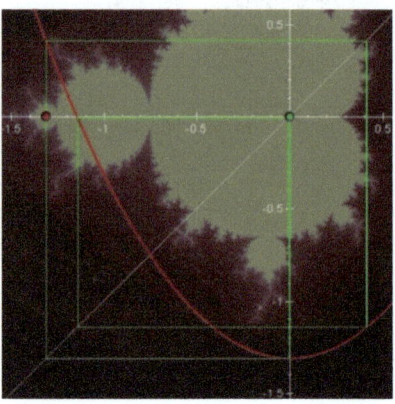

[13] allow the analysis of details from the sets of Mandelbrot and Fatou-Julia, their generalisation to higher degrees and give an approximate idea of the orbits in the complex plane or even detect small copies of the Mandelbrot set in the images associated with the Newton method (Fig. 18). In the middle image the points in black represent those values in the parameter space for which the Newton method for a cubic equation does not converge to any of the three roots.

Stereoscopy   Atractor has paid particular attention to the development of images, animations and interactive content for stereoscopic viewing, believing that, apart from the aesthetic effect, it is a great help for people, particularly children, who have difficulty in imagining certain objects in space based on their representations in the plane. Technically, two systems are used: one, using equipment produced and distributed by Atractor, allows viewing of site content [14] or printed images; the other, using two projectors with polarising filters and glasses, allows viewing of content projected onto a screen and the effect has more impact (this was the system used in the short demonstration at the Óbidos meeting). For people who can easily "see" stereoscopically without the use of aids, there are also pairs of small images such as Fig. 19.

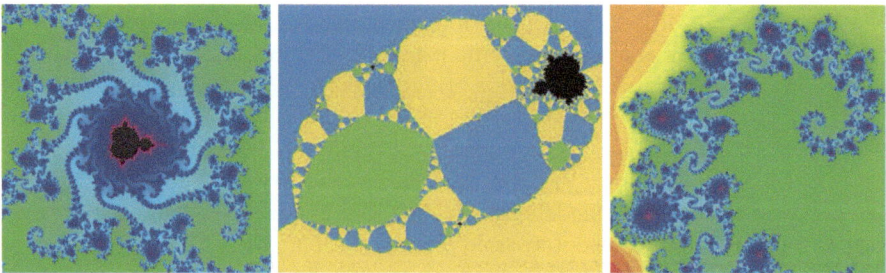

Fig. 18 More Mandelbrot sets, even in the dynamics of Newton's method for cubic polynomials (*middle*)

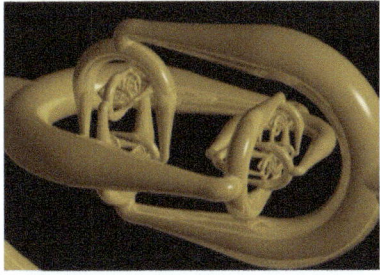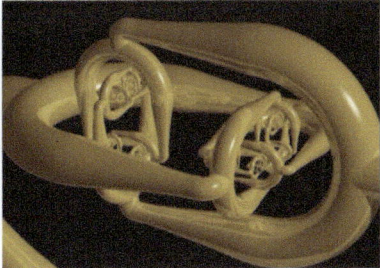

Fig. 19 Part of an Alexander horned sphere

webMathematica   This extremely powerful tool has been installed on the website, and has been used to develop content for it, which can be accessed from [15]. I will only mention the original design for some of these applications, which allows the creation of tailor-made applets with data chosen by the user. It is possible to save these applets, and use them later without an internet connection. Examples of applets of polyhedra, modified by the successive actions of various operations, were thus created remotely using *webMathematica* on Atractor's site, and can be seen in [16] (mono) and in [17] (stereo). Here are three other examples, the first two are more technical and the third one is more mundane:

1. For a smooth real function and a number n chosen by the user, an applet is created, which can then be used independently; when a point is moved along the x-axis, the graphs of the function and its Taylor polynomials up to order n on that point are shown and change as the point moves.
2. For a real rational function, chosen by the user, an applet is created, which can then be used independently; when a point is moved along the x-axis, the interval of convergence of the Taylor series of the function at that point and the graph of the function that gives the radius of convergence as a function of the point are shown (see some examples saved from *webMathematica* in [18]). See Fig. 20.
3. The user uploads an image, chooses two integers m and n, and an initial configuration; a (generalised, if $m \neq 4$ or $n \neq 4$) 15-puzzle is created on a plane, cylinder or torus, and if it is solvable the solution is found and can be saved to a solution file, which can then be used independently. To solve the puzzle an algorithm programmed with Mathematica is used (Fig. 21).

Symmetry the Dynamical Way

The DVD *Symmetry the Dynamical Way* was published[2] in September 2009, and its creation, carried out entirely by collaborators of Atractor, took several years, although there were some interruptions caused by projects on other topics. The basic

---

[2]Not only in Portuguese, but also in English, French, German, Italian and Spanish.

Fig. 20 Interval of
convergence for
$10^5/[(x + 10)[(x + 6)^2 + 1]$
$[(x - 3)^2 + 16](x - 10)]$

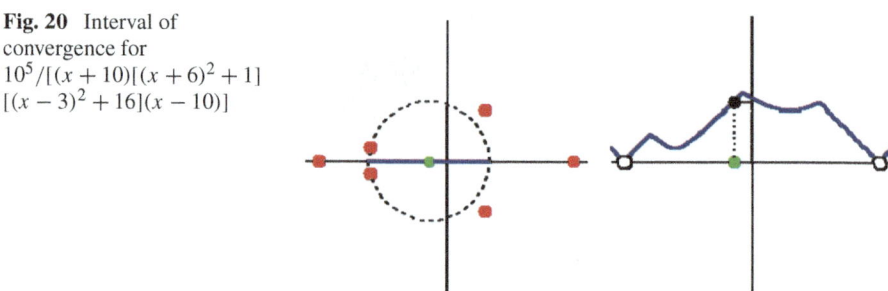

idea was to present the classification of doubly periodic patterns in the plane (wall-paper) inspired by W. Thurston's treatment using orbifolds, as far as possible in a non-technical way. The case of friezes was also handled, following ideas sketched as handwritten notes by J. Conway [20]. Presentations of orbifolds for non-specialists had been made in the past and some are available on the Internet [21], with many illustrations.

This DVD is not only illustrated by many pictures, but also by hundreds of Flash animations and interactive programs of which we will give an idea in what follows. There is also a glossary, organised in the same spirit, on ideas and results used in the text, for example, on isometries, the symmetry of a figure and the Euler number of a surface. A small program determines the composition of any two isometries in the plane and animations illustrate various topological properties.

In the DVD, after a collection of photographs of various panels of old Portuguese tiles (*azulejos*) is displayed, a presentation begins by showing animations with stamps that produce friezes or patterns with symmetries that are determined by the three-dimensional shape of the stamp. A particularly simple example is a paint roller with an asymmetric image soaked in ink. By rolling it on a wall, it leaves a frieze with only translation symmetries (Fig. 22). Another example is an equilateral triangle, which rotates in turn around each of its edges building a picture step by step and ultimately forming a doubly periodic pattern (Fig. 23). Apart from the translations defined by two translations in different directions, this figure also has the

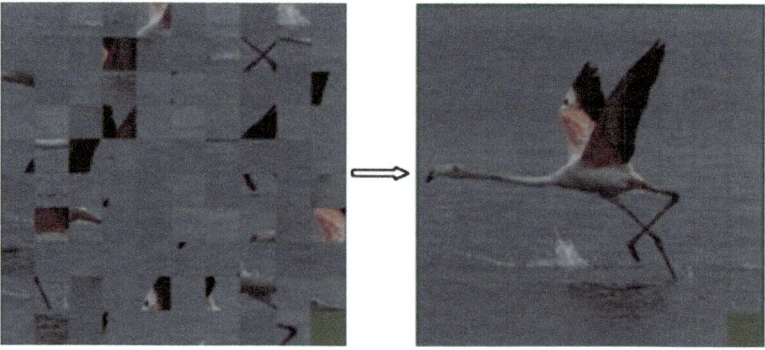

Fig. 21 15-puzzle and its solution

Fig. 22 Paint roller

Fig. 23 Equilateral triangle
pattern

following symmetries: (i) reflections about lines printed by the edges of the triangle, those having three different directions, and (ii) rotations and glide reflections obtained as composition of the previously mentioned symmetries.

More examples are given in the same spirit, but with different shapes, which also lead to patterns with symmetries that are different from the previous ones. In the chosen examples, the stamp sometimes has edges, and the pattern will have reflection symmetries, sometimes there are cone points, which give rise to rotational symmetries and sometimes there are corners on the edges corresponding to intersecting reflection lines, with their corresponding rotational symmetries (Fig. 24). And some of the stamps have all of these. The difference between the centres of rotation associated with "corners" or "cone points" is that the former are at the intersection of axes of reflection and the latter also called *gyration centres* are not.

Until this stage of the presentation, the stamps were shown stamping and the type of symmetry of the pattern obtained by each was seen. So, the reverse process naturally arises: given a pattern, how can a stamp be found which *produces* its symmetries? Some concrete examples suggest the general way to proceed—see Fig. 25.

Two new related questions then arise: (i) is it possible to go on endlessly finding patterns with different types of symmetry? Or if not, how is it possible to conclude that a finite list of patterns contains all the possibilities? (ii) When one tries to form new stamps from surfaces, to which one eventually adds edges, vertices and corners, how can one know whether they correctly stamp a new pattern?

Fig. 24 Cone point stamp
and pattern

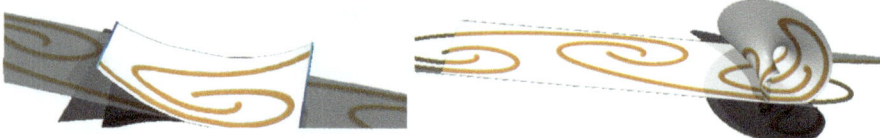

Fig. 25 Producing a stamp from a frieze

Fig. 26 Creating a stamp for a rosette

Once these questions have been stated, similar questions are considered for rosettes, where there is a much simpler answer (see Fig. 26). Indeed, it is clear that for any natural number n (>1), it is possible to produce a rosette whose symmetries are exactly n rotations around the centre of the rosette: consider an (asymmetrical) motif on a circular sector with an angle of $2\pi/n$. That rosette is stamped by a cone or a conical pad with a suitable opening, obtained by gluing the edges of the circular sector together. So there are infinitely many rosettes with different types of symmetry and there are also infinitely many *different* stamps for rosettes (Fig. 27). A natural "attempt" to adapt this method to obtain infinitely many different patterns would consist of taking "truncated cones" similar to the one used in the pattern of the image with tiles shown in Fig. 24, but this time with smaller openings. For in-

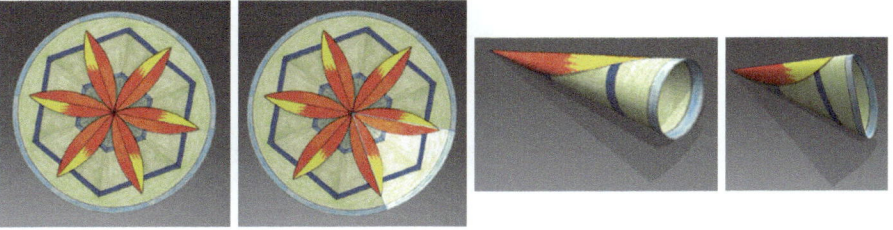

Fig. 27 One of infinitely many *different* rosettes: creation of its stamp

Fig. 28 Pattern with
rotational symmetry of
order 5?

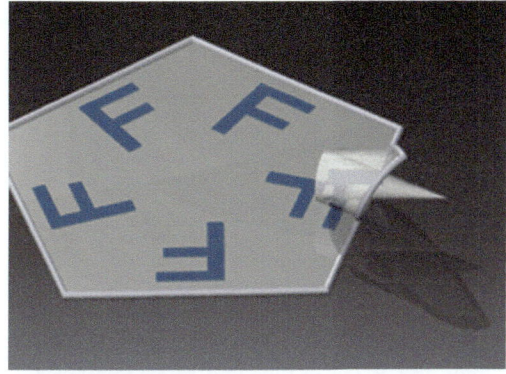

Fig. 29 Failed attempt

stance let us try to create a rotational symmetry of order 5 (Fig. 28) instead of the
previous one of order 4. The failure of this attempt is shown with an animation the
final image of which is in Fig. 29, and it shows that a stamp built in this way is
not "appropriate" for stamping the plane. This failure highlights the fact that not all
combinations of surfaces, edges, vertices and corners lead to "good" stamps for the
plane.

The next natural question is to find simple criteria to decide when a stamp constructed in the manner indicated in the previous paragraph is "good" for stamping
the plane. This question is solved by associating with each stamp constructed as

Fig. 30 Paint roller stamping a frieze

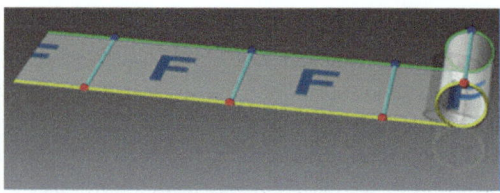

Fig. 31 Cone of order 4

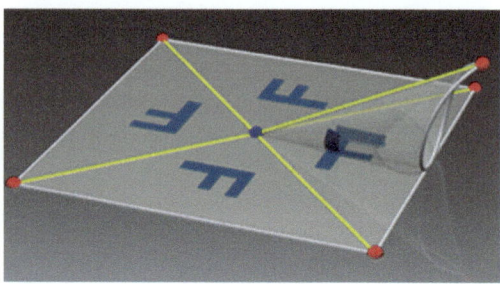

above a number called the Euler number of the stamp. The stamp is "good" if and only if its Euler number is 0. The reason for this criterion can be seen by observing a paint roller stamping a frieze (see Fig. 30): the roller, with V $(= 2)$ vertices, E $(= 3)$ edges and F $(= 1)$ face and Euler number $V - E + F$, after n turns, has stamped a figure with nV vertices, nE edges and nF faces, and therefore with Euler number $nV - nE + nF = n(V - E + F)$ (actually, to "close" the rectangle, it is necessary to add an edge). The reason why the Euler number of the stamped part does not grow indefinitely (in absolute value), despite being close to the product of the number of turns and the Euler number of the stamp, is precisely because the Euler number of the stamp is zero. This idea can be generalised, with some care, for all stamps on the plane. For instance, it is necessary to change the way that the Euler number of a stamp is calculated. Let us look at the example in Fig. 31, with a cone of order 4. When the stamp turns to stamp the square, the red vertex and the (yellow and white) edges behave like the ones on the previous roller: when the stamp turns 4 times, it stamps 4 times on the plane. But not the blue vertex. During the 4 turns of the cone, that vertex only stamps one point. If we want to apply the previous reasoning to give a useful relationship between the Euler number of the stamp and the one of the region stamped, we will have to count the vertex of the cone as ¼ instead of 1 for the Euler number of the stamp, that is, we must make a correction for that cone point of $1 - 1/4 = -3/4 = -0.75$.

An interactive application allows the user to try all combinations of different surfaces, edges (through the creation of holes), vertices and corners, to find the combinations that give a (generalised) Euler number of 0. Each value found will fill one of the 24 positions in a frame and for each of these values it is possible to choose one of five motifs and watch the formation of the corresponding pattern or frieze with the chosen stamp (Fig. 32).

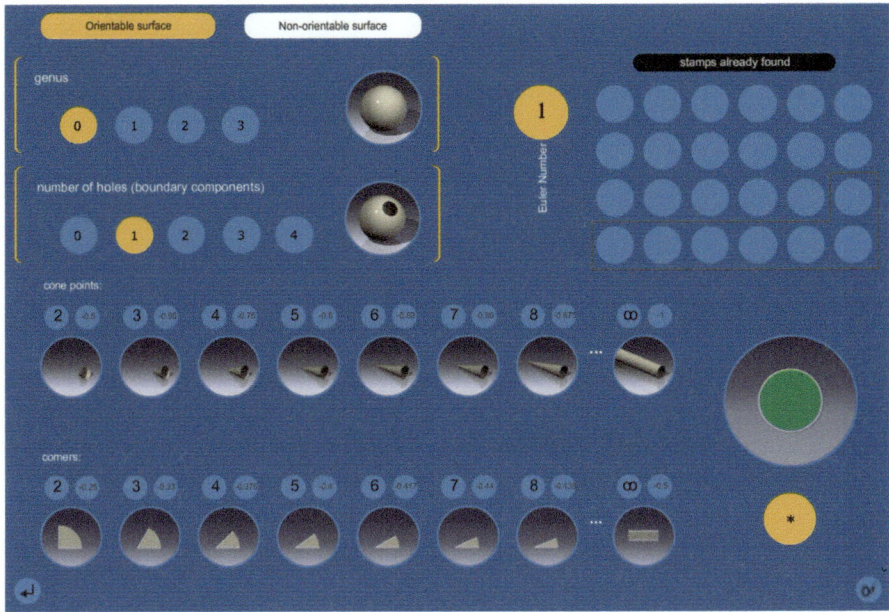

Fig. 32 Finding all stamps

Fig. 33 Local models for orbifolds

In some parts of the presentation, there are complicated questions that are perhaps best omitted on an initial visit such as the special definition of the Euler numbers of stamps. Here is a brief description of two more.

A surface can be described by the property of being like a disk in the plane in a small neighbourhood of each point. That is, a surface has disks in the plane as local models of its points. To define a stamp (or *orbifold* in Thurston's terminology) we also need to give local models for its special points: points on the edges, vertices and corners. These models are suggested by animations (see some of the images in Fig. 33).

Finally, the other *ramification* of the presentation deals with the extension of what was seen above to the study of friezes. In particular the characterisation of

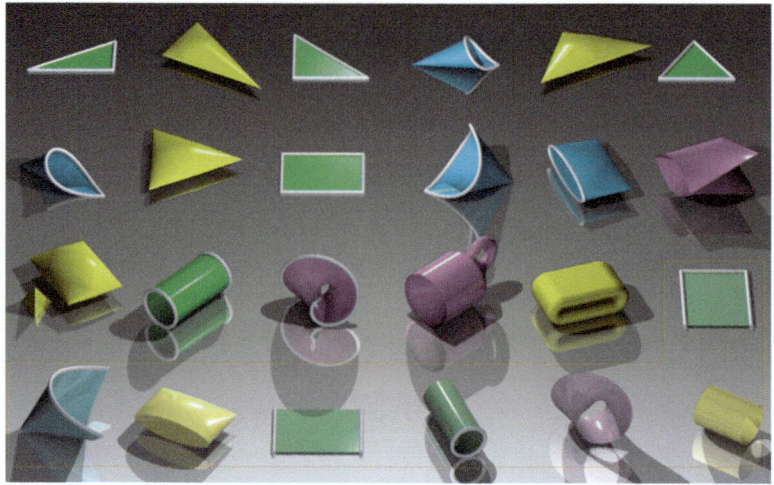

Fig. 34 24 "good" stamps: 17 for patterns and 7 for friezes

appropriate stamps (for friezes!) with the condition that they have Euler number 0, and afterwards counting how many stamps exist. Since the subject is slightly more technical, here we have included only a note[3] with very brief ideas.

Apart from the presentation described, there are other interactive sections of the DVD, which we will mention briefly. A table with all 24 "good" stamps for the plane allows the user to choose one of five motifs for each of them, and then observe how the corresponding pattern or frieze is stamped (Fig. 34).

For three of these stamps (the torus, Klein bottle and projective plane), the way that the stamp acts is not clear and it is possible to see a more detailed version of the stamp in action, by choosing "How does the stamp act?" (Figs. 35 and 36).

Another table shows five families of 17 patterns and 7 friezes, with different types of symmetry (Fig. 37). For each, the visitor is invited to (interactively) find all the symmetries; the program can automatically indicate the missing ones at any point. And at the end, an animation shows the formation of the stamp, which corresponds to the symmetries found and displays the stamp stamping the original pattern or frieze.

After the application where the user can find all "good" stamps, the next item is a generator of patterns and friezes. With one of the motifs that exist in the DVD,

---

[3]The solution consists of adding two points (at infinity) to the plane, each of which represents the point at infinity which corresponds to one of the two directions of a line that is perpendicular to the directions of the translations of the frieze. And the stamp consisting of the usual cylinder to which those two points are added is precisely a sphere. An animation shows how that (elongated) sphere, rotating around those two points, stamps an open spherical calotte, which is a model of the plane. With that rotation, if the direction is maintained, the stamp never returns to its initial position, therefore it is natural to say that the rotation is of infinite order. In the same way that it was natural for a rotation of order n to introduce the correction of $-(1 + 1/n)$, here it is natural that the correction should be -1 for each of these points.

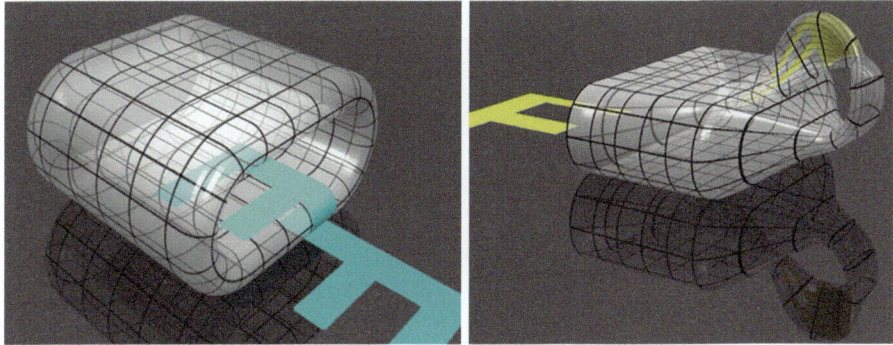

Fig. 35 Torus and Klein bottle

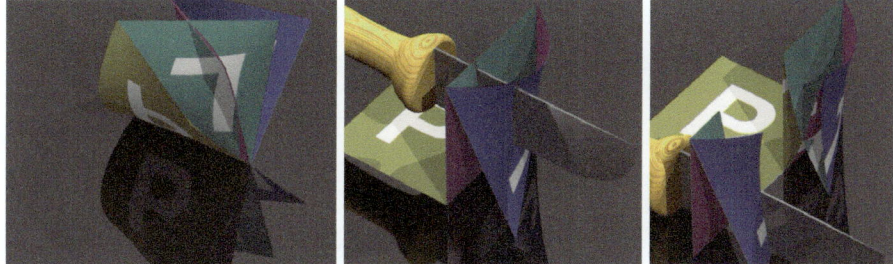

Fig. 36 Projective plane

Fig. 37 Table of patterns and friezes inspired by Alhambra

Fig. 38 Pattern and frieze generator

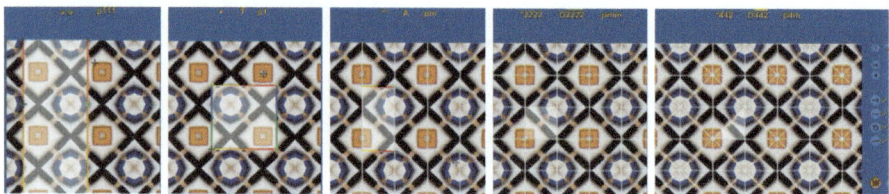

Fig. 39 Finding symmetries, stamps and motifs

or with any other that the user chooses, it is possible to obtain the pattern or frieze corresponding to each stamp. It is also possible to add drawings to the motif, using an incorporated image editor (Fig. 38).

The next program lets the user interactively try to find all the symmetries of a pattern or a frieze, and eventually find the corresponding stamp and obtain a motif (a *fundamental region*) for that stamp (Fig. 39).

It also lets the user compare the initial pattern with the one obtained with the stamp and the motifs found. All this can be done with images that are on the DVD and with images provided by the user.

In the previous programs, for each pattern (or frieze) there is a plane motif with a perfectly determined standard shape. The next program shows how, for many of the stamps, this shape is not unique. The program displays examples of motifs with

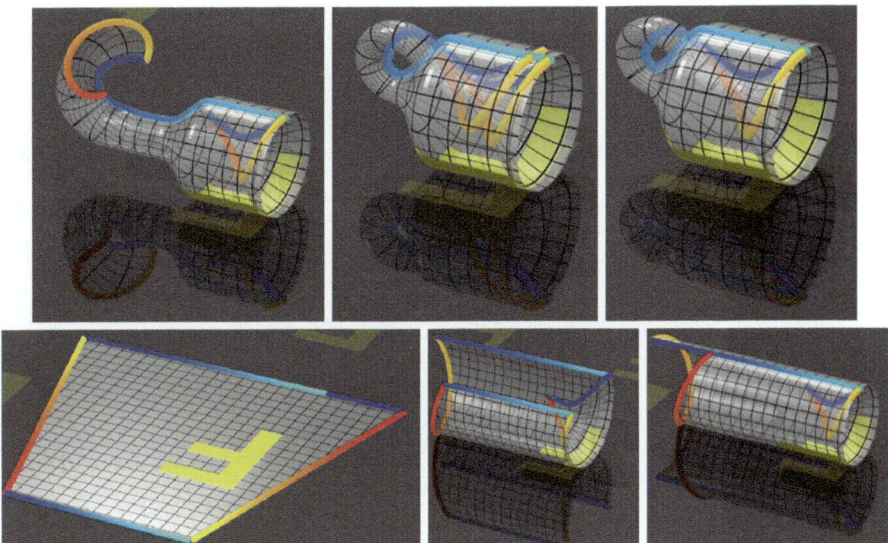

Fig. 40 Gluing a non-standard motif to build the Klein bottle stamp

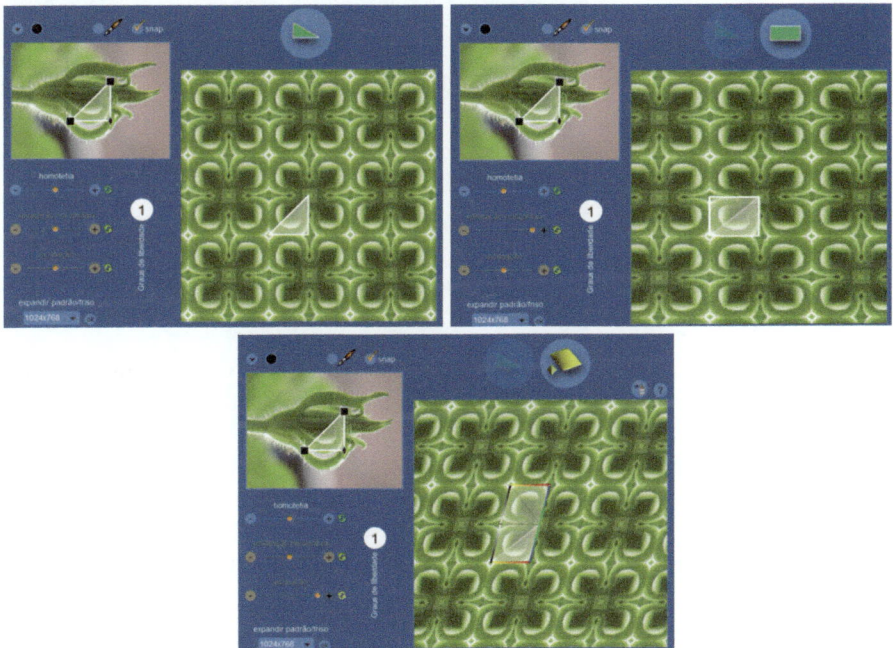

Fig. 41 Finding the degrees of freedom

non-standard shapes and highlights the fact that although these shapes are different, after the edges are glued together, the final shape of the stamps match that of the stamps obtained from the *standard* motifs (Fig. 40).

Fig. 42 Enhanced pattern generator

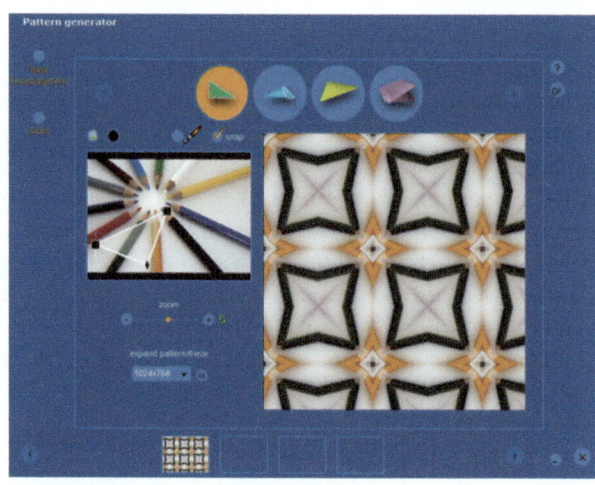

Fig. 43 Enhanced pattern classifier

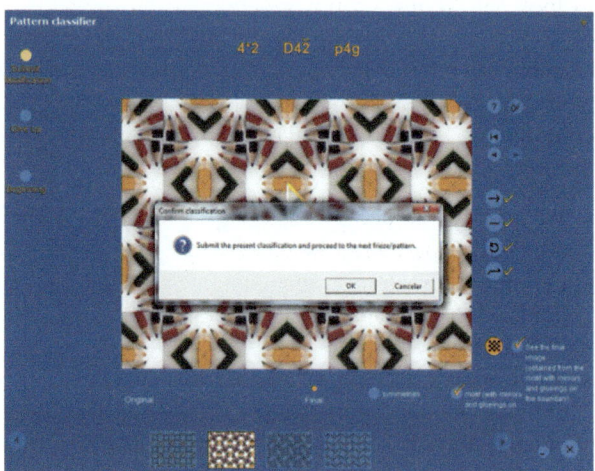

The final program highlights the degrees of freedom of a pattern (or frieze), i.e., the larger or smaller number of affine transformations that preserve every symmetry. If it is clear that a dilation of the plane transforms any pattern or frieze into another with exactly the same symmetries, there will be other affine transformations, for example a small "horizontal expansion", which may in some cases destroy some of the existing symmetries and thus lead to a pattern that requires another type of stamp and a bigger motif (Fig. 41).

A map of the DVD makes it easy to search for specific topics.

Fig. 44 A game in competition mode with four stamps and four images

GeCla (Beta Version)—Work in Progress

The development of a program using the DVD as support, for various types of activities, accessible to audiences of different ages including small children, was scheduled for immediately after its release. Compelling reasons led to this development beginning only a year later: a beta version of this program is now complete and working. It uses a more sophisticated version of the pattern generator, which saves information about the symmetries of the figure generated in an encrypted form (Fig. 42). If the new classifier is subsequently used for pattern classification, "it knows" what the symmetries are and can guide the user in classifying the pattern, according to the help level chosen, correcting errors, or just indicating at the end if the result is correct or not (Fig. 43). Teachers who want their pupils to practise can upload a collection of images onto the Internet and suggest that the pupils classify them, with the degree of help deemed appropriate. A competition version monitors the activities of two players, who alternately generate patterns and then classify the patterns generated by their opponent. At the end, a report (Fig. 44) is produced with small images of the patterns, the correct stamps, the stamps actually found and the final score of both players. The level of the game can be adapted to the age of the players by limiting the choice of stamps allowed. A bilingual (Portuguese and English) beta version of GeCla is available at [19]. A workshop on using GeCla had extremely encouraging results.

References

1. http://www.atractor.pt/div/ems.pdf/ems.pdf
2. http://www.atractor.pt/matviva/geral/modulo.html
3. http://www.atractor.pt/matviva/geral/I/I04/i4.htm
4. http://www.atractor.pt/mat/Sperner/index1-en.htm
5. http://www.atractor.pt/mat/apli_sperner/hairyBall
6. http://www.atractor.pt/mat/apli_sperner/hairyBallEsquema
7. http://www.atractor.pt/mat/conchas
8. http://www.atractor.pt/mat/logo/exemplos/coelhos.html
9. http://www.atractor.pt/mat/puzzle-15
10. http://www.atractor.pt/mat/hexlet
11. http://www.atractor.pt/mat/pendulos
12. http://www.atractor.pt/mat/din_complexa/mandel-zoom
13. http://www.atractor.pt/mat/din_complexa
14. http://www.atractor.pt/stereoP/fr-stereoP.htm
15. http://www.atractor.pt/webM/wm/materiais.htm
16. http://www.atractor.pt/webM/exemplos/poliedros.htm
17. http://www.atractor.pt/webM/exemplos/stereo.htm
18. http://www.atractor.pt/webM/exemplos/taylor.htm
19. http://www.atractor.pt/mat/GeCla/index-en.html
20. http://www.geom.uiuc.edu/docs/doyle/mpls/handouts/node39.html
21. http://www.geom.uiuc.edu/docs/doyle/mpls/handouts/node30.html

Mathematics in Action from Lisbon: Engagement with the Popularization and Communication of Mathematics

Ana Maria Eiró, Suzana Nápoles, José Francisco Rodrigues, and Jorge Nuno Silva

Abstract Although not a new subject, the Popularization of Mathematics, viewed as an activity of "sharing mathematics with a wider public" and "encouraging people to be more active mathematically", is a relatively recent topic that motivated the fifth international study (Howson and Kahane in The Popularization of Mathematics, Cambridge Univ. Press, Cambridge, 1990) of the International Commission on Mathematical Instruction and had seen important developments after the World Mathematical Year—WMY2000.

In this article we describe the engagement with the popularization and communication of mathematics that is being developed at the University of Lisbon, within the historical environment provided by its Museum of Science, which has a rich heritage. There are a significant number of initiatives open to society, from the educational community to the general public, and there has been a special collaboration with the Portuguese Mathematical Society since its foundation. Enhanced by the challenge of the WMY2000 and the project *Matemática em Acção*, developed by the *Centro de Matemática e Aplicações Fundamentais*, we describe several activities such as exhibitions, films, interactive applications and publications, that range from ruled surfaces, sundials and architecture to mathematical games, taking into

A.M. Eiró (✉)
Museu de Ciência and Faculdade de Ciências da Universidade de Lisboa, Lisbon, Portugal
e-mail: ameiro@fc.ul.pt

S. Nápoles
FCUL (Faculdade de Ciências da Universidade de Lisboa), Departamento de Matemática, Lisbon, Portugal
e-mail: napoles@ptmat.fc.ul.pt

J.F. Rodrigues
FCUL/Centro de Matemática e Aplicações Fundamentais da Universidade de Lisboa, Lisbon, Portugal
e-mail: rodrigue@fc.ul.pt

J.N. Silva
FCUL/Centro Interuniversitário de História das Ciências e da Tecnologia and Associação Ludus, Lisbon, Portugal
e-mail: jnsilva@gmail.com

E. Behrends et al. (eds.), *Raising Public Awareness of Mathematics*,
DOI 10.1007/978-3-642-25710-0_11, © Springer-Verlag Berlin Heidelberg 2012

account mathematics as an educational resource and exploring the multimedia and computational tools for communicating mathematics.

An Historic Environment

Mathematics has a strong relationship with its past and the role of mathematical objects in time and space is a very special one. The history of mathematics is a very useful tool for popularizing mathematics and for helping mathematicians and mathematics teachers build the correct image of their science which is a key for development and for facing the great challenges of the 21st century.

The Museum of Science of the University of Lisbon is located in a 19th century building in the historical part of Lisbon (Fig. 1). Its cultural heritage, however, dates from a Jesuits' school of the 17th century, with the Noviciate of Cotovia. The College of the Nobles was created in the Enlightenment, based on the reforming ideas of the Marquis of Pombal, and the Polytechnic School was created in 1837 following the liberal reforms of the 19th century. With the re-foundation of the University of Lisbon in 1911, the Polytechnic School was transformed into the Faculty of Sciences. In the 1980s, the Faculty initiated a move to new facilities providing an opportunity for the creation of the Museum of Science, in 1985 [2].

The first century of teaching mathematics, at the Polytechnic School and during the earlier decades of the Faculty of Sciences, was strongly associated with the teaching of astronomy and surveying sciences. This justified the creation, in 1875, of an astronomical observatory for training students [3], which was reconstructed in 1898 (Fig. 2). Among the professors that taught astronomy and mathematics we refer Filipe Folque (1800–1874) and José Sebastião e Silva (1914–1972), respectively.

Within this historic background and with a well-defined mission of collecting, preserving, and studying scientific instruments and scientific heritage in general, the Museum of Science aims to communicate science to broad audiences in astronomy, mathematics, physics and chemistry [4]. The Museum has established close

Fig. 1 The 19th century building of the Polytechnic School, later the Faculty of Sciences, hosting today two Museums of the University of Lisbon: Science and Natural History

Fig. 2 The Astronomical
Observatory of the
Polytechnic School (1898),
© Mark Heller

Fig. 3 Spherical triangle
calculator. Astronomical
Observatory of the Faculty of
Science of the University of
Lisbon, MCUL547

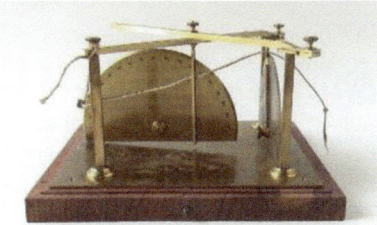

collaborations with several institutions, in particular in mathematics, with the *Centro de Matemática e Aplicações Fundamentais* (CMAF) http://cmaf.ptmat.fc.ul.pt/ and the Department of Mathematics of the Faculty of Sciences, but also it has old and strong links with the Portuguese Mathematical Society (SPM). It is important to underline that the Museum collections held important historical archives related to mathematics, including relevant information about SPM. In fact, this Society was created in 1940 in the building where the Museum now stands.[1]

From Heritage to Society

With its rich historic legacy, the Museum is increasingly trying to integrate the collection (Fig. 3), history and archives into the communication programs developed for society. Over the past 25 years the Museum has held exhibitions, educational programs and special events, including scientific conferences, workshops, and many mathematics activities.

In 1987, for the bicentenary commemorations of José Anastácio da Cunha, a Portuguese mathematician and poet, an important temporary exhibition was held at the

---

[1] The historic building of the Museum also hosted the first editorial office of *Portugaliae Mathematica*, the Portuguese mathematics research journal founded in 1937, the *Centro de Estudos Matemáticos de Lisboa* and also the first editorial office of the *Gazeta de Matemática*, both dating from 1940.

National Library in Lisbon [5]. In December 1990, the commemoration of 50 years of the Portuguese Mathematical Society at the Museum of Science included an exhibition *Efervescência Matemática 1937–1947*. This exhibition evolved into *Movimento Matemático 1937–1947*, a traveling version developed by SPM that started in 1997 at the *Museu da República*, with a catalogue published by the city of Lisbon [6].

In November 1997, an international conference on museums of science and technology, jointly organized by the Museum and the *Fundação Oriente*, was held in Arrábida, in a 16th century Franciscan monastery 30 km south of Lisbon. The proceedings [7] include two articles on the popularization of mathematics, a subject that was then creating an upsurge in museological interest.[2] Albrecht Beutelspacher revealed his pioneering project of a museum totally dedicated to mathematics [8] and Arala Chaves [9] showed the museological value of interactive exhibitions for mathematics, in particular, with the use of physical models and computer simulations.

Among the initiatives of preparation of the World Mathematical Year 2000 (WMY2000), the Museum of Science and CMAF helped organize a series of seven public lectures on mathematics and music, held at the *Fundação Calouste Gulbenkian*, in Lisbon, from January to July of 1998, that were published the next year [10]. Under the auspices of the European Mathematical Society, the Fourth Diderot Mathematical Forum on Mathematics and Music took place in December 1999 simultaneously in Lisbon, Paris and Vienna[3] [11]. In the aftermath of the WMY2000, the Museum of Science also hosted in December 2001–February 2002 the itinerant exhibition *Beyond the Third Dimension*.

In 2007, the collaboration between the Museum and SPM was reinforced and two temporary exhibitions were produced: on the Portuguese mathematician Aniceto Monteiro [12] and on the history of *Portugaliae Mathematica*, the journal founded by Monteiro in 1937. Both exhibitions provided an opportunity to develop several activities of the mathematics club. A big mathematics fair was organized for the children´s day, attracting over one thousand kids. *Symmetries and mirrors*, a temporary exhibition produced by *Atractor* http://www.atractor.pt/ was displayed in the summer of 2007 and *Experiencing Mathematics* http://www.mathex.org/, an itinerant exhibition sponsored by UNESCO, ended its six month tour of Portugal at the Museum in the summer of 2008.

Two recent exhibitions were conceived and developed locally: *Mathematical Games throughout the Ages* and *Measuring the sky to rule the territory*. The first is an exhibition where unique replicas of historical board games are on display, coupled with hands-on boards for the public to use and play. It includes games like Pentalfa, Hex, Ludus Globi, Ludus Regularis, Rithmomachia (Fig. 4), Ouranomachia,

---

[2]These proceedings also include an article on the Beijing astronomical observatory, which has more than 550 years of history and possesses unique bronze mathematical instruments of the Ming dynasty (17th century), which have recently been restored and constitute an open air museum.

[3]The Portuguese component of the Diderot Mathematical Forum was organized by CMAF and was partially held at the University of Lisbon in co-ordination with the XII National Seminar on the History of Mathematics organized by SPM.

Fig. 4 Rithmomachia in *Mathematical games throughout the ages*: an exhibition at the MCUL, Lisbon

Fig. 5 Circle of proportions, Elias Allen, London, c. 1630, MCUL501

Ludus Astronomorum and Stomachion. Still open, it was inaugurated in April 2008, when the Museum hosted the Board Game Studies Colloquium XI [13] organized by the *Associação Ludus*. During that colloquium, the first international report [14] was presented on the Portuguese Championship of Mathematical Games, co-organized by *Ludus*, SPM and APM (Mathematics Teachers Association) and sponsored by *Ciência Viva*, the national agency for science popularization. Since 2004, this championship has involved an increasing number of students aged 7 to 17, from all over the country, reaching 100,000 participants in the seventh championship in 2011.

The second exhibition started during the International Year of Astronomy and following an international conference, organized at the Museum on the *History of Astronomy in Portugal* [15], in September 2009, in collaboration with CMAF and the *Centro Internacional de Matemática* (CIM). *Measuring the sky to rule the territory* (October 2009–August 2010), was a historical exhibition on the teaching of Astronomy at the Polytechnic School, showing original astronomical instruments (Fig. 5) and documents, aimed at highlighting the connections between mathematics and astronomy [16].

Deriving from a previous exhibition about sundials, part of this exhibition explored the observation of stars for *Measuring time, measuring the world and measuring the sea* (Fig. 6), making a separate section. References to ancient Greek philosophers, as Eratosthenes of Cyrene, Hipparchus of Nicaea and Aristarchus of Samos, as well as Portuguese 15th and 16th century sea voyages were interpreted in panels and organized in that separate section. This part became a traveling exhi-

Fig. 6 *Measuring time, measuring the world, measuring the sea*: part of the exhibition *Measuring the sky to rule the territory*, at the MCUL, in Lisbon

bition, which started in Óbidos, during the RPAM workshop in September 2010, is available for loan through SPM and is shown in secondary schools and other cultural institutions.

Activities for children and young people, schools and families have also been developed at the Museum of the University of Lisbon. In mathematics these activities explore logic, geometry and topology, by using puzzles, wood constructions, strategy games and board games.[4] Typical audiences are secondary school students, the majority between 10 and 15 years old. Participation in these activities, which cover all areas of science (physics, chemistry, mathematics, astronomy and also biology and geology), reached 48,000 in 2009. It is quite impressive that between mathematics and astronomy the total number of students reached 18,000, representing 37 % of the total number of young visitors. And mathematics is one of the most popular subjects!

WMY2000 and the Project *Matemática em Acção*

During the World Mathematical Year 2000 many research institutions accepted the challenge of dedicating some effort to public awareness activities to help improve the image of mathematics. Answering this call, CMAF decided to initiate a programmatic activity on "mathematics as an educational resource" and launched a project for the popularization and communication of the mathematical sciences called *Matemática em Acção* http://wwmat.ptmat.fc.ul.pt/em_accao/. This project (Fig. 7) started essentially with two complementary goals: to enhance the interaction with schools and their teachers through the development of scientific material

---

[4]To support these activities, in addition to traditional mathematical puzzles, new materials have been produced as, for instance, an original puzzle with polyhedra and numbers exploring permutation groups invented by Jorge Rezende [41].

Fig. 7 Logo of the project
Matemática em Acção, 2000

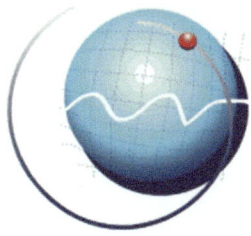

about mathematics through an innovative approach and the promotion of public awareness activities related for mathematics.

Concerning the first goal, in the year 2000 CMAF produced the Portuguese version of the video series from the California Institute of Technology *Project MATHEMATICS*, by Tom Apostol. This remarkable collection of eight didactical videos, covering topics such as the Pythagorean theorem, similarity, polynomials and trigonometry, introduces mathematical ideas by animated images, capitalizing the viewer's geometric intuition and conveying school mathematics in a rich cultural context [17]. Because these topics are integrated with the pre-university curricula, the Portuguese Ministry of Education supported all costs concerning the production and the free distribution of the Portuguese versions of these videotapes in public schools. Sessions with teachers to improve the use of the videos in the classroom were organized. These sessions showed the importance of providing scientific support to teachers concerning the deeper approaches of the topics included in the videos and to stimulate their use in a non passive way. In the following year the project *Matemática em Acção*, produced and distributed the Portuguese version of Apostol's ninth videotape, on the *Early History of Mathematics*, describing in about 30 minutes some important developments, from Babylonian calendars on clay tablets, to the invention of calculus in the seventeenth century.

Concerning the second objective and to illustrate the role of mathematics in several fields the project started to develop itinerant exhibitions, available to schools and cultural entities upon request. Starting from a virtual exhibition that was the result of an international cooperation between two mathematicians in the United States (T. Banchoff and D. Cervone) and a local team of mathematicians and computer professionals [18], and developed only through the Internet, *Para Além da Terceira Dimensão—Beyond the Third Dimension* (Fig. 8) was recreated as a physical exhibition with the collaboration of the town of Óbidos, starting on 14th October 2000. Banchoff and Cervone came to Portugal for 22nd November for the opening of the exhibition in Funchal, at the University of Madeira. Becoming a traveling exhibition it visited more than twelve institutions in Portugal and crossed the Atlantic to the *Instituto de Matemática Pura e Aplicada*, in Rio de Janeiro, Brazil, in July 2001. It is still alive in cyberspace http://alem3d.obidos.org. This virtual exhibition, named after a classic book [19] by Thomas F. Banchoff, describes twelve surfaces and includes small movies representing aspects of geometrical objects in three and four dimensions, like the "Triple Point Twist", "Necklaces", "In and Outside the Torus", "The Klein Bottle", "Math Horizon" and two animated pieces on the hy-

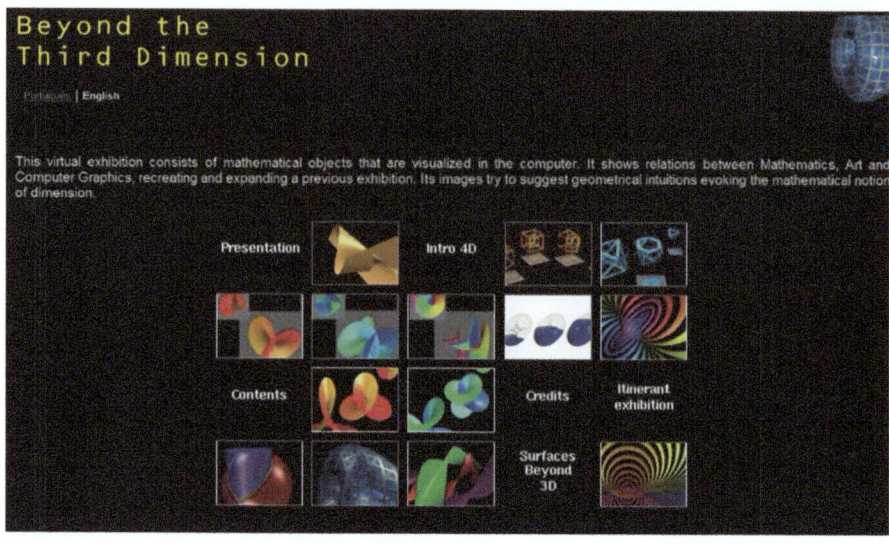

Fig. 8 *Para Além da Terceira Dimensão—Beyond the Third Dimension*

percube, "A Rotation of Cubes" and "Iced Cubes" [20]. These images suggest geometrical intuitions evoking the notion of dimension and illustrate the relationship between mathematics, art and computer graphics. They stimulate the imagination and the curiosity of young people and the public in general, showing the potential of new technology for information and communication, through a fascinating experience that relates mathematical concepts to real and abstract objects. They are one element of a virtual museum of mathematics [21], a vision that several mathematicians started to dream of in the last century and is now becoming, step by step, a reality.

Within the framework of the WMY2000, a Portuguese version of the computer-generated film on the mathematical visualization of minimal surfaces *Touching Soap Films* http://page.mi.fu-berlin.de/polthier/video/Touching/index.html, by K. Polthier and co-authors, was produced. It was distributed on a non-profit basis, for example, to the high schools. CMAF sponsored the European launch of the CD-ROM *Raising Public Awareness of Mathematics*, a project coordinated by M. Chaleyat-Maurel, V.L. Hansen and R. Brown, which took place in the town of Óbidos, on 11 November 2000. CMAF also organized the international workshop on *Multimedia Tools for Communicating Mathematics* that took place in Lisbon [22] in that year.

Two new exhibitions were organized in the town of Óbidos by CMAF, both in collaboration with SPM. The first, *As Sombras do Tempo—The Shadows of Time* [23] is a successful exhibition on sundials that began in that town in June 2002. The other exhibition, the *1st Exhibition of Mathematical Games*, opened in June 2003. This initiative and another one that started in the following year, the Portuguese Championship of Mathematical Games, were at the origin of *Associação Ludus* http://ludicum.org. This association, based at the Museum of Science also aims to promote mathematics, emphasizing its cultural, historical and recreational aspects,

Fig. 9 Ruled Surface, Fabre
de Lagrange, Paris, 1861,
MCUL 1122

focusing on mathematical games. The fascinating topic of board games was also the subject of two other exhibitions, organized by the Department of Mathematics of the Faculty of Sciences. The first, in collaboration with the City Museum of the Town of Lisbon, in 2004, with rich ethnological and archaeological components was *Pedras que Jogam* (*Stones that Play*, [24]) http://mat.fc.ul.pt/divulgacao/pedrasquejogam/ inaugural. The second, in 2006 also in Lisbon, in collaboration with the Centro Cultural de Belém was *Matemática em Jogo* and had some computer games on display http://mat.fc.ul.pt/divulgacao/mej/.

Ruled Surfaces, Sundials and Architecture

Models for designing ruled surfaces and their intersections were used in the course on descriptive geometry, created in 1859 at the Polytechnic School. The Museum of Science lodges a collection of twenty of those models, dating from the 19th century and constructed by Fabre de Lagrange in Paris (Fig. 9). This collection inspired the project *Geometry: to manipulate and to visualize*, which aimed to: (i) construct replicas of three of the models; (ii) use the replicas for the visualization of the surfaces simulated by the straight lines (cylinder, cone, hyperboloid of one sheet, hyperbolic paraboloid); (iii) model the devices using *Mathematica* and to carry out a mathematical study of the surfaces produced by the instruments.[5]

Research into ruled surfaces in contemporary architecture was developed and presented in the exhibition *The Collections of the Museum* on show at the Museum of Science from March to September 2009. As emblematic examples, we have the Cathedral of Brasília by Óscar Niemeyer (Fig. 10), which uses a hyperboloid of one sheet, and the Oceanarium in Valencia by Felix Candela (Fig. 11), which uses hyperbolic paraboloids. These surfaces are also used for roofs as by St Mary's Cathedral

---

[5]This work was carried out in partnership with the S. João de Brito School and was presented in the 4th Forum *Ciência Viva*, held in Lisbon in May 2000 http://wwmat.mat.fc.ul.pt/em_accao/superficies_regradas/.

Fig. 10 Cathedral of Brasília

Fig. 11 Oceanarium in Valencia

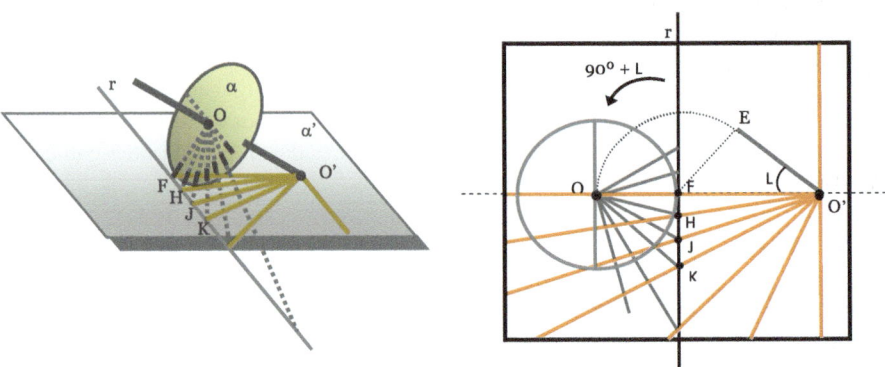

Fig. 12 How to draw the hour lines on a horizontal dial

in San Francisco (Pietro Belluschi and Pier Luigi Nervi). The Portuguese architect Álvaro Siza also used a ruled surface having a catenary as directrix for the Tent of Portugal pavilion at the EXPO 98, a building that marks the renewed east-end of Lisbon.

Sundials are a good way for students to relate concepts of astronomy and of mathematics together with the history of these instruments. They also promote an interest in architectural and aesthetics aspects, and so they are important for the rounded development of the individual.

Starting with the most basic sundial, the equatorial one, we can directly draw the hour lines of a horizontal (or vertical) dials by projecting the equally spaced hour lines of an equatorial dial (equally spaced by 15°) onto a horizontal (or vertical) plane (Fig. 12).

From a mathematical point of view, we can use basic plane trigonometry to calculate the angles between the hour lines of a horizontal dial constructed for any latitude L. Using the rectangular triangles FOO′, FHO and FHO′ we have:

Fig. 13 *As Sombras do Tempo* (The Shadows of Time)

$$\sin L = \overline{FO}/\overline{FO'}, \qquad \tan 15° = \overline{FH}/\overline{FO} \quad \text{and}$$
$$\tan \angle FO'H = \overline{FH}/\overline{FO'} = \overline{FO}\tan 15°/\overline{FO'},$$

that is,

$$\tan \angle FO'H = (\sin L)(\tan 15°).$$

Similarly, as $\angle FOJ = 30°$, $\angle FOK = 45°$, we have:

$$\tan \angle FO'J = (\sin L)(\tan 30°), \qquad \tan \angle FO'K = (\sin L)(\tan 45°) \quad \text{and so forth.}$$

For vertical sundials the formulas are analogous, using $\cos L$, if the plane of the hour lines faces south, but can be more complicated if this plane is not perpendicular to the north-south direction.

These simple properties led to the development of the already mentioned itinerant exhibition *As Sombras do Tempo—The Shadows of Time* (Fig. 13), based on the work developed with prospective teachers concerning the connections between astronomy and mathematics.[6] This exhibition was first shown in Óbidos during the interdisciplinary conference *Nexus 2002: Relationships between Architecture and Mathematics*. This conference, the fourth of a series that started in Italy, was co-organized by the project *Matemática em Acção* and contributed to the identification of the mathematical principles that are used, on one hand, as a basis for architectural design or as tools for analyzing existing monuments and, on the other hand, to see architecture as a concrete expression of mathematical ideas or, in a sense, as "visual mathematics" through history and among different cultures [25].

The interest that it raised and the multiple requests led to a new exhibition, *Sundials and Mathematics*, which included more scientific details and images of Sundials in Lisbon (Fig. 14). All those activities were coordinated within *The Shadows of Time* project http://sombrasdotempo.org/, in the framework of *Matemática em Acção*, which develops interactions between several levels of teaching using sundials. While contributing to the awareness of sundials in Portugal, this project intends

---

[6]This initiative was developed as a collaboration between the high school *António Arroio* and the Department of Mathematics, Faculty of Science of the University of Lisbon.

Fig. 14 Sundial at Sé de
Lisboa (Lisbon Cathedral)

Fig. 15 The biggest sundial
in Portugal

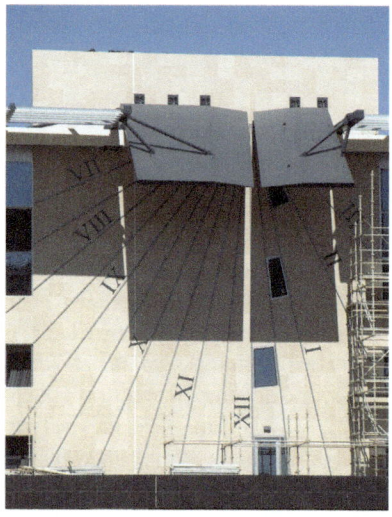

to help preserve the national heritage by encouraging the restoration of these monuments, some of them totally abandoned in recent decades [26]. It also gives scientific
support to the design of dials, as for instance, the collaboration between S. Nápoles,
N. Crato and F. Correia de Oliveira concerning the hour lines of the biggest sundial in Portugal (Fig. 15), in the front of the Tempus International building (author-
ship Tetractys-Architects). In this example of contemporary architecture, there is no
shadow projected by a gnomon to indicate the hours. It is a line of light that indicates
the hours. The declination of the wall containing the hour lines imposed the type of
the dial, an oriental one. In this case, the angle z between an hour line and the noon
line can be obtained using the corresponding angle H in the equatorial sundial with
the same gnomon and the formula $\tan z = \cos L / (\cos d \cot H + \sin d \sin L)$ where L
is the latitude and d ($= 7.78°$) is the declination of the plane of the dial.

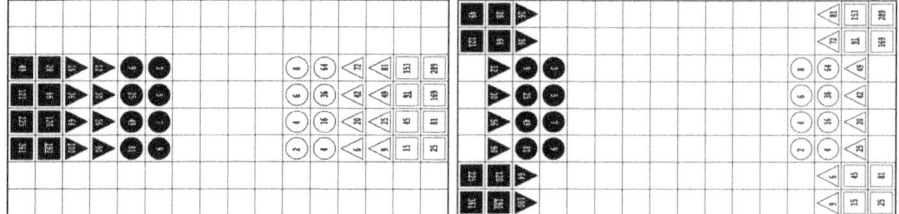

Fig. 16 Schematic representation of Rithmomachia's board

Going Further with Mathematical Games

Games and puzzles, whose practice resembles the mathematical way of thinking, in one way or another, have been known throughout history, and were surveyed recently [27]. The interaction of games and puzzles with recreational mathematics has a long tradition and shows kinship with mathematics and logic. This interaction has stimulated not only the invention of computer mathematical games such as, for example, OZ (tilings of polygons by lozenges [28]), but also mathematical challenges, like the complete solution of the problem for characterizing a class of patterns for a solitaire game formulated by Maxim Kontsevich [29]. This interaction was shown, in particular, at two colloquia on recreational mathematics (see http://ludicum.org/rm09 and http://ludicum.org/rm11), that were recently organized by *Associação Ludus*, with the support of the Museum of Science.

How can we say that one board game is more mathematical than another? The answer to this question is not easy to find. However, it is not difficult to agree that chess and go are more mathematical than Monopoly. The connections with mathematics are diverse and can be obvious or lie hidden in structure of the rules. To illustrate this fact, consider two games that deserve special attention: Rithmomachia and Hex. The first is the oldest didactical game known [30]. Created in the 11th century to teach the arithmetic of the quadrivium, it was known by every educated man in Europe until the 16th century. It was used in the classroom, first in the monasteries and later in the universities, and only when the mathematical tradition changed did this game, sharing the fate of the quadrivium, vanish. Hex was independently invented twice in the 1940s and is the paradigm of a new kind of board game: a connection game [31]. The fact that one of its inventors was the Nobel laureate John Nash, besides having been popularized by Gardner, adds to its relevance. It has been noticed that there is a profound connection between Hex and higher mathematics.

Rithmomachia unfolds on a chequered 8×16 board, where two adversaries— White (even) and Black (odd)—control their armies of 24 numbered pieces (Fig. 16).

The starting position is determined by several mathematical relations and proportions. We explain how to implement the first of the two positions shown. The second, the starting position in Rithmomachia, is easily derived from the first. We treat the disposition of the white pieces; the black ones follow a similar pattern. Each piece has a number, and below we show how the pieces are laid out in the first position, with modern notation to the right.

Fig. 17 Rithmomachia's pieces in Barozzi's manual

From the even numbers 2, 4, 6, 8 placed on circular pieces in the central cells of the sixth row, we deduce the remaining numbers. The second row, with circular pieces, has the squares of the numbers in the first:

2	4	6	8	n
4	16	36	64	n^2

The following row, with triangular pieces, gets its numbers by addition of the previous two. The next row also has triangular pieces. Each piece has a number that is the square of one more than the number on the first circular piece. The pieces of the first row of square shapes get their numbers from addition of the previous two. All these operations are illustrated below:

2	4	6	8	n
4	16	36	64	n^2
6	20	42	72	$n(n+1)$
9	25	49	81	$(n+1)^2$
15	45	91	153	$(n+1)(2n+1)$
25	81	69	289	$(2n+1)^2$

There are two pieces (91 and 190), one of each color, that are replaced by special solid objects—the pyramids (Fig. 17). These are sums of square numbers, each corresponding to one of the described shapes. In the version of the rules that we are considering, the white pyramid has six stories, numbered 1, 4, 9, 16, 25 and 36. The black one has five: 16, 25, 36, 49 and 64. Note that $1 + 4 + 9 + 16 + 25 + 36 = 91$ and $16 + 25 + 36 + 49 + 64 = 190$.

Movement is easy: discs move orthogonally one square, triangles two cells diagonally, while squares move three steps, orthogonally. Pyramids move as the chess queen, but are limited to four squares. Captures can happen in several ways. For instance, captures relating to addition and subtraction: if two pieces belonging to the same player can move to a cell occupied by an adversary piece and if the number of this piece is equal to the sum or difference of the numbers of the pieces of the first player, the piece is captured. There is no replacement.

A player wins when he places, in the adversary's half of the board, three pieces in any of the types of mathematical progression. For instance 2-15-28 is an arithmetic progression, 9-15-25 is a geometric one, and 9-15-45 is a harmonic progression. Victory can also be achieved by placing four pieces showing a combination of two

Fig. 18 Empty Hex board

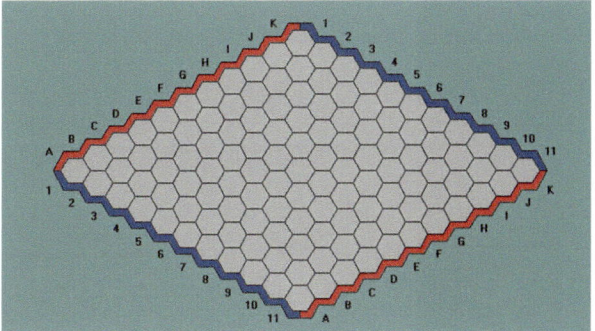

progressions, as 2-3-4-8, a combination of arithmetic (2-3-4) and geometric (2-4-8) progressions, or even three, as in 4-6-9-12.

Hex (Fig. 18) was invented twice in a short period of time first by Piet Hein, a Danish scientist and poet, in 1942, and by the mathematician John Nash in 1947. The rules of Hex are surprisingly simple: players alternate placing counters of two different colors, trying to connect parallel margins of the board. The game cannot be drawn, and this can be seen in several ways, of varying formal presentation. If we assume that the board is filled with red and blue pieces, and think of blue as water and red as one foot-high walls and bricks, then either water flows between the blue margins (blue wins) or there is a red wall preventing blue communication (red wins). Proofs using graph theory and by induction can be found in the literature. This result is known as the Hex theorem.

Nash gave a beautiful non-constructive proof that the first player must have a winning strategy in Hex. His argument—the now famous *strategy stealing argument*—goes as follows: as there are no ties, either the first or second player must have a winning strategy. Suppose it is the second. Then the first player makes his first move at random and assumes the role of the second player. The reason this trick works is that, as Hex is a connection game, an extra move, no matter how bad it might be, cannot hurt its player. Therefore, with this plan, the first player wins the game, which gives us a contradiction. We must conclude that the first player has a winning strategy. If the size of the board is large enough, nobody knows how to play perfectly, so the practice of Hex is not ruined by Nash's result.

The connection between Hex and mathematics was emphasized when David Gale, a mathematician from Berkeley, showed that the Hex theorem is logically equivalent to Brouwer's fixed-point theorem [32], a deep result in topology. Knowing this equivalence does not increase your playing strength, but shows how abstract games and mathematics are linked in a mysterious, fascinating way.

Mathematics as an Educational Resource

At the 2000 conference on *Multimedia Tools for Communicating Mathematics* organized by CMAF in Lisbon, the authors of the interactive geometry software *Cin-*

Fig. 19 Cinderella: an
interactive book of geometry
(in Portuguese)

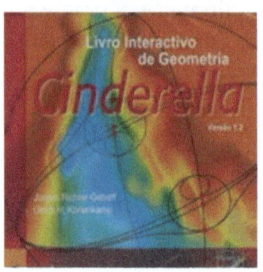

derella presented the theoretical background that is required to build a consistent
and continuous foundation for dynamic elementary geometry [33]. This helped the
project *Matemática em Acção* to develop a further collaboration with the Ministry of
Education, who sponsored the Portuguese translation and edition of the dynamic geometry software *Cinderella*, as well as its manual (Fig. 19). Every secondary school
in Portugal received in 2001 a free copy of the interactive book [34]. A virtual forum
was created on the web, containing hundreds of applets and experiences, which are
shared through the *Cinderella* Portuguese website http://cinderella.ptmat.fc.ul.pt/.
That conference also allowed the project *Matemática em Acção* to develop a collaboration with the Computer Science Department of the Faculty of Sciences of
the University of Lisbon. This collaboration meant that the concept of hypervideos
could be added to the multimedia tools for communicating and learning mathematics [35]. A first demo was developed based on the video *The Story of Pi*, by Apostol
[36], and presented in a mini-symposium at ICIAM 2003, in Sydney, and at the
National Meeting of Scientific Visualization held in Espinho, Portugal, in 2005. Finally, the complete interactive CD-ROM [37] was published in Portuguese in 2007.
Although based on the components of the Portuguese translation of the original
Apostol film, the final hypervideo has an original structure and interactive navigation, which goes far beyond the limited capability of video supporting learning and
has attached a new virtual booklet with the history of the number *Pi*. Hypermedia
gives the user greater control and autonomy, so he can explore the links to the information which is conveyed by the video and complemented by other material, aug-
menting its capabilities as a cognitive artifact. Hypermedia allows the exploration
of topics at different levels of knowledge with several levels of complexity, from
elementary topics like the calculation of areas and volumes to more sophisticated
concepts like random phenomena, approximate calculation and Fourier series. This
makes the hypervideo *A História do Pi* (Fig. 20) a useful tool for teacher training.

The use and the development of the concept of hypermedia was successfully
proposed to the University of Madeira for the training of mathematics teachers and
has led to three master's theses, including one that developed a plan for integrating
the three Apostol videos on trigonometry into a single hypervideo adapted to the
Portuguese secondary school curricula [38].

The study of the applications of Mathematics has been revealed as one of the
most efficient ways to motivate the apprenticeship of this science. A comprehensive
foundation, as well as the establishment of connections, is essential in education.
Bearing in mind the importance of providing materials and resources for teachers,

Fig. 20 *The history of Pi* in hypervideo (in Portuguese)

several have been produced [39, 40], and an interactive book is currently under development.[7] The interactive online book *Studying Mathematics with Spreadsheet* http://matematicainteractiva_dm.fc.ul.pt, includes links to computational applications and activities, designed to help teachers in almost all subjects. The design seeks to identify concepts and problems that students struggle with, and complement them with appropriate technology-based tools. The resulting tools vary in terms of their interactivity, but they are closely related to conceptual ideas presented in the textbooks and always raise questions for the students and the teacher to explore.

Final Remarks

Although the popularization and teaching of mathematics are related they are different components of the communication of mathematics, and it is clear that there are important differences between them. For example in their aims: the purpose of popularization is to raise awareness, not just to educate, and the criterion of success is not only an increase of knowledge, but also a change of attitude [1]. However, new tools and new media provide new possibilities and raise new challenges in the communication of mathematics to different audiences, from pupils to teachers, from other scientists to the general public, requiring innovative ideas and interdisciplinary collaboration.

References

1. Howson, A.G., Kahane, J.-P. (eds.): The Popularization of Mathematics. Cambridge Univ. Press, Cambridge (1990). ICMI Study #5

---

[7]This interactive book is in development through a partnership between the project *Matemática em Acção*, the Department of Mathematics, Faculty of Science of the University of Lisbon and a Basic School (EB2,3 Piscinas, Lisbon).

2. Bragança Gil, F.: Museu de Ciência da Universidade de Lisboa: Das Origens Ao Pleno Reconhecimento. MCUL, Lisboa (2003)
3. Rivotti Silva, V.: L'Observatoire Astronomique. In: Museums of Science and Technology, pp. 125–130. Fundação Oriente, Lisboa (1998)
4. Lourenço, M.C., Eiró, A.M.: O Museu de Ciência. In: Lourenço, M.C., Neto, M.J. (eds.) Património da Universidade de Lisboa – Ciência e Arte, pp. 35–54. Tinta da China, Lisboa (2011)
5. José Anastácio da Cunha (1744–1787), Matemático e Poeta, Catalogue of the exhibition. Biblioteca Nacional, Lisboa (1987)
6. Movimento Matemático 1937–1947, Catalogue of the exhibition commissioned by Ilda Perez. Câmara Municipal de Lisboa (1997)
7. Alzira Ferreira, M., Rodrigues, J.F. (eds.): Museums of Science and Technology. Fundação Oriente, Lisboa (1998)
8. Beutelspacher, A.: Towards a mathematics museum. In: Museums of Science and Technology, pp. 155–163. Fundação Oriente, Lisboa (1998)
9. Arala Chaves, M.: Expositions interactives de mathématiques. In: Museums of Science and Technology, pp. 165–188. Fundação Oriente, Lisboa (1998)
10. A matemática e a música, Colóquio/Ciências, Revista de Cultura Científica, n°24, November 1999, Fundação Calouste Gulbenkian, Lisboa
11. Assayag, G., Feichtinger, H.D., Rodrigues, J.F. (eds.): Mathematics and Music, a Diderot Mathematical Forum. Springer, Berlin (2002)
12. Rezende, J., Monteiro, L., Amaral, E.: António Aniceto Monteiro, Uma Fotobiografia a Várias Vozes. Sociedade Portuguesa de Matemática, Lisboa (2007)
13. Silva, J.N. (ed.): Proceedings of the Board Game Studies Colloquium XI. Associação Ludus, Lisbon (2009)
14. Carvalho, A., Santos, C., Neto, J., Silva, J.N.: Jogos matemáticos, a Portuguese project. In: Silva, J.N. (ed.) Proceedings of the Board Game Studies Colloquium XI, pp. 19–25. Associação Ludus, Lisbon (2009)
15. Saraiva, L.: History of Astronomy in Portugal: Institutions, Theories, Practices, pp. 32–34, CIM Bulletin n°27, Jan. 2010
16. Medir os Céus para dominar a Terra, catalogue of the exhibition, Ed. Museu de Ciência da Universidade de Lisboa (2009)
17. Apostol, T.M.: Computer animated mathematics videotapes. In: Borwein, J., Morales, M.H., Polthier, K., Rodrigues, J.F. (eds.) Multimedia Tools for Communicating Mathematics, pp. 1–27. Springer, Berlin (2002)
18. Banchoff, T.F., Cervone, D.P.: A virtual reconstruction of a virtual exhibit. In: Catalogue of the Exhibition on Sundials Co-ordinated by Suzana Nápoles et al., pp. 29–38. CMAF/Universidade de Lisboa (2002)
19. Banchoff, T.F.: Beyond the Third Dimension: Geometry, Computer Graphics, and Higher Dimensions. Scientific American Library/Freeman, New York (1990)
20. Para Além da Terceira Dimensão—Beyond the Third Dimension Catalogue of the exhibition by CMAF/Universidade de Lisboa (2000)
21. Rodrigues, J.F.: Um Museu Virtual de matemática – um desafio real para o ano 2000! Boletim da SPM, n°34, Maio 1996, pp. 77–79
22. Borwein, J., Morales, M.H., Polthier, K., Rodrigues, J.F. (eds.): Multimedia Tools for Communicating Mathematics. Springer, Berlin (2002)
23. "As Sombras do Tempo – The Shadows of Time", Catalogue of the exhibition on Sundials co-ordinated by Suzana Nápoles et al. CMAF/Universidade de Lisboa (2002)
24. "Pedras que Jogam", Catalogue of the exhibition commissioned by Adelaide Carreira et al., DM/Faculdade de Ciências/Universidade de Lisboa (2004)
25. Rodrigues, J.F., Williams, K. (eds.): NEXUS IV Architecture and Mathematics. KWBooks, Fucecchio (2002)
26. Crato, N., Nápoles, S., Oliveira, F.C.: Relógios de Sol. CTT/Correios de Portugal, Lisboa (2007)

27. Silva, J.N.: On mathematical games. BSHM Bulletin, Journal of the British Society for the History of Mathematics **26**(2), 80–104 (2011)
28. Cláudio, A.P., et al.: Implementing L'OZ. In: Silva, J.N. (ed.) Proceedings of the Recreational Mathematics Colloquium I, pp. 51–58. Associação Ludus, Lisboa (2010)
29. Duarte, P.: Solution to a solitaire game. In: Silva, J.N. (ed.) Proceedings of the Recreational Mathematics Colloquium I, pp. 161–171. Associação Ludus, Lisboa (2010)
30. Moyer, A.E.: The Philosophers' Game: Rithmomachia in Medieval and Renaissance Europe. The University of Michigan Press, Ann Arbor (2001)
31. Browne, C.: Hex Strategy: Making the Right Connections. AK Peters, Wellesley (2000)
32. Gale, D.: The game of hex and the Brouwer fixed-point theorem. American Mathematical Monthly **86**(10), 818–827 (1979)
33. Kortenkamp, U.H., Richter-Gebert, J.: A dynamic setup for elementary geometry. In: Medir os Céus para dominar a Terra, catalogue of the exhibition. Ed. Museu de Ciência da Universidade de Lisboa, pp. 203–219 (2009)
34. Kortenkamp, U.H., Richter-Gebert, J.: Cinderella: Livro Interactivo de Geometria (Versão 1.2; Trad. J.N. Silva), CMAF/Universidade de Lisboa (2001)
35. Chambel, T., Guimarães, N.: Communicating and learning mathematic with hypervideo. In: Medir os Céus para dominar a Terra, catalogue of the exhibition. Ed. Museu de Ciência da Universidade de Lisboa, pp. 79–91 (2009)
36. Apostol, T.M.: The Story of Pi. Project MATHEMATICS, Caltech (1989)
37. Apostol, T.M., Chambel, T., Nápoles, S., Rodrigues, J.F., Santos, L.: A História do Pi em hipervideo, CD-ROM Interactivo, Textos Editores, Lisboa (2007)
38. Sousa, S.: Contribuição Para um Hipervídeo na Trigonometria. Master thesis in educational mathematics (directed by J.F. Rodrigues), Universidade da Madeira, Funchal (2002)
39. Oliveira, M.C., Nápoles, S.: Using a Spreadsheet to Study the Oscillatory Movement of a Mass-Spring System, Spreadsheets in Education (eJSiE), Bond University, Australia (2010)
40. Oliveira, M.C., Nápoles, S.: Sundial's Mathematics and Astronomy with a Spreadsheet, ICERI 2010, International Conference of Education, Research and Innovation, Madrid, 15–17 Nov. 2010
41. Rezende, J.: Puzzles with polyhedra and numbers. In: Silva, J.N. (ed.) Proceedings of the Board Game Studies Colloquium XI, pp. 105–133. Associação Ludus, Lisbon, (2009)

Playing with Mathematics at *Il Giardino di Archimede*

Enrico Giusti

Abstract The article contains a brief description of the exhibitions and laboratories held at "Il Giardino di Archimede" (The Garden of Archimedes) and of the philosophy that inspires them, with special attention to three aspects of mathematical museology: history, play, everyday life.

The mathematics museum *Il Giardino di Archimede* (The Garden of Archimedes) was opened to the public on 13 September 1999. Its predecessor was an exhibition, *Oltre il Compasso* (*Beyond compasses*) dedicated to the geometry of curves and to their influence on everyday life. This exhibition was launched in 1992 and at the time of the opening of the museum it had toured a considerable number of cities, in Italy and abroad, totalling more than half a million visitors. That exhibition was the core of the museum on its opening, and it remains one of its main sections. The first location of the Garden was in the Castle of San Martino, a Renaissance villa near Priverno, a town between Rome and Naples. In 2004 the Garden moved to Florence, where it is today.

The Exhibitions

Since its foundation, the Garden of Archimedes has increased the number and quality of its sections, while remaining faithful to its initial philosophy. Currently the following sections, each constituting an exhibition on its own, are in place, though they are not all visible at the same time, due to space restrictions.

Beyond compasses. The geometry of curves (1992). This exhibition shows the generation and properties of the principal curves, from the straight line and circle to the most important algebraic and transcendental curves, together with their history and uses in technology and everyday life (see Fig. 1). The visitor

E. Giusti (✉)
Il Giardino di Archimede, Firenze, Italy
e-mail: giusti@math.unifi.it

E. Behrends et al. (eds.), *Raising Public Awareness of Mathematics*,
DOI 10.1007/978-3-642-25710-0_12, © Springer-Verlag Berlin Heidelberg 2012

Fig. 1 Cone and conic
sections

can follow any one of three theoretical paths, which are distinct but intercon-
nected. First there is a conceptual path, in which the principal notions of the
geometry of curves are described, in increasing complexity. Overlying this the
historical path shows the evolution of the concept of a curve and the progres-
sive refinement in the relevant mathematical methods. Finally a third path shows
how curves and their properties have been used at various times in science and
technology.

Pythagoras and his theorem (1999). The celebrated Pythagorean theorem and its
variations and generalisations are presented through a number of steps. By solving various puzzles (see Fig. 2), the visitor becomes acquainted with the
Pythagorean theorem and its proof, follows a generalisation of the theorem with
the squares replaced by arbitrary similar figures, takes a path leading to the
quadrature of Hippocrates' lunulae, and is finally led to more general results,
such as the well-known cathetus theorem and the lesser known theorem of Pap-
pus.

Mathematics in Italy: 1800–1950 (2000). Created during the World Mathematics
Year 2000, this exhibition follows the development of contemporary Italian
mathematics and its influence on Italian science and culture. The exhibition
consists of four sections, corresponding to different historical periods. The first
section runs from Napoleon's Italian campaign to the first congress of Italian
scientists (1796–1839). Then follows a period preceding the formation of the
unitary state (1839–1861), and a third part concerning the first 50 years of the
unified Italy, from 1861 to the First World War. The final section covers the Fas-
cist period, the Second World War and the birth of the Italian Republic, ending
in 1950.

Fig. 2 The Pythagorean theorem

A bridge over the Mediterranean Sea. *Leonardo Fibonacci, Arabic science and the revival of Mathematics in the West* (2002). The year 2002 marked the 800th anniversary of the publication of Fibonacci's *Liber Abaci*. The exhibition celebrates Fibonacci's role in the transmission of Arabic mathematics to Western Europe, and the impact of the new mathematics on Italian society (see Fig. 3). After a short illustration of Arabic civilisation and science, the core of the exhibition concerns the life and works of Leonardo Fibonacci, with particular emphasis on his *Liber Abaci*, whose content is illustrated by several panels. There follows a section dedicated to the Abacus schools, a phenomenon unique for its scope and diffusion. Special attention is devoted to the diffusion of the abacus culture in Florence, where more than 20 schools were active between 1300 and 1500. A final panel illustrates the rediscovery of Fibonacci in the nineteenth century.

A short history of calculus (2005). This exhibition started a new line of products by the Garden of Archimedes: pocket-size (15 panels of 70 × 100 cm) and inexpensive exhibitions illustrating important moments in the history of mathematics, suitable for display in schools and universities. More than 20 copies of the exhibition were produced, and are currently used to give the proper historical setting to subjects taught in regular courses. In this vein, the Garden of Archimedes produces light versions of the Pythagoras and Fibonacci exhibitions, ready for printing and also available in languages other than the original Italian.

Ancient mathematics at the post office (2008). The development of mathematics from antiquity to the Middle Ages is seen through the stamps produced by different countries and celebrating important mathematicians and their discoveries. The exhibition covers the mathematics of ancient civilisations: Egypt, the Middle East, China and India; it also includes Greek mathematics and Arab mathe-

Fig. 3 A bridge over the Mediterranean Sea

matics and its transmission to the West. Like the preceding ones, this exhibition is available in small format.

Helping nature. *From Galileo's* Meccaniche *to everyday life* (2009). The simple machines of Renaissance mechanics as described by Galileo in his *Meccaniche*, have been materialised as several large-scale objects. Several common tools show the usefulness of these machines in everyday life. The panels carry quotations from Galileo's book *Le meccaniche*, each illustrating one of the classical machines: the lever, the inclined plane, the pulley (Fig. 4), the screw, the capstan and the Archimedes screw. These instruments are reproduced in large scale, so that visitors can play with them while experiencing their properties.

Weapons of mass instruction. *Mathematical games, puzzles and pastimes* (2010). Various games and pastimes illustrate several mathematical methods, ranging from the induction principle to the proof by contradiction, from topology to binary notation, from solid geometry to probability. Thus the tower of Hanoi introduces the principle of induction (Fig. 5), the game of Hex models a proof by contradiction, while the game of Nim gives the occasion for a simple treatment of binary notation and several variations of the Königsberg bridges lead to the world of topology.

Most of the exhibitions include a catalogue, or rather a book in which the underlying ideas are treated in more detail, but always in an informal way.

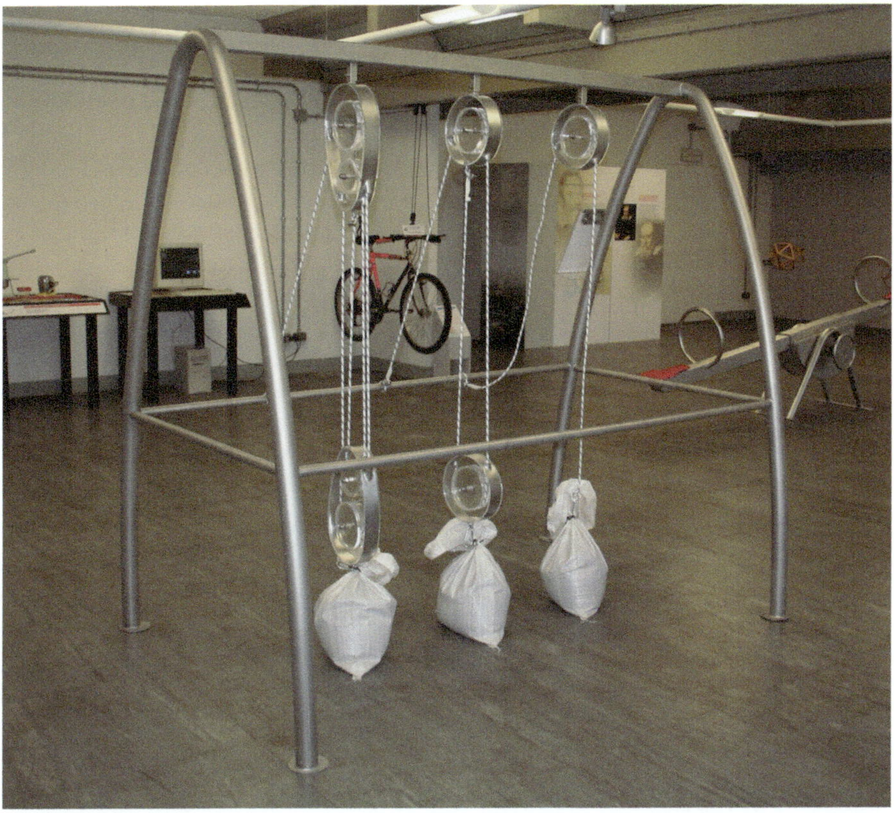

Fig. 4 Pulleys

Fig. 5 Tower of Hanoi

This short description might already have provided some insight into the general philosophy that underlies the activities of the Garden of Archimedes.

In the first place, there are three main directions of research, each with its own importance and methods, but by no means mutually exclusive:

1. Mathematics and history. Mathematics is never seen as given once and forever; on the contrary, we tend to stress the importance of understanding the historical process and the successive modifications of mathematical concepts and theories from antiquity to the present. Moreover, we have not forgotten the social aspects of mathematics, in particular the role of mathematicians in the cultural and social history of their nations.
2. Mathematics and games. For us, mathematics is not a dry land of figures and formulas, as might appear from the point of view of school learning, but the source of amusement and play. Therefore, we try to present modern mathematical ideas, such as the induction principle, the form of a game that can be in two ways: as a form of play in the first place, but also as a source of deeper insights and as a model for concepts which are otherwise difficult to grasp.
3. Mathematics and everyday life. Mathematics is not a closed world, but interacts continuously with our lives in ways sometimes hidden and difficult to perceive, but nonetheless present even in the simplest objects of common use.

A second character of the exhibitions is their interconnection. A single exhibit never stands by itself, separated from the others. On the contrary, they are all organically linked by a number of paths, which the visitor takes in succession. In this way, the path and not the object is the minimal unit of an exhibition, which is composed of a sequential series of paths, all linked by a common idea. Thus, for instance, the exhibition *Oltre il compasso* consists of three subsections: *Straight lines and circles*, *Conic sections* and *Transcendental curves*, each being in turn composed of one or more paths. This structure permits a detailed presentation of the main properties of the curves in question, from the three different points of view specified above: history, play and everyday life.

The third and perhaps most important point concerning the communication of mathematics at the Garden of Archimedes is that we always try to speak "mathematics" as opposed to speaking "about mathematics". In other words, we never—well, almost never—describe a mathematical result without giving at least a hint of how this result was obtained.

Take for instance the Pythagoras exhibition. It begins with three puzzles in which one can use the same pieces to build either the square of the hypotenuse or those of the catheti. Then follows a proof of the theorem, based on two different arrangements of four right-angled triangles inside a square whose side is equal to the sum of the catheti (at a deeper layer of understanding, one can use the same exhibit to introduce the formula $(a + b)^2 = a^2 + b^2 + 2ab$; the next step might be to show that Pythagoras' theorem depends on the fact that the sum of the angles of a triangle equals two right angles, and therefore on the parallel axiom). In the next section we substitute the squares with arbitrary similar figures: hexagons and stars. The validity of Pythagoras' theorem for these figures is justified by the fact that the area of similar figures is proportional to the square of corresponding sides (one can insert here a digression on how Socrates doubles the square in Plato's *Meno*, and possibly on the problem of doubling the cube). We consider rectilinear figures due to the use of puzzles, but it is possible to experiment on the same result for arbitrary shapes by weighing the figure on the hypotenuse against those on the catheti. This is done for the quadrature of Hippocrates' lunulae.

The same approach is followed in the Galileo exhibition *Helping nature*. On the one hand, the visitor can experiment with the proof of the law of the lever, which materialises the proof given by Archimedes and again by Galileo. But more important is the fact that by using Galileo's principle (the product of the force of its displacement equals that of the resistance by the length travelled),[1] one can easily calculate how much force is needed in the various machines: lever, capstan, pulley and screw.

Thus we introduce the visitor to the very core of mathematics: proofs. Of course, in many cases proofs are only sketched and only the general idea is given: we never insist on technical details. But the idea of how to prove is there; even better, the idea that mathematical statements must be proven and not simply trusted.

As an example, consider the proof of the law of the lever. Our proof begins from Galileo's statement that it is impossible to cheat nature, and that machines can only "help" nature. One of the laws of nature is that it is impossible to overcome a greater force by a lesser one. How then can we move a heavy weight by means of a small force? Assume that we want to move a weight of, say, 100 kg with a force of 1 kg. One possibility is to break the weight into 100 pieces of 1 kg each, and then move them one by one. In this case, to displace the weight by 10 m, the 1 kg force must travel this distance 100 times, for a total of 1,000 m. But what if we cannot break up the weight? Enter the machines, whose only task is to allow us to move the weight in one single movement without breaking it, using our small force of 1 kg. The machines can do that, but they cannot reduce the distance that must be travelled by the small force: it will always be 1,000 m. This is true for every possible machine we can invent: if we want to move a weight of 100 kg using a force of 1 kg, then the distance travelled by that force will have to be 100 times that travelled by the weight. This general principle holds for the lever, the pulley and the capstan, and establishes the rule governing their action: the product of the force by the distance is the same for the moving force and the resistance. Thus the visitor can calculate how much force is needed to lift a weight in different situations.

Once again, the exhibition can be seen at different levels, and one can find a physical realisation of Archimedes' proof of the law of the lever, in which the mathematical passages are replaced by the action of a machine. Archimedes' (and Galileo's) proof goes as follows:

1. In Fig. 6, the cylinder CDFE is in equilibrium when suspended from G, due to the symmetry of the configuration.
2. If we cut thread GX, and substitute it with AC and BD, the cylinder remains in equilibrium (Fig. 7).

---

[1] For the sake of brevity I am rather imprecise here. However, in the exhibition panels this is explained using the exact words of Galileo.

Fig. 6 Cylinder hanging
from one string

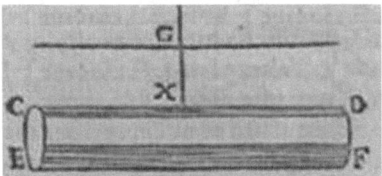

Fig. 7 Cylinder hanging
from two strings

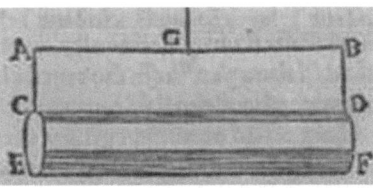

Fig. 8 Cut cylinder hanging
from four strings

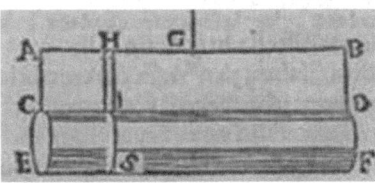

Fig. 9 Cut cylinder hanging
from two strings

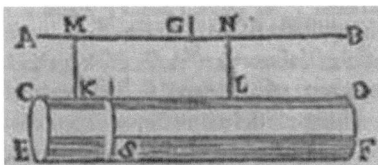

3. Now cut the cylinder in IS, adding threads HI (Fig. 8). Since nothing has been moved, equilibrium persists.
4. In each of the two parts CISE and ISFD replace the two threads at the extremities with the central threads MK and NL (Fig. 9). Nothing changes. But now the two weights CISE and ISFD hang from distances GM and GN, that are easily seen as being inversely proportional to lengths ES and SF, and therefore to the weights of the cylinders CISE and ISFD.

I already mentioned that some exhibitions are pocket-sized, and can be easily shown outside the museum. This is now true also for many of the exhibitions that were originally designed to be shown at the Giardino di Archimede; for most of these there is a light version, suitable for export. Both these and those designed specifically for diffusion are available for sale at a moderate price.

Fig. 10 Sumerian calculi

The Workshops

More recently, the activities of the Giardino di Archimede have expanded, and now include a number of workshops. When compared with the exhibitions, the workshops are characterised by a much greater involvement of the participants, who are requested not only to follow the guide and operate the machines, but to participate directly in the mathematical activity. The interaction with the material and the tutor is much stronger, and includes solving problems and answering questions; in short, doing mathematics.

On the other hand, the general philosophy is always the same, and the same ingredients are present; only the recipe is slightly different. In the first place, the workshops are inspired by history, and the reference to methods and practices of the past is always present, possibly even more than in the exhibitions themselves. Actually, since the workshops are for the general public, including children from five and up, they require little or no mathematical knowledge, and therefore the mathematics is at the lowest possible level. The historical setting adds interest to a subject that might otherwise seem rather dry, and avoids reducing the workshops to a sequence of exercises.

The core of the workshops offered by the Giardino di Archimede concerns the art of counting and computing in various civilisations, under the general title "At the beginning of counting" (All'inizio del conto). This includes:

1. *Numbers and calculations with the ancient Sumerians* (age 5+)

    One of the most ancient tools for computing consists of small stones or pebbles. In ancient Mesopotamia, using clay *calculi* whose different shapes carry different values, a "concrete" representation of numbers was obtained, that gradually evolved into a sort of symbolic writing, while calculi continued to be used as calculating machines (Fig. 10).

    The laboratory discusses the Sumerian system of representation of numbers and the transition from calculi to writing, and the use of the calculi for computing, ranging from simple sums to complex division.

2. *The hieroglyphs of the ancient Egyptians* (age 5+)

Fig. 11 Egyptians
hieroglyphs

From the beginning of the third millennium BC, the ancient Egyptians had special symbols for numbers, which appear on monuments and inscriptions. This numbering was founded on a strictly decimal base and involved the combination of symbols corresponding to the values 1, 10, 100, 1,000, 10,000, 100,000, 1,000,000 (Fig. 11).

In the laboratory one can write numbers like the ancient Egyptians did, and can experiment with their computing techniques, in particular multiplication by successive doubling.

3. *Computing boards* (age 5+)

These simple and efficient calculating devices are found in various forms from antiquity to the Renaissance and are mainly used in accounting. They consist essentially of a board divided into rows or columns on which chips are placed. Depending on the row (or column) in which they are located, tokens take different values.

In the workshop the boards are used to record numbers and to discover how to compute with chips, without any written calculation.

4. *Chinese counting boards* (age 5+)

Counting boards have been in use in China from ancient times until relatively recently. They consist of a board divided into squares, on which numbers are represented by sticks (originally made of ivory or bamboo).

In the workshops the children quickly learn how to count with sticks, and then to perform more complex operations.

5. *Rods for multiplication and division* (age 8+)

At the beginning of the seventeenth century, John Napier invented his famous sticks that facilitate multiplication without any knowledge of multiplication tables, a prelude to the first mechanical calculating machines a few decades later.

In the workshops, Napier rods are used to multiply, together with rods invented by Genaille and Lucas, a variant discovered in the early nineteenth century (Fig. 12). Similar rods are used for division.

6. *Japanese Soroban* (age 9+)

A close relative of our abacus and probably descending from the same ancestor that was widespread in the ancient world, the Soroban—as well as the Chinese version called the suan-pan—is a powerful computational tool in the hands of the skilled and experienced, who can compete in speed with modern calculators (Fig. 13).

Fig. 12 Genaille-Lucas rulers

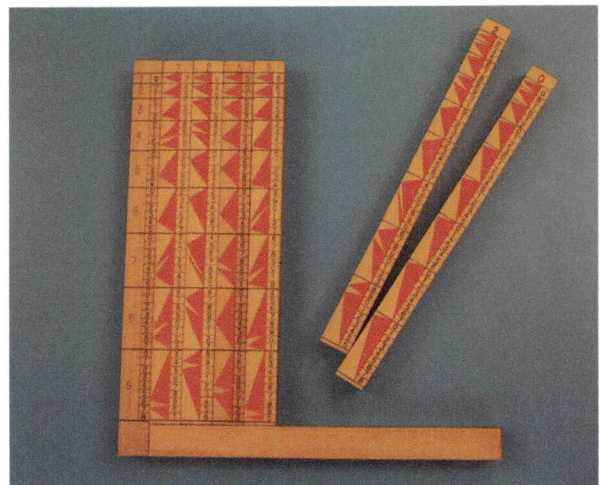

Fig. 13 Japanese Soroban

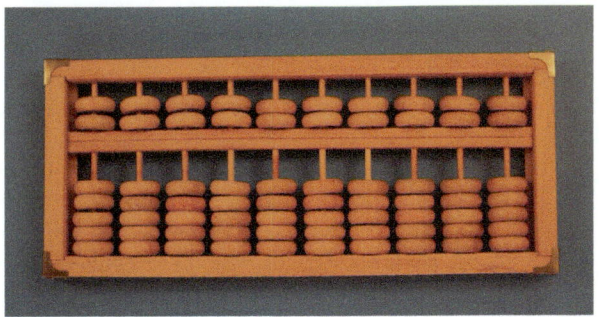

In the workshops, after becoming familiar with the Soroban, pupils can use it for computing, from additions to multi-digit multiplication.

7. *Various techniques of multiplication* (age 9+)

The arithmetic algorithms that we learn in school are not the only ones possible, and many others have been used over the centuries. In particular, multiplication can be performed in many different ways, some of which are somewhat easier and more entertaining than ours. In the workshops, pupils learn to multiply using a number of different techniques, from those of Indian mathematicians to their Arabic variants, from the abacus masters to those of the Russian peasants.

In addition to the above, the Giardino di Archimede offers other workshops on various topics:

1. *Paths, strategies and geometry in play* (age 5+)

The workshop explores the different metamorphoses of the problem of the seven bridges of Königsberg (Fig. 14), introducing the simplest topological ideas, and experimenting with games with the same topological content.

2. *Deciphering secret messages: an invitation to cryptography* (age 6+)

Fig. 14 Königsberg's islands

An introduction to some simple techniques for deciphering hidden messages, but also a reflection about some of the characteristics of our language, such as the different frequencies of the letters of the alphabet. By trial and error, we reconstruct the patient investigative work of those who, in the past, tried to steal the secrets of others.

3. *Creases, folds and ... explanations. The mathematics of origami* (age 5+)

Origami allows the discovery of the mathematics hidden in a piece of paper. Using the Japanese art of paper folding several models are constructed, which explore the properties of plane and solid figures, but also the arithmetic of fractions.

4. *From medieval algorithms to early mechanical computing* (age 14+)

Three workshops describing the progression from medieval abaci to the first adding machines through Napier rods and other mechanical instruments:
1. Computations in medieval abacus schools
2. From Napier bones to the first mechanical calculators
3. Analogue adding machines and the slide rule

5. *Leonardo Fibonacci, the Liber Abaci, and the revival of Mathematics in the West* (age 14+)

The story of the passage to the West of the Indo-Arabic number system offers many insights in the transmission and integration of knowledge and its interplay with the socio-economic context. The *Liber Abaci* is an excellent example of how different cultures can integrate and produce an innovation that becomes a universal heritage.

The material used in most of these workshops is available for purchase. Currently, some of the workshops of Il Giardino di Archimede are the core of the project PitAgorà (innovative training project for teachers and trainers in mathematics), supported by the European Union within the framework of the Leonardo da Vinci programme, Action "Multilateral projects: Transfer of Innovation".

Il Giardino di Archimede, Via San Bartolo a Cintoia 19a, 50142 Firenze, tel. +39-055-7879594, fax +39-055-7333504, www.archimede.ms

IMAGINARY and the *Open Source* Math Exhibition Platform

Andreas Daniel Matt

Abstract   The traveling exhibition IMAGINARY by the Mathematisches For-schungsinstitut Oberwolfach explores interaction and participation in math commu-nication. We present the key experiences made from over 40 exhibitions in several countries and introduce the *open source* exhibition platform. Using this platform ev-erybody will be invited to copy and exchange material, stage their own exhibitions and add new ideas. Through a selected case study of activities we show the future potential of IMAGINARY and the platform designed to engage a broad audience and make them part of the math communication and math popularization process.

Introduction

The Mathematisches Forschungsinstitut Oberwolfach

Oberwolfach is a small village in the German Black Forest synonymous with the internationally renowned research institute in mathematics. Every year the institute hosts about 2,500 mathematicians. They gather in a non-formal atmosphere to ex-change ideas about current research topics in the form of weekly workshops and special programs (see the chapter in this book by Gert-Martin Greuel). In 2008, the institute at Oberwolfach started its own math outreach program on a broader level with events, publications, the exhibition IMAGINARY and the museum of mathe-matics and minerals MiMa [1, 2]. Many of the mathematicians visiting the institute take an interest in these activities and contribute to them. The institute and its partici-pants serve as the link to current research topics and as the main hub to communicate its outreach activities through different universities in many countries.

A.D. Matt (✉)
Mathematisches Forschungsinstitut Oberwolfach, Oberwolfach, Germany
e-mail: matt@mfo.de

E. Behrends et al. (eds.), *Raising Public Awareness of Mathematics*,
DOI 10.1007/978-3-642-25710-0_13, © Springer-Verlag Berlin Heidelberg 2012

IMAGINARY

IMAGINARY is a traveling exhibition that presents visualizations, interactive installations, virtual worlds, three-dimensional objects and their theoretical background from algebraic geometry, singularity theory and differential geometry in an attractive and understandable way. The exhibition was originally designed for the German Year of Mathematics 2008 and supported by the Federal Ministry of Education and Research (BMBF) from 2008 to 2009 [3].

The exhibition has visited more than 40 cities in Germany, Austria, France, Ukraine, Switzerland, Spain, Poland, the United Kingdom and the USA. Talks, workshops and teacher training programs have also been held in Argentina, Colombia, India, Israel, Pakistan and Portugal. The free exhibitions have taken place in publicly accessible areas, such as universities, railway stations, bank buildings, research institutes, theme parks or fairs, with the joint organization of local partners including universities, research institutes, advertising agencies and city councils. The local partners have assisted in media work, supervision of the exhibition and logistics.

The title of the exhibition expresses the main concept: to show the abstract or *imaginary* property of mathematics through *images*. The core element of the exhibition is its interactive mathematical installations. The idea is to allow users to experiment with mathematical objects and create new ones. In this way the underlying mathematics is discovered by the visitors on their own and evokes motivation and curiosity. Various programs are available at the exhibition and on the Internet, which can be downloaded for free used at home or in school. The programs have been developed for IMAGINARY or specially adapted by mathematicians and artists from different institutions. Details of the IMAGINARY exhibits can be found later in this chapter.

More than 180,000 individuals have visited IMAGINARY, among them over 900 school classes on free guided tours. Additionally about 120,000 visitors have seen the exhibits at special exhibitions and other events, for example on a science ship or at science days. In September 2009, the exhibition received the award *Land der Ideen 2009* presented by the Deutsche Bank under the patronage of the President of Germany [4].

Key Experiences in Math Communication

Since 2008 we have gained a lot of experience. We were surprised by the unexpected interest of the general public and their desire to explore and understand mathematical content. Our approach was to be as close as possible to the public by offering individual guidance and collecting feedback. Since the inception of the exhibition we have constantly changed its contents of the exhibition according to this feedback. This has helped us to take a few important decisions at the right moment. In the following we present a selection of these key experiences and decisions.

Fig. 1 The IMAGINARY exhibition at Leibniz University Hannover, 2010. In the center you can see the picture cube and around the cube there are a series of interactive installations

The Artistic Packaging

We decided to present the mathematical content in a more artistic than scientific style. For that reason we chose high-quality prints displayed in a walk-in cube structure with dimensions $4 \times 4 \times 4$ meters, which could be placed in any room of an appropriate size, see Fig. 1. The logo, texts and installations follow simple, aesthetic design guidelines with an emphasis on a combination of forms and formulas, strong colors, clear shapes and white spaces. It was important to have this cube construction as an eye-catcher and central point of the exhibition, while the interactive stations tempt the public to stay and experiment. For the interactive exhibits it was necessary to offer non-standard interfaces: instead of mouse or keyboard input we used 68-inch touch screens, a special mouse and for one station three-dimensional glasses. The artistic packaging of the contents and the use of new or lesser known technologies proved to be the perfect combination for generating a positive first impression on our audience.

From Interaction to Participation

Science communication in museums or at exhibitions, as communication in general, is becoming more and more interactive and user generated. Interaction can start with the simple involvement of the users, for instance the possibility of influencing the exhibit by pressing a button. At IMAGINARY we decided to offer tools where users can create something new, beyond the imagination of the exhibition producers with limitless possibilities. This kind of interaction we define as *creative interaction*.

As an example take the exhibit SURFER, where users invent algebraic equations, with or without any prior knowledge, which are used to create pictures, i.e. algebraic surfaces. The variety and quality of these pictures, all created by visitors, is stunning and we receive novel and new creations every day!

Creative interaction is also a step towards participation at IMAGINARY. First, we started adding user-created pictures to our exhibition. We allowed visitors to print their images directly at the exhibit and to place them on open display in a user gallery. Since these pictures turned out to be appealing to other users we then decided to offer user galleries on our website and also to organize exhibitions consisting only of pictures by our users. Users started to interact and a community was created, a key ingredient for our new open-source idea.

The Human Touch

One way to transmit the joy of mathematics to a broader public is through direct communication with a dedicated individual, mathematician or trained exhibition tutor. Explanations at an individual level help to reduce prejudices towards mathematics. Our exhibition demonstrators are generally young students of mathematics or its didactics who are curious and interested in geometry. Their task is to catch any visitor at the exhibition and guide them by providing information on the exhibits and also talking about their own motivations and interests. Sometimes the demonstrators started a tour by stating that they did not understand everything about the exhibition and that they were surprised how much they could explore together with the visitors.

Another appealing human feature is humor. Imagine what could happen when you change the algebraic equation of a heart by only one exponent? See Fig. 2 for the result. When they are shown during a guided tour, such examples produce laughter and interest.

Many of the mathematicians behind IMAGINARY have attended some of the exhibitions to give individual explanations and presentations on their current research. The authors of the exhibits are always very welcome and their interaction with the audience helps us to receive feedback.

Free Software and Open Approach

All the programs behind the interactive installations of the exhibition are available to download for free at the IMAGINARY web site. Free software is highly appreciated

Fig. 2 The *Herz*, heart, given by the equation
$(x^2 + \frac{9}{4} \cdot y^2 + z^2 - 1)^3 -$
$x^2 \cdot z^3 - \frac{9}{80} \cdot y^2 \cdot z^3 = 0$
becomes *Herz in der Hose*, heart in your boots (literally trousers), when you change the last cube into a square

by the visitors. It adds a sustainable component to the exhibition: users can not only prepare themselves before coming to the exhibition, but also continue experimenting at home and involve friends and family after their visiting. This open approach has helped IMAGINARY to reach users in countries where no exhibitions have been staged.

Wakening and Satisfying Curiosity

We have observed that the exhibition can be a starting point for evoking curiosity and motivation so that visitors continue exploring and reading about the underlying mathematical contents. Interaction and participation by the visitors raises this interest. Visitors are especially attracted by current research and intrigued by difficult mathematical problems. How to disseminate the latest research is an important topic in science communication [5]. We communicate for example the open problem of the maximum number of singularities for algebraic surfaces, see world record surfaces in the chapter by Gert-Martin Greuel for further details. The problem is illustrated with beautiful images and in principle even the visitors could solve it. A user of our program SURFER, Torolf Sauermann, a German surveying technician, once claimed to have found a new world record surface, raising the record for septic surfaces from 99 to 100 singularities. It then turned out that he had made a counting mistake and had only found another surface with 99 singularities. He thus equaled the record, see Fig. 3. This is an example of the possibility that through math communication interesting scientific results can also be found by laymen, hobby mathematicians or the broader general public.

The background articles provided at the IMAGINARY web site give an introduction and insight into the mathematical topics covered by the exhibition. Several experts were invited to write these articles for different levels of mathematical education. The IMAGINARY programs have been downloaded more than 300,000 times in the last three years and the background articles more than 150,000 times.

Fig. 3 World
record-breaking singularities.
Left: The Labs Septik with 99
singularities. *Right*: The
Sauermann Sunflower, also
with 99

Competition and Community

An exhibition is not only an exhibition: there is also a series of media and marketing activities attached to it. The web site and social networking tools are important for attracting visitors before and after the exhibition, and also for including and connecting with the general public outside the current exhibition venue. During our second exhibition we decided to start online picture art competitions using our program SURFER with the cooperation of the media, for example *Die Zeit* or *Spektrum der Wissenschaft* [6]. The competitions were very successful and we received several thousand entries in just a few weeks. The side-effect of such competitions was that users became connected to each other. The competitions were used to showcase individual work. We decided to always immediately display any entry submitted, so competitors were aware of each others images.

Through the competitions we received more than 40,000 pictures by over 5,000 users. Many of these users are also in direct touch with us and thus form one part of our community. The other part is formed by the several hundred IMAGINARY demonstrators, i.e. students who have worked at the different exhibitions. Some of them are now teachers and use the programs in school; others keep in touch and help us at new exhibitions to train other tutors. More information on the importance of an active community can be found later in this chapter.

The *Open Source* Platform

Motivation and Idea

> *To teach is not to transfer knowledge but to create the possibilities for the production or construction of knowledge. Paulo Freire [7]*

Our approach to communication is to enable audiences to be part of it, to become communicators themselves. Since our first IMAGINARY exhibition, schools and individuals have enquired about the possibility of copying the images and using our programs at their own exhibitions, events or at home. With the agreement of the authors of our exhibits we handed over the digital content to universities, museums and individuals for copying and started to prepare installation manuals, further descriptions and content packages. The demand for our content has increased and since property in open source is configured fundamentally around the right to distribute, not the right to exclude [8], a logical next step for us was to develop an *open-source* math exhibition platform.

Through such a platform not only can software including source code be provided easily, but also design templates, web site infrastructure, digital data and production details of pictures or sculptures, technical plans, guidelines for funding and media work and contact information of the companies and mathematicians involved in the project.

Our first motivation was to facilitate our work by having a more automated system to provide already existing content, but then we realized that open source projects usually involve a dynamic structure and an active community and offered possibilities for creating and updating our content. We were aware of open-source software from the operating system *Ubuntu*, which we were using for our exhibition and museum, and also of software products like the word processor *OpenOffice.org* or the web content management system *Plone*, which we use for the web site of our institute. Open-source projects are mainly software based, but there exist hardware-based approaches, for example the Rapid Prototyping Printer *RepRap* [9], a free desktop three-dimensional printer capable of printing plastic objects, including many parts of itself. There are other platforms such as *i2Geo.net*, a network enabling math educators throughout Europe to publicly share interactive geometry constructions [10], and the European online database for scientific traveling exhibitions *extrascience.eu* with its portal for open science resources [11]. The portal emphasizes interoperability structures and has a toolbox for registering science exhibits. Another interesting project is *exhibit commons*, where Internet users can create and interact with science exhibits in museums [12].

In the following sections we discuss the key ingredients of an open source exhibition platform and give an overview of the IMAGINARY contents we are planning to share through such a platform.

The Key Ingredients

The following ingredients need to be considered before starting an open-source project. We will not discuss funding, the importance of having a strong main partner and dedicated organizers, which are essential requirements for starting such a project. In our case Oberwolfach and its IMAGINARY coordinators plan and implement the platform, while sponsors, foundations or incubators are asked to support it financially. While writing this article we received two years project funding from the Klaus Tschira Stiftung, Germany [13].

Legal Issues

An open-source platform working with a series of exhibits created by different individuals in various countries needs a clear agreement on the copyright and terms of use of all content offered. Open source does not mean free or only unpaid work and the distribution rights have to be properly and professionally defined. We are considering licensing our exhibits under the creative commons license *CC by-nc-sa*. This license lets others remix, tweak and build upon our work non-commercially, as long as they credit the authors and license their new creations under identical terms [14]. Another license option that grants rights to modify and re-distribute would be the GNU public license [15].

The issue of non-commercial use is a controversial topic [16], for example when offering exhibits at a science museum with entry fees. We are thinking of allowing non-profit organizations to generate income from the exhibits to cover expenses. Another issue to consider is that some of our content is already protected under similar or other licenses, some even commercial. Thus some content might not be included in the platform in order to keep a simple license structure.

Community Creation

One goal of the open-source platform is that the distribution, maintenance and modification of its contents are carried out mainly by the community. Creating an active community is therefore the key to decentralizing the project, i.e. to have many contributors, controllers and marketers within the community working on the project. Note the difference between a network of contacts and a community of self-active, committed individuals.

IMAGINARY created or became connected to several communities that overlap: there are a few thousand *math artists*, interested in creating images and exploring their mathematical context, and about 300 demonstrators at the exhibitions, who guide visitors and know about the technical and mathematical contents. Then we have the schools, i.e. about 1,000 teachers with their students, and the mathematicians interested in raising awareness of mathematics in general and others interested in participating and showing their own work. Another group are the experts in certain fields, such as architects, designers or computer scientists with special knowledge applicable to our contents, for example, a graphical user interface programmer who offers his help to improve the interface of our programs, a multi-media expert rebuilding the whole exhibition as a virtual one, or a film-maker using our programs to make math tutorial videos.

All of them have their own reasons for participating and it is important to be aware of them. In general the members of these communities identify with the concept of IMAGINARY, they like it and see its purpose in society. They can see that the project is open and dynamic and feel part of a bigger, developing initiative. To keep a community involved, communication is vital. One has to be active and quick to respond. Social reputation inside the community or within wider social systems helps and the contributions of the community members have to be clearly visible. The community will only be motivated to contribute if they understand and can influence the concept, that is the strategy, ideals and philosophy of the project. Another issue is to ensure that the values of the concept in general and the shared content in particular do not interfere with the personal expression of the authors. The project depends on contributions from the community. Problems can arise if individual self-fulfillment comes before the project's goals or purpose.

The motivation of the community is maintained through the constant exchange of ideas and feedback and through the high quality of the project, for example being a market leader in its niche, and also through community activities, for example the distribution of project T-shirts or other giveaways, the formation of user groups, the organization of non-formal community meetings, conferences, talks or yearly conferences.

Quality Control and Maintenance

To guarantee the quality of the contributions, maintenance and support, we plan to use management and coordination tools in combination with a clear community structure. Responsibilities can then be distributed and the community split for example into core developers, a framework team for organization and quality control and other groups for translations or marketing.

We think that it is important to start a project with a sound base of high-quality material, in order to set the standard. A good community will develop a strong system of self-evaluation and self-monitoring using web or IT tools. The technology is important in facilitating control and approval of new contributions or updates. Here too, a good initial structure, for example templates for each type of contribution, is crucial.

Platform and Technology

Online services like web platforms, databases, management tools or software repositories are constantly changing and evolving. Primary tools have to be chosen. These include documentation formats and the template structure of the content. As an initial step for IMAGINARY we plan to have a web portal offering our material so that it can be enjoyed online or offline, and where it can be reproduced virtually or physically but also changed and adapted and placed back onto the portal. Later completely new material could be added, for example images or programs for other exhibitions or topics. The platform has to be simple to use and responsive to the activities of the community.

There are many technologies available to create such a platform, and typically open-source projects also use open-source tools. The platform itself could then run under an open license, so that other organizations could use its structure to host and create their own math exhibition.

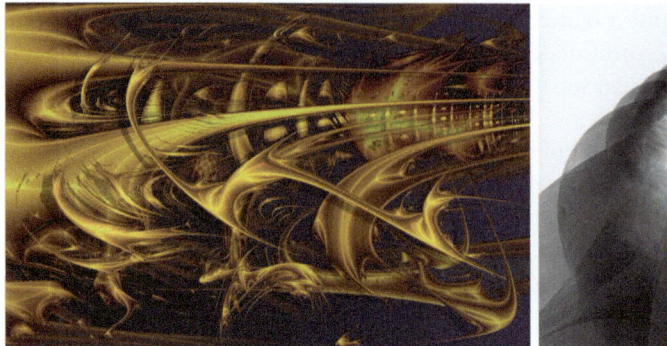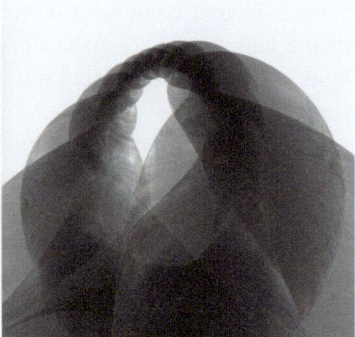

Fig. 4 *Left*: Lyapunov Play by Luc Benard. *Right*: Twizzle Torus by Nicholas Schmitt

Contents of the IMAGINARY Platform

In this section we give an overview of the current IMAGINARY exhibits and items
we are planning to offer through the *open-source* platform. There will be packages
of several items for different exhibition budgets and spaces, and the items will also
be available individually. At the moment most of the content is available in German,
English, French, Spanish, Catalan, Basque and large parts also in Ukrainian, Polish
and Portuguese. The translations were made by our exhibition partners or directly
by members of our community. We plan to add all the available sources needed to
reproduce the items, for example source files to recreate images or LaTeX/InDesign
files to reproduce text panels.

Images

The exhibition consists of a gallery of about 80 mathematical images from algebraic geometry, differential geometry, projections of four-dimensional polyhedra
and artistic interpretations of mathematical objects. Each image is provided with
explanatory texts and if possible instructions on how the image was generated together with source code and references to the software used to create the image. The
explanatory texts are offered in different versions to cater to a variety of knowledge
levels, occasions and needs. They are available in several languages and formats. Information on the printing and mounting of the images is given later. Figure 4 shows
two images from our collection.

The authors of the pictures are Herwig Hauser (Austria), Luc Benard (Canada),
Étienne Ghys and Aurélien Alvarez (France), Jos Leys (Belgium), Oliver Labs, Ulrich Pinkall, Steffen Weissmann, Nicholas Schmitt and Tim Hoffmann (Germany)
and Richard Palais (USA).

Fig. 5 *Left*: Three-dimensional prints of algebraic surfaces. *Right*: Sculptures Calypso and Spitz by Herwig Hauser

Sculptures

The German companies *Voxeljet Technology* and *Alphaform* manufacture three-dimensional models made from layers of glued plastic powder using stereolithography. Both companies have accepted the challenge of printing a selection of the algebraic surfaces from the exhibition in three dimensions. It was difficult to find appropriate data formats for the models. The institute FORWISS at the Passau University has implemented various techniques for the exhibition to translate algebraic surfaces into printable data. Twelve sculptures with a diameter of about 25 cm are shown at the exhibition, see Fig. 5. We plan to offer the digital data needed to reproduce them on local three-dimensional printers and also the contact details for the companies, who are selling the sculptures.

Programs and Interactive Exhibits

The interactive stations of IMAGINARY are all based on software programs that were developed or adapted for IMAGINARY. The programs are used at the exhibition through touch screens or other non-standard control devices. At home they can be used on any computer and controlled with the mouse. Some programs have different versions: a reduced exhibition version with less options and a special kiosk mode and an extended version with more options for home use. Here are the four programs we use at IMAGINARY:

- SURFER is a fast ray tracer of real algebraic surfaces with a simple and intuitive user interface [17]. On a huge touch screen visitors can play with formulas and forms. One does not have to be a mathematician to simply change a pre-defined formula and observe in real time the new amazing figures (all points in space that solve the given equation) that appear. The figure can be turned in space, colored or changed through parameters. Who can create an interesting surface, a funny one, a pointy one or one that resembles a real object? The program features sample galleries with explanatory texts on singularities, known surfaces and fantasy surfaces. The program was developed by the Mathematisches Forschungsinstitut Oberwolfach (MFO) in collaboration with the Technical University Kaiserslautern. The developers are Gert-Martin Greuel, Christoph Knoth, Oliver Labs, Andreas Matt, Henning Meyer, Christian Stussak and Maik Urbannek. In Fig. 7 you can see the program in use at an exhibition in Berlin.

- Cinderella is interactive geometry software for the visualization and simulation of data and algorithms [18]. One of the authors, Jürgen Richter-Gebert from the Technical University Munich, prepared a series of interactive applications for IMAGINARY using Cinderella. The Java-based applications communicate various mathematical topics such as simulation, chaos or symmetries in a playful way. A user can experiment with and easily create geometrical constructions or virtual physics simulations; the left image in Fig. 6 shows an example of a simple robot simulation.

- The programming environment jReality allows virtual modeling of mathematical objects [19]. For IMAGINARY a special application has been developed to explore seven objects, mainly from differential geometry, in a virtual physical environment. The application looks like a modern three-dimensional computer game. Users move through a virtual world to interact with the objects using a special control device. jReality is a project by the DFG Research Center MATHEON and the IMAGINARY application was developed by Ulrich Pinkall and Steffen Weissmann. The control device of a jReality station can be seen in the right image of Fig. 6.

- Martin von Gagern's program Morenaments allows the user to paint symmetrical patterns in one of the 17 space groups in the Euclidian plane. It allows them to create aesthetic designs and ornaments and especially attracts young people. The program is available under the GNU General Public License [20].

- 3D XplorMath is a comprehensive program for visualizing mathematical objects. The program has a series of galleries of different categories of interesting objects, ranging from planar and space curves to polyhedra and surfaces, to ordinary and partial differential equations, and fractals [21]. As a special feature the objects can also be viewed using three-dimensional glasses. The program was developed by an international consortium organized by Hermann Karcher and Richard Palais.

Films

Étienne Ghys, Aurélien Alvarez (France) and Jos Leys (Belgium) produced the film *Dimensions*, which explains the mathematical concepts of the fourth dimension for a broad public [22]. It is available under the creative commons license type CC by-nc-nd, which is a non-commercial license where no derivative of the film is allowed. The film MESH by Konrad Polthier and Beau Janzen on discrete mathematics has also been shown at the exhibition [23]. For the open-source project we will include a selection of free films.

Fig. 6 *Left*: Exhibit Cinderella in Stuttgart, 2008. *Right*: Exhibit jReality in Leipzig, 2008

Fig. 7 The interactive SURFER station in Berlin, 2008

How to Organize an Exhibition?

We can provide a series of documents for exhibition and event management. Organizing an exhibition includes media work, logistics, technical work on site, training of demonstrators, fundraising and a lot of organization [24]. Checklists, technical plans, training manuals, mounting instructions and best-practice documents, for example on opening or closing ceremonies, communication with schools or web design are included. For example we describe several options for printing and mounting pictures, from laminated print-outs, to plastic banners, textile prints or plexiglass and aluminum versions. The images can then be mounted using gallery systems, on individual structures or pin boards. To facilitate media work, we offer template files in common desktop publishing formats for posters, flyers and invitation cards and examples of school invitations, press releases, media reports and general advice on working with journalists [25]. In collaboration with our partners we plan to provide special offers for printing or technical equipment and share our list of companies and also exhibition tutors or coordinators in order to support other exhibitions.

Case Study of a Spreading Exhibition

If we are in education then its purpose is that our students should be better than us. Raza Kazim [26].

One result of math communication is that our audience surprises us with its curiosity and creativity and continues the communication process. New ideas arising by our users or visitors, be it newly created algebraic surfaces or other ideas on how to present or play with IMAGINARY objects, surpassed our imagination. We sometimes call IMAGINARY a spreading exhibition instead of a traveling exhibition, because it seems to have a life of its own which spreads to other countries. In this section we present a selection of activities that emphasize the need for an open-source approach of IMAGINARY and shows its potential. The selection includes open science days, science museums, workshops, talks, films and media campaigns. This case study also gives ideas for different presentation formats and for items we might include in the open-source platform. In this section we mention the key people involved, to give them their credit and also to show our appreciation for their effort and work.

Schools, Universities and Museums

During our first exhibition in Germany a school enquired about the possibility of reproducing the exhibition for an open school day. They reprinted our images, laminated them and even built a simple form of the IMAGINARY picture cube. Since then we have received many requests and several schools have carried out their own exhibitions. Universities have also used our programs for education and public events, for example at the Girls' Day 2008 in Berlin. In Fig. 8 you can see images of a school exhibition and a university event.

In fall 2009, Sergiy Ovsiyenko, math professor at the Taras Shevchenko National University in Kiev, Ukraine, and his team independently organized an IMAGINARY exhibition in the Museum of Russian Art in Kiev. They found a sponsor for the technology, translated and adapted our software programs and for the first time IMAGINARY was shown in a big art museum.

Science museums in Europe and elsewhere started to show interest in IMAGINARY. The first permanent IMAGINARY installation will open in July 2011 in

Fig. 8 *Left*: Girls' Day presentation at the Technical University Berlin using IMAGINARY software. *Right*: Open school day at the Otto-Hahn-Gymnasium Saarbrücken with their own small IMAGINARY exhibition

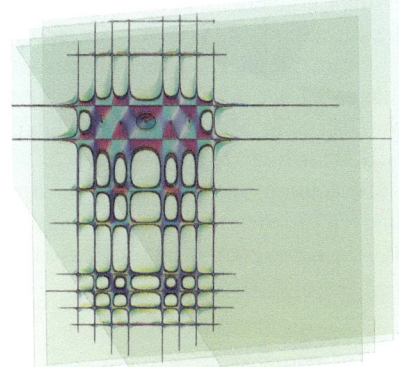

Fig. 9 *Left*: A SURFER picture reproduced as a quilt by Hiltrud Heinrich. *Right*: SURFER art by Kurt Ballay

the German Museum in Munich with three interactive stations. The programs have been adjusted by the curator and designers of the museum to serve for a museum environment with many visitors and no tutors on site.

Artists and Artwork

Several artists use our programs, mainly SURFER, to create pictures. The German artist Hiltrud Heinrich has already exhibited her SURFER pictures at different museums and events and has even organized her own math art exhibition. Other artists have explored the possibilities of the program including its visualization errors. See Fig. 9 for two examples.

Animation and Music

In July 2010 a four-day workshop on "math art and music" was organized for 10–12-year-old school kids as part of the *KinderUniKunst* project in Vienna. The aim was to make animations with SURFER and then add music and sound. Austrian musician Bernhard Fleischmann and two students from the math department of the University of Vienna conducted the workshop. The results were four short movies. The novelty for the participants was that being a mathematician meant to be the cameraman and director of a movie. At the close of a creativity fair in Vienna the musician and a mathematician performed live: together algebraic surfaces and music were created on the spot and linked together. Performing at this closing ceremony would not have been possible without the music.

In February 2010, José Francisco Rodrigues during a visit to the MFO had the idea of creating an artistic film on cryptography in high definition using SURFER animations as a core element. The final film LPDJLQH D VHFUHW was produced by the Centro Internacional de Matemática (CIM) in Portugal in collaboration with the animation artists Armindo Moreira and Victor Fernandes of Casa da Animação in Porto, Stephan Klaus of Oberwolfach and the film-maker and mathematician Bianca Violet in Berlin. It was presented at the opening of the congress "Raising Awareness in Mathematics" in Óbidos, September 2010. The film can be downloaded for free [27].

By adding other art forms such as film and music, IMAGINARY has reached an even broader audience. New ideas, for example adding algorithmic music to the SURFER program might increase its audience even more.

Popularization

In 2009 a German advertising company started an AIDS awareness campaign showing a picture of a typical student's room [28]. We saw the final campaign when it was printed as a big banner in the Cologne subway and were surprised: in the picture between the skateboard and the rock poster there is a print of an algebraic surface with its equation, taken during an IMAGINARY exhibition.

This is an example where content spreads on its own and is integrated into media and daily life. Similarly, we sent digital poster files to many of our users and they sent us pictures of IMAGINARY posters printed and mounted in living rooms of private houses, offices, school classes and libraries. Each of these small exhibitions helps to promote mathematics; the communicators are the people themselves.

RSME-Imaginary 2011–2013

"This is not a lemon" was one of the first sentences uttered by Sebastia Xambó, coordinator of the 13-cities IMAGINARY exhibition tour for the centenary celebra-

Fig. 10 This is not a lemon and that is not an apple, but the meeting point of the concrete *visualization* of mathematical objects and the *imagined* reality. The equation of *Citrus* is $x^2 + z^2 = y^3 \cdot (1 - y)^3$. The picture by Magritte is a photograph taken from the *Medford Art Goes to School* project [29]

tion of the Royal Spanish Mathematical Society (RSME) when he saw the IMAGINARY poster. He was comparing the logo image *Citrus* to the surrealist painting *"Ceci n'est pas une pomme"* by René Magritte and thus stated that the leitmotiv of IMAGINARY is the meeting point of the *imagined* reality and the concrete *visualization* of mathematical objects, see Fig. 10.

Right from the beginning the team of RSME-IMAGINARY added their own philosophy and ideas to the exhibition: a new storyline was found to connect the pictures with real objects and the explanatory texts were rewritten in a more entertaining way. The design of the exhibition was updated and simplified using the same type of technology for all stations. As well as the traveling exhibition, a semi-permanent museum installation has been devised, featuring algebraic images and sculptures with the interactive station SURFER, which was shown in the Cosmo-Caixa museums of Madrid and Barcelona.

The example of taking IMAGINARY contents, changing them and adapting them—much more than just translating everything into Spanish, Catalan and Basque—shows the potential of an open source concept. RSME-IMAGINARY was inaugurated in January 2011 in Madrid and Salamanca and will be shown until 2013 in 13 cities in Spain. It is so far the biggest IMAGINARY exhibition tour after the German year of mathematics. And this time, everything has been organized by someone else!

Conclusion, Future and an Invitation

We have presented our key experiences of math communication with the IMAGINARY exhibition and a case study of activities that were mainly conducted by or with our users and visitors. In the coming years we want to develop an international, extendable platform for the attractive communication of mathematical resources, which will enable the independent continuation of IMAGINARY and math exhibitions in general. Following our experiences and the first steps that we took to open the exhibition to a broader public we are positive that such an open-source platform will not only help to promote this exhibition, but also serve as an example for other exhibitions and activities in this field. We believe the decentralized approach and the ability to copy and change the exhibits will ensure it reaches an even broader audience.

We would like to invite you to be part of IMAGINARY and our future open-source project. We would be happy to assist you in organizing an exhibition or event at your institute, school or at home and we are looking forward to new ideas and contributions.

More information: www.imaginary-exhibition.com.

Acknowledgements The project "IMAGINARY—open source Plattform für interaktive Mathematik-Vermittlung" will be funded by the Klaus Tschira Stiftung. I would like to thank the following people for providing ideas and help while writing this article: Gert-Martin Greuel, Anna Hartkopf, Stephan Klaus and José Francisco Rodrigues for their valuable feedback and Armin Stroß-Radschinski for his expert advice on the open-source concept.

References

1. IMAGINARY. http://www.imaginary-exhibition.com
2. MiMa, Museum for Minerals and Mathematics. http://www.mima.museum
3. Greuel, G.-M., Matt, A.D.: IMAGINARY through the eyes of mathematics. Travelling Exhibition Catalogue (2009)
4. 365 Orte, Eine Reise zu den besten Ideen Deutschlands. Dumont (2009)
5. Chittenden, D., Farmelo, G., Lewenstein, B.V.: Creating Connections: Museums and the Public Understanding of Current Research. AltaMira Press, San Francisco (2004)
6. Pöppe, C.: Das Ausgedachte in Sichtbares umgerechnet. Spektrum der Wissenschaft (2008)
7. Freire, P.: Pedagogy of Freedom: Ethics, Democracy, and Civic Courage. Rowman & Littlefield, Totowa (1998)
8. Weber, S.: The Success of Open Source. Harvard University Press, Cambridge (2004)
9. Sells, E., Smith, Z., Bailard, S., Bowyer, A., Olliver, V.: RepRap: the replicating rapid prototyper-maximizing customizability by breeding the means of production. In: Handbook of Research in Mass Customization and Personalization, vol. 1, pp. 568–580 (2009)
10. Libbrecht, P., Kortenkamp, U., Mercat, C.: I2Geo: a Web-Library of Interactive Geometry. Towards a Digital Mathematics Library. Grand Bend, Ontario, Canada, July 8–9th, 2009, pp. 95–106 (2009)
11. Portal for Open Science Resources. http://www.osrportal.eu
12. LaBar, W.: Exhibit commons: using the internet for a new exhibit paradigm. In: Museums and the Web (2006)
13. Klaus Tschira Stiftung. http://www.klaus-tschira-stiftung.de
14. Creative Commons. http://www.creativecommons.org
15. GNU General Public License. http://www.gnu.org/licenses
16. Commons, C.: Defining 'noncommercial': a study of how the online population understands 'noncommercial use'. Creative Commons Wiki (2008)
17. SURFER—Visualization of Algebraic Surfaces. http://www.imaginary-exhibition.com/surfer
18. Richter-Gebert, J., Kortenkamp, U.H.: The Interactive Geometry Software Cinderella. Springer, Berlin (1999)
19. jReality. http://www.jreality.de
20. Morenaments. http://www.morenaments.de
21. 3D XplorMath. http://3d-xplormath.org
22. Dimensions—Une Promenade Mathématique. http://www.dimensions-math.org
23. Janzen, B., Polthier, K.: MESH (Springer VideoMATH) (2007)

24. Christensen, L.L.: Hands-on science communication. IAU Special Session, 2 (2006)
25. Göpfert, W., Lang, V.: Medienkompetenz: Wissenschaft Publik Gemacht. Klaus Tschira Stiftung GGmbH (2006)
26. Kazim, R.: Matters Relating To Educational Redesign. Lectures By Raza Kazim, pp. 1–2 (2005)
27. LPDJLQH D VHFUHW—a film of art and mathematics on elliptic curves and cryptography (2010). http://www.cim.pt/?q=LPD-UHW
28. AIDS campaign: Mach's mit. http://www.machsmit.de
29. Medford Art Goes to School project. http://web.me.com/kacyang/Medford_AGTS/Home.html

Part III
Popularisation Activities

www.mathematics-in-europe.eu

Ehrhard Behrends

Abstract In this article we describe how the popular mathematical web page www.mathematics-in-europe.eu of the European Mathematical Society was created: from its "prehistory" to the end of September 2010 when it was officially launched.

The RPA Committee of the EMS

The European Mathematical Society (EMS) was founded in 1990 to represent mathematicians throughout Europe working at research institutions, universities and other institutions of higher education. A number of subcommittees exist to support the work of the EMS: education, electronic publishing, meetings, etc. During the World Mathematical Year 2000 a new committee was founded: the committee for raising the public awareness of mathematics (RPA committee for short). RPA activities are very important for the EMS since in nearly all countries the public understanding and appreciation of mathematics are far from satisfactory. And as a consequence, politicians and other shapers of public opinion generally have only a very vague idea why this field is of vital importance, and furthermore, many students hesitate to study sciences in which mathematics is essential.

Since 2000 the activities of this new committee have included round-table discussions and competitions to find the best published article that raises the public awareness of mathematics.

At the beginning of 2009, new members and a new chair were elected by the executive committee (EC) of the EMS. It was the desire of the EC that a new RPA activity should be established: a popular mathematical webpage under the auspices of the EMS.

E. Behrends (✉)
Fachbereich Mathematik und Informatik, Freie Universität Berlin, Berlin, Germany
e-mail: behrends@math.fu-berlin.de

E. Behrends et al. (eds.), *Raising Public Awareness of Mathematics*,
DOI 10.1007/978-3-642-25710-0_14, © Springer-Verlag Berlin Heidelberg 2012

Fig. 1 *First row*: Nuno Crato (Portugal), Krzysztof Ciesielski (Poland); *second row*: Thomas Bruss (Belgium), Robin Wilson (UK), George Szpiro (Israel and Switzerland), John Barrow (UK), Betul Tanbay (Turkey), Steen Markvorsen (Denmark), Ehrhard Behrends (Germany), Mireille Chaleyat-Maurel (France), Franka Brueckler (Croatia); from left to right

Preparations for the Webpage: Contents

The first meeting at which ideas concerning the structure of such a popular webpage were discussed took place in May 2009 in Brussels, Belgium (local organizer: Thomas Bruss; see Fig. 1).

The most important decisions were:

- The name of the webpage will be www.mathematics-in-europe.eu.
- It will be crucial to find a sponsor in order to be able to realize this project.
- All of the contents will be provided in English, but contributions written in any other European language will also be included (together with an English translation).
- For the technical realization the content management system (CMS) Joomla will be used. It is based on PHP and uses MySQL database.

Many ideas concerning the contents, the design, and the structure were collected. Here is a survey:

The home page   The home page will link directly to the main parts of www. mathematics-in-europe.eu: news, information, popularization activities, competitions, math help and miscellaneous. There are further links to

- The mission statement (what do we have in mind?)
- Welcome messages (who are the intended visitors of this webpage?)
- EMS (information on the European Mathematical Society)
- Our sponsor (Munich Re)
- Languages (our policy is that everything can be found in English, other languages are welcome)
- Contact (how to contact www.mathematics-in-europe.eu?)
- Imprint (who is responsible?).

Also there will be a "news" section on the home page as well as building blocks that change depending on the date or at random:

- The "top five inspirations of the month" (some interesting web sites)
- The quotation of the day
- Historical reminder
- The popular article of the day.

Information   *What is mathematics?* This is a collection of various viewpoints. Some articles already exist, others will be re-used from the EMS newsletter. They will have a short commentary (easy to understand, demanding, philosophical, ...).

The mathematical landscape: This will include information on *key concepts* and *mathematical areas.*

Mathematics and ... : This will include articles on the philosophy of mathematics. As far as the history of mathematics is concerned we will use several articles from the newsletter and we will have a separate section "The mathematical calendar".

Research: A list of links to European research institutions has been prepared. Franka Brueckler will write something about the millennium problems. It will also be important very soon to have articles concerning the following:

- There are numerous examples of contemporary mathematical research that concern important problems of real life.
- Sometimes there is "mathematics inside" places where nobody would have expected it: computer tomography and the Radon transform; data transfer via the Internet and cryptography, data compression, ...

Popularization activities   We intend to have a summary of the national popularization activities and to provide help for colleagues who want to start RPA projects.

Competitions   There are a collection of links to the most important mathematical competitions, and also relevant information on the most prestigious prizes in mathematics (the Abel prize, etc.).

Math help   One important section will be "Mathematics in other languages", with a mathematical dictionary of (as many as possible of) European languages. Also we want to collect a database of ideas for school teachers.

Mathematics as a profession   This section will contain mainly descriptions of typical professions and interviews with mathematicians who work in the "real world".

Miscellaneous   This page will link to a "virtual museum" with stamps, banknotes, etc. and the "mathematical tourist".

Our Sponsor

The contact with the reinsurance company Munich Re (see Fig. 2) started in the nineties when one of the members of the executive board was very active in the

Fig. 2 The logo of Munich
Re

Fig. 3 Some of the pictures
proposed by Iser and Schmidt

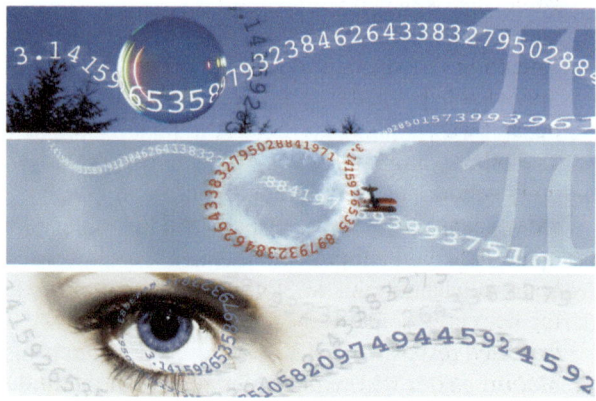

German Mathematical Society (DMV). Munich Re later became the sponsor of the popular German web page www.mathematik.de, and very soon the idea arose of trying to create a similar cooperation at the European level.

Negotiations started in the summer of 2009 and came to a successful conclusion in December of this year when a sponsor agreement between Munich Re, the EMS and the DMV was signed. A considerable amount of money will be provided to prepare and to run www.mathematics-in-europe.eu; the secretary of the DMV will be responsible for administration.

Creation of the Webpage: Graphic Design

The quality of the design of a webpage is highly important, and therefore it was clear that professional help would be indispensable. We asked the graphic agency Iser and Schmidt (Bonn, Germany, www.iserundschmidt.de) to present a proposal. Our rather vague idea was that there should be a series of pictures with "some relation with mathematics", so that each day a new one could be chosen at random.

Iser and Schmidt presented such a series where "circles in nature" are combined with the first digits of the number π. For some members of the committee circles and π were considered as being maybe too trivial for a webpage run by a professional mathematical organization, but the majority was convinced of the need to use well-known motifs for a popular site.

The committee agreed to use six of the twelve pictures proposed by Iser and Schmidt, some of which are shown in Fig. 3.

On the basis of the graphical and structural decisions a complete layout was designed by the agency. A "typical" home page for www.mathematics-in-europe.eu is shown in Fig. 4.

Fig. 4 An example of a home page for www. mathematics-in-europe.eu

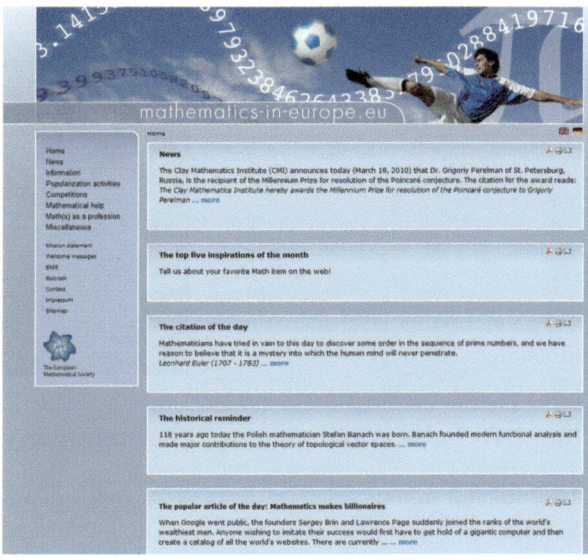

Realization

As soon as the layout was available the transformation to a "real" web site could begin. This was carried out by Inbion (www.inbion.de), a small enterprise specialized in the administration of webpages. As soon as this frame was prepared we—the members of the RPA committee and many collaborators—started to provide "real" content. We can only mention a selection here:

- The president of the EMS, Ari Laptev, wrote a greeting.
- Many popular articles were collected to be used in a random order as the "popular article of the day".
- Colleagues from all over Europe helped us with the mathematical dictionary.
- More than a hundred authors gave us permission to use their articles that had been written for the newsletter of the EMS. In this way very interesting material on various aspects of mathematics was available: history and philosophy, Abel Prize winners, research, . . .
- A large number of introductory texts had to be produced: a site notice, welcome messages, introductions to the subpages, etc. They all had to be translated into English, and it has to be stressed that David Kramer—a professional translator— was of invaluable help for nearly all of these texts.
- . . .

Needless to say that most of the contributions had to be transformed to a file format suitable for the web. Of great help was a team of students at the Freie Universität Berlin, they also collected mathematical citations as well as "historical reminders" for the 366 days of the year (see Fig. 5).

Fig. 5 The team of students
from the Freie Universität
Berlin

The Official Launch of www.mathematics-in-europe.eu

After several months of hard work www.mathematics-in-europe.eu was officially launched on September 26, 2010, in Óbidos, Portugal. Óbidos is a charming town situated one hour by car to the north of Lisbon. In the "International Year of Mathematics 2000" (WMY2000) it was the site of the creation of the international exhibition "Beyond the Third Dimension" (see http://alem3d.obidos.org/en/) and of a meeting of the EMS WMY2000 Committee. This committee has launched there the CD-rom "Raising Public Awareness of Mathematics", done in the framework of an EC project.

In September 2010 Óbidos hosted again an international workshop: "Raising the Public Awareness of Mathematics" (organizers: E. Behrends, Berlin; N. Crato, Lisbon; J.F. Rodrigues, Lisbon; see http://c2.glocos.org/index.php/RPAM/rpam2010). The opening was on September 26, 2010, it took place in connection with a "mathematical afternoon" organized by the Portuguese Mathematical Society (SPM) in cooperation with the town of Óbidos. At this event mathematical films and lectures for a general public were presented (see Fig. 6).

Later, one could participate in a reception for an itinerant mathematical exhibition ("Medir o Tempo, o Mundo, o Mar") on the use of geometry to measure the universe and help astronomical navigation, jointly organized by the SPM and the Museum of Science of the University of Lisbon. The exhibition and a reception took place at a local art gallery.

And it was at the occasion of this public awareness event that our web page www.mathematics-in-europe.eu was "officially" launched. And the fact that many members of the EMS/RPA committee were present in Óbidos was used to discuss in a separate meeting the next steps in connection with the realization of this website.

The Present State of www.mathematics-in-europe.eu

When writing this article (January 2011) many of our aims have already been realized. In particular one should note:

Fig. 6 *Left*: The opening ceremony with the mayor of Óbidos and the president of the Portuguese Mathematical Society. *Right*: The organizers of the RPA conference in Óbidos (Behrends, Rodrigues, Crato)

1. We had in mind a *mulitlingual* webpage. In fact, the menu and the introducing texts (mission statement, welcome messages etc.) are now available in 14 European languages: English, Croatian, Danish, Finnish, French, German, Italian, Norwegian, Polish, Portuguese, Russian, Spanish, Swedish, and Turkish. The "clever" webserver recognizes from where the query comes: a user in Spain finds the menu not in English when he or she enters www.mathematics-in-europe.eu but in Spanish etc.

2. Every day another "first impression" is offered: not only the picture, but also a citation of the day, a historical reminder and a popular article of the day change. And everyone is invited to follow a recommended link to an interesting web page ("Top five inspirations").

3. Our *mathematical dictionary in the European languages* provides help in 15 languages

   - How do you pronounce x^2 in Swedish? No problem: "x i kvadrat".
   - And how is the Spanish "conjunto vacío" called in Italian? They call ist "insieme vuoto".
   - An here are the Polish words that contain "funk" together with their Croatian translations (see Fig. 7).

4. There exist about 300 articles and sub pages with information on the various aspects of mathematics: research, philosophy, history, ...

5. What are the most important facts concerning mathematics in the European countries: societies, research institutions, public awareness activities, ... Admittedly this part is far from being complete, but we hope that we can present soon a more representative survey.

6. Regularly the "historical calendar" reminds us to important events in the history of mathematics: "Principia Mathematica" published 100 years ago, etc.

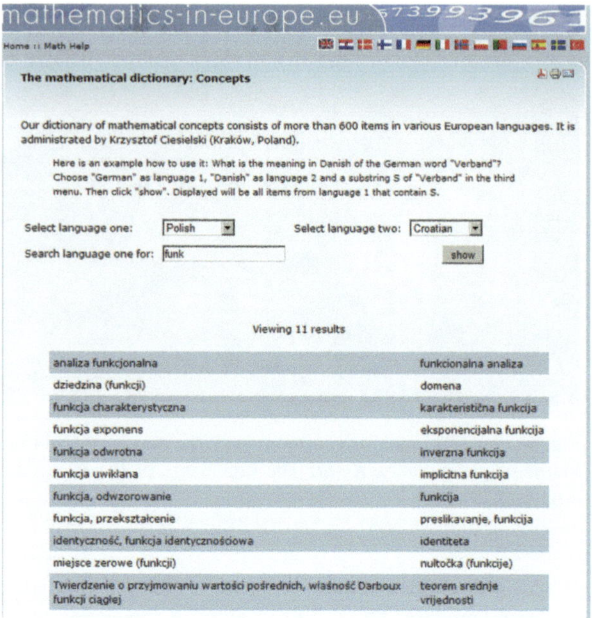

Fig. 7 The dictionary

The Next Steps

It doesn't come as a surprise that a big amount of work remains to be done. At present we work on "Miscellaneous" were we realize something like a virtual exhibition will be shown: mathematical stamps, coins, the mathematical tourist etc. Other gaps concern "mathematics and art", "mathematics and music", and more information on examples of mathematical professions, to mention a few. Even more severe is the fact that we have not yet started to collect attractive material for the European teachers. An explanation might be that all members of the RPA committee work at universities and the experience with teaching at schools is rather limited. However, for this year a collaboration with experts is planned, and there is good hope that a second sponsor will help us with this aspect of www.mathematics-in-europe.eu.

The details how one could organize the next steps as effectively as possible will be discussed by the members of our committee at the occasion of our next meeting in Zagreb, Croatia.

The Maximin Principle in the Popularisation of Mathematics: Maximum Effect with Minimum Costs

Franka Miriam Brueckler

Abstract Limited means can boost the development of new ideas and of finding new ways of presenting old ones. The example of popularisation of (not only) mathematics in Croatia is a corroboration of the previous statement. In Croatia there is a long tradition of mathematical journals publishing popular articles more or less regularly, tens of mathematical books for a general audience have been published in recent decades and the schools programme on national television has mathematical broadcasts. Still, a real change from the communication of mathematical topics to the popularisation of mathematics happened in the last 15 years. In that time the author and several other enthusiasts developed a mainly interactive and live communication-oriented approach to the popularisation of mathematics. As the financial means for the organisation and presentation of mathematical and other popular science workshops were and are quite scarce, most activities depend on the enthusiasm and original ideas of individuals and on finding presentation ideas that need only cheap materials. In this paper the author presents some of her experiences in the popularisation of mathematics in Croatia and ideas for workshops that can be organised with a very low budget.

A Tale of a Zigzagging Mathematician (or: How One Can Become a Populariser of Mathematics, at Least in Croatia)

Once upon a time there was a small country in Europe. After some time of independence and a long time of dependence and coexistence with other nations in various states and political systems, it became independent again. Soon after a young mathematician finished her studies in the capital.[1] The usual routine: she was good in maths at school, studied mathematics because "it was easy enough" and because it

---

[1] Zagreb.

All the pictures in this paper are from the author's personal archive.

F.M. Brueckler (✉)
Department of Mathematics, Faculty of Science, University of Zagreb, Bijenička 30, 10000 Zagreb, Croatia
e-mail: bruckler@math.hr

E. Behrends et al. (eds.), *Raising Public Awareness of Mathematics*,
DOI 10.1007/978-3-642-25710-0_15, © Springer-Verlag Berlin Heidelberg 2012

promised a profession independent of political circumstances, finished on time and with good grades .... The expected consequence was that she should stay at the faculty as an assistant. But this mathematician noticed something as her graduation was approaching: mathematics *was* beautiful, but nobody except the mathematicians noticed or could explain why. Even worse, she rarely noticed a mathematician try to do something about it. Was it so unnatural to wish to be able to tell something of her profession to the people she met? If she just waited to meet the usual fate of a good mathematics student, it seemed probable—even certain—that she would never have the chance of normal communication about her job with her family, friends, people she just met. This seemed quite a terrible prospect. So, shortly before graduation she decided she did *not* want such a future. She started working as a journalist, graduated, continued working as a journalist .... And discovered: the other side also was not without faults. Many outsiders didn't really care about having the correct information about anything, let alone about mathematics, particularly if receiving the information involved personal (brain) activity. This left our mathematician very frustrated and disappointed. Every communication needs at least two sides, and her experience seemed to show that in the case of communication between "a mathematician" and "a non-mathematician" both sides preferred not to communicate at all. And after one year of being a journalist she decided that if she had to choose one of the two sides as "her side", it would be the mathematical one. And she went to her faculty and got the assistant job. The next two years were all that she had expected in her first projection of the future: she learned what it means to "do mathematics" and never, ever, talked about what she was really doing to anyone other than her colleagues.

And then one spring she also started teaching mathematics in another city,[2] a city in the part of the country that is famous for good food. Even mathematicians eat,[3] and one dinner changed everything. A professor asked her to draw posters to announce the faculty's mathematical colloquia. Wow! Drawing, her hobby since she was a kid, and freedom to interpret the titles in a humorous way. Next year she was asked to help her best friend and colleague to prepare the mathematics department's presentation at the university fair. Hmm .... What should they do? They were advised to put a few mathematical books on the table, and decided to put some basic information about the mathematics courses on the wall of the stand. Our mathematician created the posters, but somehow was dissatisfied. Was this all? The university fair was a complete shock to her. The next stand belonged to the chemistry department and a "crazy" guy and his team not only demonstrated experiments—they made people *do* them. There were explosions, colour changes, *interesting* smells .... This stirred our mathematician's fighting spirit: if they can do it, we can too! But how? It was much easier for chemists to devise an attractive presentation. However, if you really want to do something, you will. In short: next year, next fair, our

---

[2]Osijek.

[3]Our mathematician's personal experience was that mathematicians were always able to suppress the urge to eat when absorbed with a theorem or its proof—a trait she didn't find in herself and was starting to think this implied she was not really a mathematician.

mathematician answered the chemists' challenge.[4] The first mathematical hands-on activity presented at the university fair was very simple: solving recreational maths problems on a portable board. And this time she instructed all the mathematicians at the stand to actively try to involve visitors in talking about mathematics. After some part of a chemical experiment landed on her head, she answered with a typical mathematical weapon: paper (aircraft). And suddenly there was humour, and people were attracted: the mathematical stand was almost equally well visited as the chemistry stand. Our mathematician was now proud that she was a mathematician because not only was mathematics beautiful, but she found there were ways—fun ways—to talk about it and to involve people in mathematical activities. And years went by, the mathematical stand developed into an annual ever-changing place of mathematical hands-on activities: various recreational problems found in books and on the Internet were adapted into versions suitable for presentation in a small space and big crowd, see Fig. 1.

There was only one problem: several of her colleagues considered (and openly said so) that these activities were nice, but of a "lower" kind, and it was not worth spending so much time on them. If some thought otherwise, she never heard. Mathematicians, people who should want to tell others how nice a job they have and why it was a useful one, not only did not understand her, but even considered that this was a nice hobby, but of no professional consequence for *real* mathematics. Was it really something she had been doing only for her own fun? After all, no man is an island, and she felt that if you insert "group of people sharing a profession" instead of "man", the sentence should still be true. Was something wrong with her?

The Year of Mathematics 2000 came and our mathematician went to the European Congress of Mathematics in Barcelona. She found a round-table discussion *How to Increase Public Awareness of Mathematics?* This sounded interesting, and was even more so. Now she learned that there was a term for what she had been doing—the *popularisation of mathematics*, that it was important for the mathematical community and that there exist ($\exists$) mathematicians who think so. However, she found herself at a crossroads: either to continue with popularisation and risk that although it took a significant fraction of her working time it would never be considered a professional, but a personal, task, or to leave the whole thing and return to the standard life of a professional mathematician.

Of course, the reader can guess what happened: she continued to popularise mathematics. She wrote popular articles, but felt that you can't beat face-to-face communication. The media, filled with prejudices, were (and are) not easy to approach. From the set of popular maths activities only one was left: public lectures. She gave some, and with each one she tried to involve the public in active participation more and more. At some point she noticed that the most attractive[5] of the popularisation activities carried out by the chemists (many of them now her friends) were called "workshops" and consisted mostly of hands-on activities, similar to

---

[4] Although they had no idea that they had challenged a mathematician.

[5] By incident (or not?) these were also the ones where the public learnt the most.

a 6th Fair, 2002   b 13th Fair, 2009

c 14th Fair, 2010   d 15th Fair 2010

Fig. 1 Pictures from presentations of the Department of Mathematics of the University of Zagreb at the University Fairs

those at the university fair. And as she tried to develop more activities, the fact that her state was not a very rich one and consequently not very financially supportive for science implied she had to be extra creative.

And now our story approaches the end. Our mathematician continued developing mathematical workshops, finding students and colleagues[6] willing and able to present mathematical topics in an unusual, often humorous way, and willing to sacrifice at least a little of their time for this. A few of them even do it regularly and developed their own activities. Even better, she found she was not the first to do active and systematic popularisation in her country—in another city[7] there is a well-organised "Golden Section Club" involved in lots of popularisation activities. As usual, when you create opportunities, more open. Now the country has regular (yearly) Science Fair Weeks in the spring in several cities, teachers of mathematics and science often ask for popular maths lectures at their meetings and in schools, and our mathematician also organises regular popular mathematics lectures for sec-

---

[6]So far it seems that students are easier to involve than colleagues, surely because they—the colleagues—have more duties.

[7]Rijeka.

ondary school children at a high school. Even the reactions of the mathematical community are more favourable (or maybe she just became better in filtering out the negative comments). Ah, yes, the financial problem still exists, but our mathematician found that this had a good side effect: it stimulated the development of more creative ideas. Still, it would be nice to have more funding. And so, something that seemed just a diffuse idea close to impossible, transformed into a colourful and exciting reality, with constant challenges and occasional obstacles (but who wants only sunshine without rain?). So our mathematician lived happily (hopefully: ever) after ....

The story has two morals: it makes sense to do something you believe in, even if in the beginning it only makes sense to you. And the popularisation of mathematics is a challenging field with a variety of methods, suitable for each type and style of communication and presentation, and many require little money.

How to Set up a Low-Cost Mathematical Workshop?

As mentioned in the story above, the workshop style is a particularly effective popularisation method. The presenter not only sees and feels the effect immediately, but also—provided the workshop is prepared well—the audience learns (and remembers) more than by other methods because they actively participate and discover mathematics. Most of the materials needed are cheap: paper, ropes, scissors, rulers, pencils, adhesive tape, glue, dice, playing cards, dominoes, matches, pebbles, (non-professional) footballs, candies, ... In the following we shall describe a few workshop projects. The three that are described in detail are chosen as the ones where the author had particularly good experiences, and also because of their additional advantages. The first, mathematical origami, has the particular advantage of being adaptable to very different levels of previous knowledge. The second, mathematical magic, is easily performed spontaneously and is particularly good for helping people with little mathematical knowledge to discover basic mathematical facts. The third, gambling and probability, is an example of how to introduce more advanced topics. The author claims no right on the originality of the activities in the workshops, but does so on the originality of (at least some of) the presentations.

The Mathematical Origami Workshop

So far the most popular mathematical workshop given by the author is a very simply prepared and extremely low-cost one: mathematical origami, see Fig. 2. Since 2007 the author has conducted the workshop 13 times, once presented it for an educational programme on national television and on several occasions presented some of its parts spontaneously at popular science events, in coffee shops and schools. The basic routine is to show people how to make a few simple modular origami polyhedra models, and then let them work by themselves using photocopied diagrams,

Fig. 2 The mathematical
origami workshop at the 2010
Science Fair in Zagreb

offering them help and explanations. We usually let everybody take home what they
made—note that this is at the same time good advertising (people are happy to take
something home with them), educational (they can dissemble the model at home
and try to make others of the same kind) and there is no additional cost (since the
used papers cannot be reused). For readers who are not familiar with the terminology: modular origami is an origami technique where two or more sheets of paper are
folded into units (modules) that are assembled in a puzzle-like manner into the final
model (without gluing and without cutting the paper, except possibly to reduce the
paper to the proper format), see Fig. 3. There are far more models of polyhedra that
can be made by the modular origami technique than by classical origami (where
models are made from one sheet of paper). Most models are not overly "touchy"
about exactness and slight misfoldings usually have no significant effect on the final
model.

The Mathematical Origami Workshop Recipe

Ingredients   Paper, approximately 250 A4 sheets and 250 square sheets (not
necessarily special origami paper); photocopied diagrams for making the models;
paper-clips (to help hold the modules together while assembling), ruler and scissors
or paper knife (optional).

Time   The preparation time for this workshop is practically zero, except for the
time needed to buy paper and arrive at the specified workshop location. If doing it
for the first time or after a longer period, you should plan a few hours for learning
or relearning how to make the models you plan to present.

The suggested time for the workshop is about one hour, but the workshop can
be adapted to as little as 15 minutes (making only one, at most two models) and

Fig. 3 Modular origami cube

of course also to longer durations. In one hour people have enough time to make three to five models, depending on their dexterity (note that children under eight to ten usually are not very skillful in making precise folds; a solution is to prefold the paper for them and leave them the task of assembling the models as a puzzle; generally, I recommend this workshop for ages from 10 up).

How many people?   The amount of paper suggested under "Ingredients" is enough for a workshop with twenty to thirty people. You should have one workshop leader in your team per ten participants that are expected to be present simultaneously.

Equipment/Environment   The ideal environment for the workshop is a room with a few large tables. It is best if the workshop leader(s) can freely move around and come into direct contact with the participants.

Directions   After a few words of introduction, the participants are taught step by step how to fold their first model. The steps are shown and the folds people make are checked. The final assembly should be left as a puzzle, but it is often good to help someone by holding the model for them while they assemble. The procedure is repeated with a second and third model. After the basic models have been made, we usually leave the choice of further models to the participants. For this

Fig. 4 Modular origami skeletal octahedron

we use photocopied diagrams, but we give suggestions (e.g. warnings if a model is particularly complicated to fold) and regularly check/supervise the procedure. Usually each of the workshop leaders knows one or two models more (than the basic three) by heart and we found it very practical for each of us to "specialise" in different additional models, so that we can divide people into subgroups according to the models they choose.

The usual starting models in the author's version of the workshop are the cube, the skeletal octahedron and the regular tetrahedron. For the cube one needs six equal units made from square sheets of paper, which are folded and assembled as shown in Fig. 3. Similarly, the skeletal octahedron consists of six equal units made from square sheets of paper, which are folded and assembled as shown in Fig. 4 (every module has two opposite triangular parts over and two under the neighbouring modules). You need two A4 sheets per tetrahedron (or one cut in half into two A5 sheets). The steps are shown in Fig. 5.

Variations   The workshop can be made more "programmed" by only showing some, say five, previously selected models. In this case, particularly when the number of participants is known in advance, it is a good idea to prepare "packages" of paper needed per participant. Also, one can present a lecture-workshop combination, by giving a short lecture on origami geometric constructions (see below).

Notes   There are many advantages of this workshop. First, since modular origami is based on making several identical units, it is easy to adapt it to group work (each participant in the group can make some of the units and then the whole group can assemble the model). Secondly, the workshop is communication-intense since it is necessary to work *with* the participants, and this offers a simple way of showing a different picture of mathematics and mathematicians than is usually perceived. Finally, but most importantly, the workshop can easily be adapted to various levels of mathematical background and ages: smaller children and people with little mathematical background enjoy the artistic aspect and learn the names of some polyhedra. On a higher primary school level one can introduce more mathematical terminology when giving instructions (e.g. "fold the perpendicular bisectors of the sides") and introduce symmetry considerations. Secondary school children enjoy more com-

Fig. 5 Modular origami regular tetrahedron

plicated models and can also be given various additional problems (e.g. "how big should the paper be to get a tetrahedron of double the height of the previous one?").

For more mathematically interested participants, there is also a huge topic on origami geometric constructions as an alternative to classical ruler-and-compass constructions. Namely, there exists a set of origami construction axioms and the style of mathematical thinking is the same as for Euclidean constructions, but origami constructions are more powerful and allow us to solve cubic algebraic equations (e.g. the classical problems of doubling the cube and trisecting an arbitrary angle can be solved by origami constructions). There are even connections between origami and graph theory. This particular feature of being adaptable to all levels of mathematical understanding is a very big advantage of this workshop, since the others usually are either not adaptable for smaller children, or the more advanced aspects cannot be presented in a popular and workshop manner.

Further Reading   The book *Mathematical Origami: Geometrical Shapes by Paper Folding* by David Mitchell (Tarquin, 1997) is a perfect introduction to the topic; the three models discussed are taken from this book. There is also a wide range of web pages with diagrams or instructions for making origami polyhedra, e.g. http://www.origami-resource-center.com/modular.html. There is also another special modular origami technique created by Mitsonobu Sonobe, and the units are accordingly known as Sonobe units; with this technique one can create a wide variety of very attractive polyhedra, see e.g. http://downloads.akpeters.com/previews/ Mukerji-Excerpt.pdf. For those interested in various mathematical aspects of origami, a good start is the page http://mars.wnec.edu/~th297133/origamimath.html and the paper *Origami and Geometric Constructions* (http://www.langorigami.com/ science/hha/origami_constructions.pdf) by Robert J. Lang. There are also several web-pages giving ideas for additional problems connected to mathematical origami, e.g. http://math.serenevy.net/?page=OrigamiHome.

The Hocus-Pocus Mathematicus Workshop

Mathemagic is a term that first appeared as the title of a book by Royal Vale Heath in 1953. Mathemagic, or mathematical magic, refers to magic tricks founded on mathematical principles. In contrast to usual magic tricks, mathemagical ones are completely fair (i.e. there are no secret switches of items in the performance); the most "unfair" thing a mathemagician does in the performance is to discreetly check the lowest card in a pack. Also, mathemagical tricks mostly do not need any special equipment, but can be performed with everyday objects: playing cards, dice, ropes and so on; some can even be performed without any equipment at all. Since the items are generally small, they are parlour tricks. The mathematical background is generally simple (arithmetic, elementary algebra, logical thinking), but there are some tricks relating to non-trivial topics in mathematics (the so-called class of topological tricks and some others). In the author's workshop version there is an additional difference from classical tricks (and announced at the workshop): in contrast to common magicians, the performer explains how and why a trick works.

The Hocus-Pocus Mathematicus Workshop Recipe

Ingredients   One or more standard decks of 52 cards; a few dice; pencils and paper; a pocket calculator (optional); a few thick ropes of about one metre in length; pebbles, matchsticks or similar small items; a paper roll and adhesive tape (for Möbius strips); optional other items (e.g. special number cards) depending on planned tricks.

Time   The workshop can be made very short (e.g. a spontaneous performance in a train), but generally the optimal time is 30 to 45 minutes to cover about twenty tricks. Most of the tricks need one or two minutes to perform, but some require more time. The preparation time is practically zero, except if learning or relearning is needed.

How many people?   The number of performers should be large enough that, depending of course on the workshop environment, each performer at a table is easily visible to everyone in a subgroup. Generally, if only one person leads the workshop, the number of spectators should not be above 25.

Fig. 6 The Hocus-Pocus Mathematicus workshop at the 2010 Science Fair in Zagreb

Equipment/Environment   The ideal environment for a workshop is a room with a large table for the items, see Fig. 6. Mathemagical tricks are not easy to perform on a stage or when far away from the audience, since for many tricks it is necessary that the audience sees the relatively small items. Thus, it is best that the performer stands in the middle surrounded by the workshop participants. An alternative is to hold the workshop in a e.g. classroom or coffee shop with enough performers for each table. It is also possible to split the performing team so that each takes a table and performs some of the chosen tricks, and the audience circulates from table to table.

Directions   The instructions are simple: learn some tricks from the books or from the Internet, think about why they work (this is usually not described with the tricks) and perform them. As the idea is to present mathematics, and not the tricks by themselves, one should give a short explanation on why a trick works after it has been performed one or more times. Even better, one can stimulate the audience to try to explain the trick by themselves after showing them how it is done. For topological tricks we suggest a slightly different approach: after performing *all* of the planned topological tricks, a short explanation is given of what topology is, as well as comments on its applications. As there is quite a large amount of literature on mathemagical tricks (see below), we shall describe here only three of the author's favourites:

Gergonne's trick is a mathemagical classic. The standard version is performed with 27 cards that are dealt face up in three columns with nine cards each. A spectator is asked to choose a card and name its column. The performer collects the cards by columns, putting the named column in the middle, and deals the cards again (by rows!). The procedure "ask the spectator for the column where the card is, collect the columns putting the chosen one in the middle, deal by rows" is repeated twice. Then the card that was chosen by the spectator will be in the centre, and the performer can choose how to reveal it.

The trick uses a simple elimination rule (the first step reduces the number of possible cards to 9, the second to 3 and the last to 1). There is an extension of this trick in which the chosen card can be made to land on any chosen position

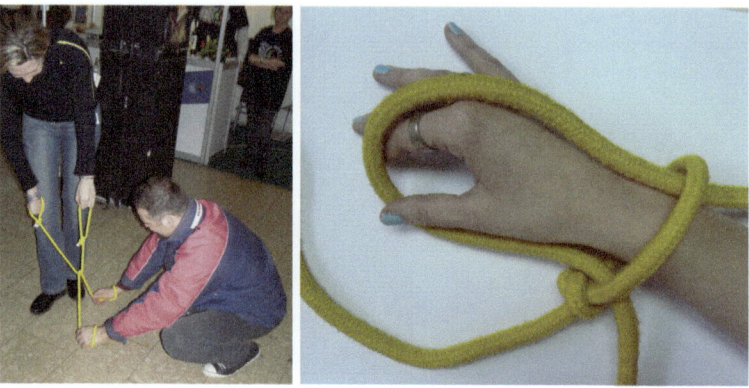

Fig. 7 The tangled couple trick

after a sequence of three steps almost identical to the ones described; this extension is based on numbering the positions 1 to 27 in base 3, see e.g. Martin Gardner's *Mathematics, Magic and Mystery* (Dover, 2003).

The tangled couple is the most fun trick in the topological class. Two people are tied together in a special way (see Fig. 7), but in a topological sense they are not connected at all. They can try to prove this themselves by finding a way to separate without cutting or untying the ropes (the solution is also shown in Fig. 7). It is best to leave them for a few minutes to try, as it is usually very entertaining to watch them trying,[8] but it is quite rare that somebody who hasn't seen the trick before finds the solution.

How many coffees (cokes, juices, etc.) do you drink per day? is an arithmetical trick based on simple properties of multiplication, addition and subtraction. To make sure the calculations are correct it is advisable to give the chosen spectator a pencil and paper, or a calculator. The instructions are best presented as a story.[9] I usually choose the following story: "What do you like to drink most? OK, write down how much of this you drink per day. Now, a good friend came by celebrating a reward he got and invited you to have as many of your favourite drinks as you drink daily, so multiply your original number by 2. Now, another friend came by and bought you 5 more—add 5. What you have now is obviously nearer to your weekly than daily number of your favourite drinks. I want you to estimate your yearly number of drinks; as there are 52 weeks in a year, one should multiply by 52, but as I only want a rough guess, let's make it easier—multiply by 50. Now, have you already had your birthday this year? If yes, add 1761, otherwise add 1760.[10] Now, subtract the full

---

[8]If performing in a mixed setting, like a science fair, tangled couples are an excellent eyecatcher and often attract other bystanders to the mathematical presentation.

[9]The original version, easily found on the Internet under the name "Your Age by Beer Math", involves beer, but as the trick is often performed with children, this is obviously inappropriate.

[10]This only works in the year 2011. It is an easy mathematical task to find the corresponding numbers for other years.

year when you were born, all four digits, and tell me the result." The last two digits of the result are the person's age,[11] and the other digit(s) show the number chosen in the beginning (e.g. a final number 415 indicates that the person is 15 years old and drinks 4 of their favourite drinks per day).

Variations One can "detach" the topological tricks and combine them with other popular topological topics (e.g. the four-colour theorem) into a topological workshop.

Notes Mathemagical tricks are a very attractive way to approach people with mathematical topics. Since the background is generally simple, they are most interesting for children and for adults who have (possibly long ago) finished school without learning much mathematics. It is very important to make sure that you as the performer really understand why and how a trick works, because in the author's experience is that it is quite common for people at the workshop to ask questions (even of the type: what would happen if ... ?). For most tricks there are two good ways to disclose the mathematics behind them: by performing the trick one or more times, then disclosing the procedure and (a) explaining the mathematics involved or (b) leaving it to the spectators to find the explanation themselves. The (b) version is only applicable if there is enough time and the audience is generally willing to cooperate (e.g. it does not work well when the audience consists of school children who were *told to* come to the workshop, as they are often harder to animate).

Further Reading Besides the books by M. Gardner and R.V. Heath, there are several other good books on mathematical magic. The author's personal favourite (because it contains several tricks not described in the other books) is William Simon's *Mathematical Magic* (Dover, 1993). There are also many mathemagical tricks described on the web, see e.g. http://www.numericana.com/answer/magic.htm or http://www.cut-the-knot.org/arithmetic/rapid/magic.shtml.

The Wild West Math (Gambling and Probability) Workshop

It is obvious that basic probabilistic and statistical concepts can be introduced via gambling games. Still, many are far too complicated to be discussed mathematically in a popular workshop. The gambling games that existed in the Wild West are quite a good choice since for most of them the probabilistic analysis is elementary. The author prefers the following games: roulette, blackjack, faro, poker (straight[12]), high dice, hi&lo, chuck-a-luck and grand hazard. The rules for these games can easily be found on the web.

---

[11] Note that if performing it with an adult it is better to choose a male spectator since the person's age is disclosed to the public.

[12] The popular game Texas hold 'em poker didn't exist in the times of the Wild West. Draw and stud poker did exist, but are too complicated to analyse for a popular workshop.

Fig. 8 The Middlemath
Saloon introduction to
probability at the 2009
Zagreb University Fair

The Wild West Math Workshop Recipe

Ingredients   One or more standard decks with 52 cards; a roulette wheel (a cheap children's play version will do);[13] three (or more) dice; a dice cup and a dice tray (optional); pencil and paper; a pocket calculator (optional).

Time   The workshop ideally lasts about 60 minutes, but anything above 30 minutes will do. As with the two other workshops, the preparation time is minimal (just the time needed to set up the gambling tables), provided you do not need to learn the game rules and probability behind the games. Of the three workshops, this one requires the most learning time before the first performance.

How many people?   The workshop ideally works with at most 10 people per workshop conductor.

Equipment/Environment   In the minimal version (one conductor, small group of participants) one table for the roulette wheel and enough space to roll the dice and play card games. The participants should have, at least some of them, somewhere to sit, see Fig. 8. As for the other workshops, this one also doesn't work well if the workshop leader is only a demonstrator in front of the audience. If a larger workshop with more leaders is organised, it is best to have one table with stools per leader and so that each table is used for one game (or group of games), and then let the participants circulate.

Directions   Generally, the workshop works best by going through the games sequentially. Each game is first played, then the leader asks questions and introduces the necessary probability terms, trying to stimulate the spectators to obtain the conclusions by themselves. Some more complicated facts can be described at the end.

---

[13] An American version with both 0 and 00 is common, but to the author's knowledge it is not possible to obtain a cheap version, and even the "real thing" is not easy to get.

The best strategy is to start with the game that offers the simplest introduction to probability: roulette. Typical questions to ask are to find the probabilities of winning for some of the various types of bets. Note that you'll probably use the European version with only one zero, so it is natural to ask the audience to compare the odds in the American and European versions. Other typical questions are: is winning more probable if one sticks to the number(s) chosen, or is it better to change for each spin of the wheel? Why? Determine the expected value of the winnings per unit bet (for one sort of bet)! If the opposite of this value is called the house edge, is the house edge larger in European or American roulette?

After roulette, hi&lo is the next simplest. Players bet on the sum of values on two dice (they can bet that the sum will be larger than, smaller than or equal to 7, or they can bet on a concrete pair). Typical questions would be to get the audience to calculate the probability of rolling a 7 and to roll a pair of ones, to compare the probabilities of rolling a sum larger or smaller than 7, to calculate which odds offered would be fair (in hi&lo, the offered odds are 2:1 for the bets "under 7" and "over 7", and 4:1 for "exactly 7" or a bet on a concrete pair). It is also possible too let the audience discover the probabilities experimentally, by recording the sums obtained in a number of rolls. Chuck-a-luck and grand hazard can be used to extend the learned facts to three dice, and high dice to introduce conditional probability. The analysis of craps, the most attractive of the dice games to play (and the only one of these dice games still played in casinos today), combines all the basics of probability that were previously introduced.

Faro, the classic Old West card game, takes the conditional probability questions further, but in another setting (while in high dice and craps the number of dice rolls needed to end a game is not known in advance, faro is played until the card pack is spent). Blackjack and poker are the most popular, but also the most probabilistically complicated of the games, so we are limited to simple questions like: what is the probability of being dealt a blackjack (an ace and one of the cards 10, J, Q, K)? What is the probability of being dealt four-of-a-kind? Is it more probable to draw (exactly) a pair or no pair?

Variations   Obviously, one can choose other gambling games (or lotteries) and not stick to the "Wild West" theme. Also, more complicated gambling games like blackjack, poker and craps can be a good topic for separate workshops.

Notes   If calculations require more time than a visitor is willing to spend, the author usually encourages them to estimate the result or to compare it with some known probabilities, and to try to explain how the correct one should be calculated. If time is short, one can even give most answers, but should take care that at least the simpler questions are answered by the spectators. It is also interesting to note that some of the games are not pure games of chance, but also involve strategic aspects (this is particularly true for blackjack and poker). As some of the games are optional since they don't introduce any really new probabilistic term or topic, one or more of them can be omitted in advance if time is scarce, or omitted "on-the-go" if analysing some other game took more time than planned. These "omittable" games are grand hazard, chuck-a-luck, faro and poker. For additional effect you can use decorations and special clothes, e.g. such as those that can cheaply be bought during carnival time.

Further Reading   There are several books on probability and gambling games, e.g. R.A. Eppstein's *Gambling and Statistical Logic* (Elsevier, 2009) and E.W. Packel's *The Mathematics of Games and Gambling* (MAA, 1981). There are also many web pages and articles on probabilities in craps (e.g. http://mathforum.org/library/drmath/view/56534.html), blackjack (e.g. http://www.bjmath.com/) and poker (e.g. http://mathforum.org/library/drmath/sets/select/dm_poker.html). The author has written a short analysis of the mentioned less-known dice games, *Wild Wild Math (Gambling and Probability in the Wild West)*, for the http://www.mathematics-in-europe.eu web page.

A Few More Ideas

There are many more mathematical topics that can be presented on a low budget, but are still very effective. Ideas can easily be found in the literature and on the Internet, and with a little imagination most can easily be adapted for a workshop. The list of such ideas is not infinite, but long enough for all practical purposes. Besides the three workshops mentioned and their variations, the following have also been organised and performed by the author: "How to divide a cake?" (the problem and algorithms for the cake-cutting problem; we use candies and a pizza); "Let us play" (mathematical games such as nim, hex, set,[14] tic-tac-toe, etc.; we use pebbles, photocopied grids, pencil, set-game cards); "Soccer/Football Maths" (geometry, symmetry, probability and statistics; we use a football, a picture of a football field, pencil and paper, a pocket calculator); "Symmetries and tilings" (classical tiling topics and connections with crystallography; we use pictures of various tilings, objects with various symmetries); "The Rubik's Cube Group" (introduction to group theory; we use a few Rubik's cubes and fifteen-puzzles); "Rabbits and Pine Cones" (Fibonacci numbers in nature; possible outdoor workshop; we use plants where the number of leaves, petals or another feature is a Fibonacci number and also plants that do not involve Fibonacci numbers); "Flexagons" (we use photocopied grids for making various flexagon models, scissors, glue, coloured pencils); "Magic Squares and their Relatives" (magic and Latin squares and other "magical" grids; we prepare worksheets with tasks and use playing cards for demonstrating Latin squares, self-made squares and cubes with numbers); "How to hide a message?" (basic cryptography, including classical methods and visual cryptography; we use tables with various codes and transparencies to demonstrate (2, 2)- and other threshold schemes, pencil, paper and pocket calculators); "How a mathematician draws and colours a map?"

---

[14] A card game designed by Marsha Falco in 1974, which uses a deck of 81 specially designed cards.

(scales, projections and the four-colour theorem; we use photocopied examples of maps, also maps without names showing only borders and some geographical features, rulers, empty paper, coloured pencils and prepared models for demonstrating various projections). All of these workshops can be divided into smaller ones with a more focused topic (e.g. a workshop only on the game of nim) or adapted to lectures with interactive components. Also, one can combine parts from different workshops to present a specific mathematical area (e.g. some mathemagical tricks, Rubik's cubes and tilings as an introduction to algebra).

Last, but not Least ...

Perhaps the preceding text gives the impression that it is easy to set up an attractive popular maths workshop. However, this is not true—the author only claims that it is not expensive and that it is fun. Professor Vagn Lundsgaard Hansen formulated the following statement, which I present as a theorem:

Theorem 1 (Hansen) *The popularisation of mathematics is not easy.*

Note that by "popularisation" we mean "good popularisation": effective and interesting, with simplification but no misinterpretation or distortion of mathematical facts. The "proof" of Hansen's theorem is easily achieved by contradiction: suppose the popularisation of mathematics is easy ... (I leave the rest of the "proof" to the reader). As popular mathematical workshops are a way to popularise mathematics, we immediately obtain the following:

Corollary 1 *It is not easy to organise and conduct a popular mathematics workshop.*

As a consequence of Hansen's theorem, one can "prove" the preceding corollary directly. Workshops mean active participation and the term "active" refers to drawing, making models, experimenting and similar hands-on activities. Many interesting mathematical topics have no natural hands-on interpretation. This is particularly true for most theoretical ("pure mathematics") topics, and it requires a good amount of imagination and creativity to interpret them in a visual and experimental way suitable for a workshop. Still, there are more than enough applied topics, which are usually more concrete and their presentation is thus easier to transform into a workshop. Also many topics suitable for children have already been transformed into hands-on activities and are described on the Internet or in papers and books, so the inventory of workshop ideas should suffice for some time for most communities.

But there are other reasons why popular mathematics workshops are not a trivial activity. The fact is that conducting a workshop means—in contrast to writing a popular text, much more so than giving a TV or radio interview on a popular math topic and also significantly more than giving a popular public lecture—to be able to adapt easily and quickly to changing circumstances and various types of public and their comments. The popularisation of mathematics is—and should be—(in the main) carried out by professional mathematicians and maths teachers, many of whom are not used to such demanding forms of communication. A choice of a suitable topic, creative interpretation and good preparation (of the whole team) are the minimum conditions for a good popular mathematics workshop. But even the most attractive topic prepared diligently and interpreted originally, even using expensive or fancy props, will not suffice. Imagine the following situation: a mathematician develops a workshop on mathematical magic. He decides to dress in a magician's robe with a nice hat, buys nice cards and big dice, prepares well, knows all the tricks by heart and is also prepared for many possible questions because he thought of all the mathematics behind the tricks he wants to show in his workshop. And then our mathemagician comes to the appointed place, stands in front of the audience, and says (possibly mumbles): "Well … err, let's do some tricks. (pause, shy smile, pause) I thought you could, well, maybe, help me? (pause, half-turn to the table where his cards are, turning his back to the public for several seconds) …", then he performs the trick and the spectator involved miscounts the required number of cards and our mathemagician gets angry: "Oh, no, you spoiled the trick, you should be able to count correctly I think?" I guess the point is obvious.

So, I wish to end this text with the following statement, which I believe to be an axiom of all interactive versions for the popularisation of mathematics:[15] the most important ingredients of a good popular mathematical workshop cost nothing—verbal (and non-verbal) communication skills and humour.

Acknowledgements I wish to thank my non-mathematical friends, geologist Professor Tihomir Marjanac, chemists Professor Ivan Vicković, Professor Nenad Judaš, Professor Tomislav Friščić, Dr. Krešimir Molčanov and Dr. Vladimir Stilinović, and many others, whose university fair presentations were a source of inspiration, to my mathematical friends, Professor Šime Ungar, Professor Vedran Krčadinac, Željka Bilać and Maja Kurek for their ideas, support and contribution in the activities, and to all the other assistants and students of mathematics without whose cooperation the workshops would have failed, or even not existed. To all the named and many of the unnamed friends: thank you for believing in me.

---

[15]The "axiom" is formulated under the assumption that popularisation is carried out by a professional mathematician. Otherwise it should be extended to also include the proper understanding of the mathematics presented.

WMY 2000: Ten Years on …

Mireille Chaleyat-Maurel

Abstract In this article I recall briefly the main features of the operation_WMY 2000: a year for mathematics. Then, I give some examples of the continuation of this event like the travelling exhibition Experiencing mathematics. The last paragraph contains some thoughts for the future of raising public awareness in mathematics.

WMY2000: A Year for Mathematics

The celebration of the World Mathematical Year has taken place all around the world with a great number of events, but this article focuses mainly on the main themes of this celebration as all the details can been seen on the WMY2000 server (http://wmy2000.math.jussieu.fr).

Framework

On May 6th 1992 in Rio de Janeiro (Brazil), the then chairman of the International Mathematical Union (IMU), Professor Jacques-Louis Lions, declared the year 2000 to be World Mathematical Year. See the Declaration in the Appendix.

This enterprise, placed under the aegis of UNESCO and the Third World Academy of Science had three different axes:

– The great challenges of the 21st century
– Mathematics, keys for development
– The image of Mathematics.

The three aims have received a lot of attention from mathematicians with a special emphasis on the third theme which can be called Raising Public Awareness of Mathematics.

M. Chaleyat-Maurel (✉)
University Paris Descartes, Paris, France
e-mail: mcm@math.jussieu.fr

E. Behrends et al. (eds.), *Raising Public Awareness of Mathematics*,
DOI 10.1007/978-3-642-25710-0_16, © Springer-Verlag Berlin Heidelberg 2012

215

Fig. 1 Logo of the
WMY2000

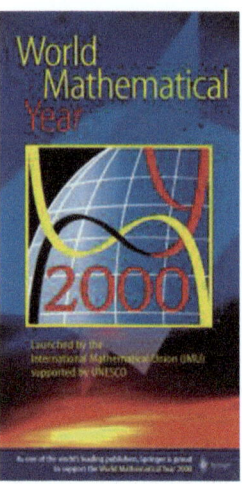

Origins of the Operation

'The idea of declaring the year 2000 to be World Mathematical Year is, without any doubt, one of the numerous initiatives of Jacques-Louis Lions', Gérard Tronel, one of his closest students, told me.

This project arose probably around 1990, but it is difficult to know why and how it came at this period. Of course, it is influenced by J.-J. Lions' involvement in the IMU, but it corresponds also to the vision he had of the globalization of mathematics and the problems it would generate.

Mathematicians had anticipated this globalization with the creation of the International Union at the end of the XIXth century. But all the relationships of J.-J. Lions developed with many colleagues from Asia, Latin America and Africa had convinced him that the future mathematicians would also come from developing countries.

From the beginning he understood the power of computers and networks for the diffusion of information and the education of elites and he saw very clearly that a worldwide operation would demand a precise definition of the objectives and a broad adhesion to them. Jacques-Louis Lions was a visionary, as attested by his written papers where one can recognize the interest he had for the great problems of society as, for instance, the salvaging of Venice, pollution control, the forecasting of weather and even the global change of climate!

The activities of J.-J. Lions were oriented towards the future, but he was also very attached to the mathematicians of earlier generations. For instance, he had a deep admiration for Henri Poincaré and he insisted that the seat of the Newsletter was located at the Institute Henri Poincaré.

Follow up of IMU

Let us quote, here, an interview with David Mumford, then President of IMU, which took place in May, 1995, during the IMU Executive Committee meeting in Paris.

Q: How does IMU feel, and more precisely, what do you think about the WMY 2000 initiative?

D.M.: I think it is an excellent initiative and I think it offers a great opportunity to try and improve the image of mathematics and to try and give some inspiration to the mathematical community. This should be done in terms of several factors like Hilbert looking ahead to where we stand in mathematics and what the major challenges are and, at the same time, looking at our relations with the applications of mathematics, looking at the issues in mathematical instruction and looking at the hopes of involving the Third World more intensely in mathematical activities.

Q: Specifically on these three topics which have been defined for the WMY, have you a personal view on what the challenging problems in mathematics will be in the 21st century?

D.M.: I personally feel that a very important issue is to restore the free interchange of ideas between pure mathematicians and applied mathematicians. During the 19th century, you see that most mathematicians were both. Fourier series, for instance, were inspired by applications. But there has been a divergence, especially in the US, although I think to some extent in other countries too. I hope that the present emphasis on the usefulness of science in general (if the public is going to put money into scientific research, then it will be useful), will not lead simply to pushing applied subjects at the expense of work on pure subjects, but rather to a sense of common purpose where pure researchers can find inspiration in applications and use their theoretical ideas to grapple more effectively with applied problems. I think there has always been an interchange, but that will really be a major challenge.

Sponsorship of UNESCO

After the support in 1992, of the then General Director of UNESCO, Federico Mayor, S. Raither (the person in charge of mathematics at UNESCO) suggested that a resolution be put to the General Assembly. In November 1997 a resolution was presented by Luxembourg. It was backed by 15 countries and adopted by the Plenary of the General Conference. See the text in the Appendix.

Newsletter WMY2000

In 1993 it was decided to publish a newsletter (Newsletter WMY2000) which would gather, and redistribute widely, information on initiatives taken by the international mathematical community for the preparation of WMY 2000.

Nine issues have been published and distributed (7 000 copies for each). Their content is on the web-site.

The following institutions supported the Newsletter: IMU, ICSU, Comité National Français des Mathématiciens, Collège de France, École Polytechnique, IHÉS, The mathematics department of the University Pierre et Marie Curie (Paris, France).

The distribution received the help of the ICTP (Trieste, Italy) and of the CIMPA (Nice, France).

Web-site

Thanks to the Institute of Mathematics of Jussieu (Paris, France), a web-site has been developed with the following address: http://wmy2000.math.jussieu.fr.

It contains all the information and documents concerning WMY2000: pictures of posters, postcards and stamps, reports of conferences and announcements of events.

A great number of people accessed the site and many of them decided to contribute to WMY2000 after seeing the projects.

Most of the Mathematical Societies have established a link to the WMY2000 web-site.

Publicity

The main publicity came from the web-site and the Newsletter, but we also took advantage of the big International Congresses (ICM94 in Zurich (Switzerland), ICM98 in Berlin (Germany) and 3ECM in Barcelona (Spain)) to advertise for WMY2000 with a special booth where documents on the World Mathematical Year could be found.

Two posters were widely distributed (one with a design of Marie-Claude Vergne (who has created the WMY2000 logo) and one offered by Springer Verlag). We would like to thank Springer, which also provided a page for us in each of its 2000 Newsletters where we could announce some of the WMY2000 events.

Many papers on WMY2000 were published in various European and International Newspapers.

Associations

All the International Mathematical Societies have contributed to WMY2000, either by special conferences and events or by their annual meeting, with sessions devoted to the aims of WMY2000. Some of them have created a special Committee.

For example, the book Mathematics: Frontiers and Perspectives published by the IMU, PACOM'2000 (the fifth Pan African Congress of Mathematicians) in Cape

Town (South Africa), the first World Congress of the Bachelier Finance Society in Paris (France), Mathematics and 21st Century in Cairo (Egypt), the Latin American Congress of Mathematics in Rio de Janeiro (Brazil), etc.

Countries

More than 40 countries have participated in celebration of the year in different ways.

Argentina, Canada, France, Germany, Italy, Portugal, Spain and UK contributed to the year very actively with good access to the general public and the politicians.

We regret the small participation of Asian Countries (Year 2000 is not remarkable for them!) and of Eastern and Developing Countries (they have other priorities ...). But let us mention the conference Macau 2000: Mathematics and its Role in Civilization, which was held January 11–14, in Macau (China).

Fortunately, some countries like Algeria and Senegal have organized fine events, despite their difficulties.

Projects

Conferences

Several international conferences have been held during year 2000 (see the list on the web-sites with some reports). In some cases, they would have taken place independently of this special occasion; nevertheless, they have contributed to make WMY2000 more visible and very often they have organized round tables or talks about the aims of WMY2000.

For example, the Conference in Los Angeles organized by the AMS and a special day in Paris devoted to the Clay Prizes.

Lectures for the General Public

Mathematicians have become more aware of the need for communicating mathematics to the general public and such events have taken place in many countries (Algeria, Argentina, Australia, Canada, Denmark, Finland, France, Germany, Greece, Italy, Norway, Portugal, Senegal, Spain, UK, ...).

Books

Some books have been published especially for WM2000. In addition to the book of the IMU, others include Development of Mathematics: 1950–2000, published by

Fig. 2 Stamps from Argentina and Spain

Birkhäuser under the direction of Jean-Paul Pier (Luxembourg) and the book from
the publishing house Springer, Mathematics unlimited-2001 and beyond.

Stamps

Stamps related to WMY2000 were issued in the following countries: Argentina,
Belgium, Croatia, Czech Republic, Italy, Luxembourg, Monaco, Slovakia, Spain
and Sweden (see the web-site for some of them).

Posters

Following the poster competition arranged by the European Mathematical Society,
posters or series of posters have been produced in Belgium, Denmark, France, Germany, Italy, Luxembourg, Portugal, Spain and UK. Some of them were in the Public
Transportation Systems (Buses and Undergrounds) or presented in Science Museums like the series, Mathematics in the daily life, produced by a team in Paris within
a European Contract (RPA-MATHS) related to the European Science and Technology Week.

Other Activities

The event, launched in Obidos (Portugal), in year 2000, was a project of the European mathematical Society, in conjunction with the University of Bangor, within the
framework of the European Science Week of the European Commission where we
have produced, posters and a leaflet on the theme—mathematics of the daily life—
and a CD-Rom with beautiful images of mathematical objects and superb sculptures
by the sculptor John Robinson, inspired by mathematics.

Fig. 3 Poster from the
subway in Brussels

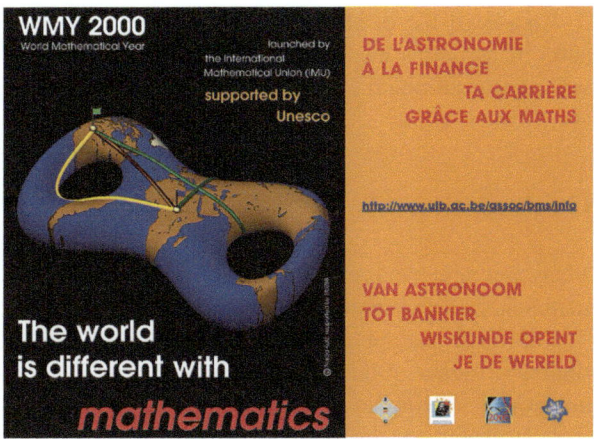

Fig. 4 A postcard published
in France

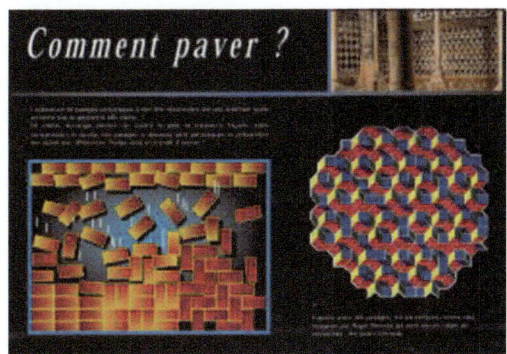

Postcards inspired by the posters have been printed in Denmark, France, Germany and Portugal.

The tramway in Torino (Italy) was emblazoned during one month with the WMY2000 logo.

Among many other activities (see the list on the web-site), let us quote mathematical exhibitions, articles in Newspapers and some (very few ...) Radio and Television Programmes.

Conclusion

From this big and successful operation, it is clear that for some years, many mathematicians are convinced that communicating mathematics to the general public is necessary and very helpful for the community.

The production of posters with mathematical themes has been of considerable interest to schools, where teachers use them to motivate and inspire pupils.

An e-mail list of about 900 mathematicians, all volunteers, interested in these sorts of activities has been established. They ask for ideas and material and are willing to continue the work.

Continuation

Many mathematical societies now consider that to raise the public awareness of mathematics is an important task and they have plans to continue this work.

The EMS has appointed a new Committee (RPA Committee) to ensure a continuing presence at the European Science and Technology Week and to encourage activities involving the general public.

Concerning the image of mathematics, I think that the different actions have contributed to encouraging the learned mathematical societies to make efforts in this direction. It can take time ... but, for instance, this year the French Mathematical Society have nominated a member of the council to be in charge of raising public awareness and a special page has been created in its Web site for the public at large. There is also the beautiful French web site managed by the CNRS—Image des Mathématiques. Let us also mention the European Mathematical Society which is launching its web site—mathematics-in-europe.

The introduction of posters in subways showed us the limitations of this sort of action for the general public: we dreamed of workers who, in the morning, would be interested in the Fibonacci sequence ... Instead of this, some teachers pulled off the posters to put them in their classrooms! But, even now, I receive requests from teachers from around the world, asking for posters.

For developing countries ... It is less clear, but I shall discuss this in the context of the exhibition—Experiencing Mathematics.

World Mathematical Year 2000 has inspired several other initiatives; among them:

– The International year of Physics in 2005: physicists have copied us!
– The year of mathematics in Germany: a great success chaired by Ehrhard Behrends.

Experiencing Mathematics

Among the follow up of WMY2000 sponsored by UNESCO, there is the travelling Exhibition—Experiencing Mathematics.

After the WMY 2000, the person from UNESCO in charge of basic sciences, Minella Alarcon wanted to promote international cooperation in mathematics with a very visible operation. She had close contacts with researchers in the Philippines and Japan already involved in the popularisation of mathematics.

Fig. 5 The exposition "experiencing mathematics"

We proposed to her the building of an international traveling exhibition, putting together mathematical objects and posters. So, the exhibition—Experiencing Mathematics—was born, designed by Centre Science with the support of several different institutions, among them the International Mathematics Union (IMU) and its commission for education (ICMI) and the European Mathematical Society.

The exhibition is mainly addressed to those between the ages 10–18, but also to their teachers and parents. It has been conceived with three main objectives:

– to raise public awareness and interest in mathematics, to demonstrate not only that mathematics is indispensable and everywhere but that it is interesting, challenging and fun as well
– to demonstrate that mathematics is within everyone's reach, that, contrary to what is generally assumed, a good grasp of basic mathematical concepts can be achieved by the majority, and that important mathematical ideas can be made widely accessible.

All the experiences are conceived so that teachers will be able to use them in their classrooms. A tool for their training!

The exhibition was first presented at the 10th International Congress in Mathematics Education, in July 2004 in Copenhagen (Denmark).

After this, mathematicians and mathematics teachers from different countries asked for it, and thus it began to travel around the world. Particular attention was paid to presenting it in developing countries, by arranging the circulation of the exhibition by continental regions: Southern Africa, Latin America, Asia, West Africa, which was more economic for the countries petitioning and enriching for the exhibition.

For the past 6 years (2005–2010), the exhibition has been presented in 90 cities of 32 different countries in Southern and West Africa, China and East Asia, Latin America and Europe. It has welcomed more than 1 200 000 visitors, of whom about 70 per cent were young visitors, and more than 20 000 teachers. There are now in circulation, three copies of the exhibition with many improvements concerning the understanding, the strength (so many visitors . . .) and the weight (for the travels!).

Everywhere, the presentation was preceded by training sessions for teachers and animators lasting 3 days on average. These presentations have been well covered by the media (newspapers, radio and TV, buzz on many many web-sites) and was an opportunity for organizing specific activities linked to mathematics (workshops, lectures, competitions).

Some characteristic examples were:

– in Namibia (2006), 50 000 pupils and teachers in 12 cities in 3 months;
– in Madrid, for the ICM2006, nearly 40 000 visitors in 3 months (with queues outside in the summer holidays!) and students from more than 100 academic establishments;
– in 2007, Bangkok, Laos (4 cities), Viet-Nam (2) and Cambodia (4), Singapore, 120 000 young visitors with 4 specific training sessions for teachers around the usefulness of Mathematics in Cambodia sponsored by IMU, ICMI, the CIMPA and UNESCO;
– in India (2008), nearly 100 000 visitors in 4 towns (Delhi, Calcutta, Bangalore, Mumbai) with 2 specific training sessions for teachers organized by the local Unesco;
– in Pakistan (2008), 50 000 young visitors in 3 cities with the National Science Foundation;
– in Latin America (2008–2010): 8 countries—in Brazil, 10 cities in 7 months— and at the end in Chile—for the second time—in January 2010, 2 000 teachers for two weeks of mathematics training sessions (with the Chile University);
– and last, more difficult but with success, in West Africa, in 2009–2010, in 3 countries (Senegal, Benin, Burkina Faso) with the help of the local mathematicians.

All these presentations were carried out with the logistic and practical support of the local Ministries of Education, of Researchers, the Mathematical Societies, the Embassies, the Science Foundations of each country and the Science Museums of each city.

To present all the news of the exhibition circulation, of the maths news and the partners, a web site was conceived: www.MathEx.org.

A surprising thing for us is that it is easier to find people interested in the exhibition in developing countries than in Europe or North America. So, if you have some contacts in these areas, please, tell me!

To complement the exhibition, a virtual exhibition in 4 languages (English, French, Portuguese, Spanish) was conceived in 2008 with the financial support of Southern Africa Unesco (Mozambique, Namibia and, especially, Angola).

Including more than 30 virtual and interactive experiences, 300 activities are proposed as mathematical experiences. The virtual exhibition presents examples of the educational uses of the themes of the exhibition with printed documents for teachers, in order to reinforce its educational impact, especially in developing countries.

These virtual tools are conceived so that teachers will be able to use them in their classrooms without the aid of computers but with only the usual materials like paper, string, cardboard and glue and experience mathematics with their students. It could also favor the development of material showing how the themes of the exhibition can find resonance in different cultures. It was distributed on 800 CDroms, and is freely accessible on the Web: www.ExperiencingMaths.org.

It was presented in specific training sessions in Angola (Inide) in 2008 and 2009 (more than 150 CDroms were distributed in Angola) and training sessions were proposed in 5 regions of Angola.

Others were given in all countries where the actual exhibition was presented, especially in Chile in January 2010 to more than 500 teachers for the 200th anniversary of independence of Chile and in Spain in September 2010 for their XIII CEAM congress.

You can see a presentation in the Spanish mathematical web site Vol. 4, no. 3 (June 2009) www.matematicalia.net/index.php.

Now, I want to tell you something about a new project and a new web site:

What careers in maths? What jobs use maths?

The public in general, numerous pupils and even mathematics teachers are very poorly informed about opportunities offered to the graduates in mathematics at any level and generally believe that only research and teaching are accessible to them. So, it is necessary to (prompt teachers and?) inform parents and pupils early enough in their schooling.

To do that, in some places mathematical events linked to the exhibition have been organized, like showing mathematical movies, conferences of professional mathematicians and even theater plays related to mathematics.

This new project aims to offer new tools to young people, their teachers and their parents, in the shape of:

– an original interactive web site with games, reports, interviews (written or filmed);
– panels or flyers in the exhibition with some portraits of young professionals using a great deal of mathematics in their jobs, but also a variety of professions with different levels of maths qualifications and illustrating these, paying the necessary attention to different kinds of consideration such as gender, regional balance and also not limiting this to well developed countries.

It could be craftsmen, mechanics, office workers, technicians, engineers and managers who profit in their job from their mathematical background. This could show young people (secondary and high school students), as well as their parents and their teachers, the diversity of possible jobs after a scientific education.

We also plan to take advantage of having the exhibition in a specific region to interview well-known mathematicians of the area. They could report on the specific situation and the need of mathematics in their country.

Some Recommendations

First, as my old friend Vagn Hansen, who is here, has said, if a subject becomes invisible, it may soon be forgotten and eventually it may even disappear.

Therefore, the communication of mathematics to the general public is not only obliged to go on, but it is also necessary that it increases.

For example: it is necessary to explore the new media and communication techniques: the web, of course, Twitter, blogs, etc, without forgetting that it can be dangerous to replace the old ones (papers, . . .) by new ones without checking that they are fit for purpose. For instance, replacing regular mails by a blog which nobody connects to . . . It is best to add than to replace!

Another example is that it would be better to collaborate with the bodies devoted to the popularization of science. After all, we are pursuing the same aims! In Europe, I think of Euroscience; in France to Revoluscience, and so on.

In addition, there are new fields for mathematicians like—'computational biology', 'computational medicine'. We could join the corresponding learned societies in communicating on these subjects.

It is necessary to find new areas of mathematics to interest the public at large. I must confess that in many science museums, the mathematics section very often shows the same tricks: the Galton board, the brachistochrone, and so on ...

We should make more of extended ideas like discussing trees, which are simple to describe and which are related to many fruitful domains like:

– biology,
– theoretical computer science,
– web,
– etc.

I finish with the subject of using the visual approach to mathematics, like poster campaigns (we know that people do not read ...). First it is necessary that the images are beautiful to fight the common idea of the austerity of mathematics. They should be conceived like an entry portal in order to capture the attention, to raise interest and give the desire to learn more: to go to a Science Museum where there are hands on displays, with a brochure to take away related to the exhibition, to go to the web site of the project.

That is the way that mathematics will become more visible.

Appendix

I. The Declaration of Rio de Janeiro on Mathematics

On May 6th, 1992, in Rio de Janeiro, during the celebration of the 40th anniversary of the world-wide reputed Institute of Pure and Applied Mathematics (IMPA), Professor Jacques-Louis Lions, President of the International Mathematical Union (IMU) declared in the name of this Union, that the year 2000 will be the World Mathematical Year.

WMY 2000 is set under the sponsorship of UNESCO (Professor Federico Mayor), of the Third World Academy of Sciences (Professor Abdus Salam and Professor Carlos Chagas, who took part in the declaration of Rio de Janeiro), of the French Ministry of Research and Space (Professor Hubert Curien), of the Brazilean Academy of Sciences (Professor Israel Vargas) and of the Swiss Federal Counsellor (Dr. Flavio Cotti), the next International Congress of Mathematicians being organized in Zürich in August 1994.

The declaration of Rio de Janeiro sets three aims:

– The great challenges of the 21st century,

– Mathematics, keys for Development,
– The Image of Mathematics.

1. First aim: the great challenges of the 21st century. During his conference in Paris in 1900, David Hilbert listed a series of the main problems that the then ending century had to challenge. The American Mathematical Society suggested in 1990, at the last General Assembly of IMU in Kobe (Japan), that first class mathematicians, to be represented within the Turn of the Century Committee, organize the efforts to envision what the great challenges of the year 2000 would be. This Committee is chaired by Professor Jacob Palis Jr, IMPA (Brazil), Secretary of IMU.

2. Second aim: Mathematics, keys for Development of Pure and Applied Mathematics are one of the main keys of the understanding of the world and of its development.

That is why it is essential that countries which are members of UNESCO be gradually encouraged to reach a level enabling their admission to IMU, of which there are 50 nations for the time being. Therefore, the second aim of the Declaration of Rio de Janeiro is that most countries which are members of UNESCO reach such level by the turn of century.

That implies great additional efforts in the fields of Education, of Training, and—a very sensitive point for countries that face difficulties in having currency resources—of access to Scientific Information.

Such efforts which have already been widely undertaken, will be confirmed and raised by the two main commissions of IMU: ICMI (International Commission on Mathematical Instruction), which is chaired by Professor M. de Guzman from Madrid and whose Secretary is Professor M. Niss from Denmark, and the CDE (Commission on Development and Exchange), whose president is Professor M.S. Narasimhan from Bombay and whose Secretary is Professor P. Bérard from Grenoble, France. Both commissions are linked with UNESCO which was represented in Rio de Janeiro by Professor A. Marzollo, responsible for mathematics.

3. Third aim: the Image of Mathematics. The Declaration of Rio de Janeiro sets as its third goal, which is also of great importance, a systematic presence of mathematics in the "Information Society" thanks to examples and applications which will be scientifically exact and open to the largest number.

That will be developed in connection with such efforts which have already been undertaken by many countries that are members of IMU. The declaration of Rio de Janeiro on Mathematics announcing the World Mathematical Year 2000 was warmly supported not only by all the mathematicians present in Rio and who had come from all continents, and of course many of Brazil's most eminent mathematicians, but also by professors in other subjects too, and especially Professor Carlos Chagas, former President of the Pontifical Academy of Sciences.

II. The UNESCO Resolution

In its November 11, 1997 plenary meeting, the UNESCO General Conference followed the recommendations of Commission III and approved draft resolution 29

C/DR126 related to the World Mathematical Year 2000, allocating 20,000 US dollars to this series of events. The following 15 countries co-sponsored the draft resolution: Belgium, Benin, Brazil, Colombia, Cote d'Ivoire, Denmark, France, Ireland, Luxembourg, Philippines, Netherlands, Russian Federation, Spain, Thailand, Uzbekistan.

The General Conference

- Considers the central importance of mathematics and its applications in today's world with regard to science, technology, communications, economics and numerous other fields;
- Is aware that mathematics has deep roots in many cultures and that the most outstanding thinkers over several thousand years contributed significantly to their development, and numerous other fields;
- Is aware that the language and the values of mathematics are universal, thus encouraging and making it ideally suited for international cooperation;
- Stresses the key role of mathematics education, in particular at primary and secondary school level, both for the understanding of basic mathematical concepts and for the development of rational thinking;
- Welcomes the initiative of the International Mathematical Union (IMU) to declare the year 2000 the World Mathematical Year and carry out, within this framework, activities to promote Mathematics at all levels world-wide;
- Decides to support the World Mathematical Year 2000 initiative;
- Requests the Director General to collaborate with the international mathematics community in planning the World Mathematical Year 2000 and to contribute during 1998–1999 funds of 20.000 US dollars from the Regular Programme and Budget in support of preparatory activities.

Some Remarks on Popularizing Mathematics or a Magic Room

Krzysztof Ciesielski and Zdzisław Pogoda

Abstract In this chapter, we first discuss activities for raising the public awareness of mathematics in Poland. Then, we write about the personal experiences of the authors concerning these activities. In the last and main part of the chapter we describe how the authors present mathematical terms and ideas to the general public, with a few examples.

Introduction

1. In Poland, many people are involved in raising the public awareness of mathematics. Probably three forms of activities are most frequently: writing about mathematics, speaking about mathematics and organizing different mathematical competitions. Here, we consider the first two.

There are a number of journals of popular mathematics, which are published in Poland. First of all, we should mention here the monthly *Delta*. This monthly was founded by the Polish Mathematical Society and the Polish Physical Society. In the beginning, it was devoted to mathematics and physics. The first issue was published in 1974. Later, the journal started popularizing also astronomy and (recently) informatics. In recent years the monthly has been published by the University of Warsaw. Since the beginning, Marek Kordos has been the editor-in-chief. Many outstanding Polish scientists have written articles for *Delta*, for example Karol Borsuk and Andrzej Schinzel.

Some years ago, a new magazine, the quarterly *MMM—Magazyn Miłośników Matematyki* (*The Magazine for Fans of Mathematics*) started being published. It presents mathematics in a rather elementary way, a large part of each issue is devoted to interesting elementary problems and puzzles.

K. Ciesielski (✉) · Z. Pogoda
Mathematics Institute, Faculty of Mathematics and Computer Science, Jagiellonian University, Łojasiewicza 6, 30-348 Kraków, Poland
e-mail: Krzysztof.Ciesielski@im.uj.edu.pl

Z. Pogoda
e-mail: Zdzislaw.Pogoda@im.uj.edu.pl

E. Behrends et al. (eds.), *Raising Public Awareness of Mathematics*,
DOI 10.1007/978-3-642-25710-0_17, © Springer-Verlag Berlin Heidelberg 2012

Also, there are magazines for mathematics teachers in schools. The oldest is a 64-year-old monthly, *Matematyka* (*Mathematics*). It presents mathematics in a simple way, mainly to interest teachers.

These three journals are probably the best known in Poland.

It must be noted, that there are two famous journals that popularize science generally and from time to time they include mathematics as well. They are *Wiedza i Życie* (*Knowledge and Life*) and *Świat Nauki* (the Polish version of *Scientific American*). For many years a very good monthly *Problemy* (*Problems*) was published, which popularized science at a high level, but publication stopped about 20 years ago, because of economic changes and financial problems.

From time to time, articles about mathematics are published in other magazines and even daily newspapers. Usually they are connected with sensational news, but sometimes readers may find a good article written by a professional mathematician.

Several books of popular mathematics have been published in Poland as well. However, the character of such books has changed. Up to 30 years ago, the majority of books were addressed to people who knew something (or even quite a lot) about mathematics, mainly in the series *Biblioteczka matematyczna* (*The small mathematical library*). For many years, there were only a few books available for people not having a suitable mathematical background. It should be noted that a famous book by an eminent Polish mathematician Hugo Steinhaus, *Kalejdoskop matematyczny* (*Mathematical Snapshots*) was published in 1938, in 1954 and in 1956 after World War II, and then we waited more than 30 years for the next edition! However, in the last 20 years the situation has changed. Many popular mathematical books have been published and many have been translated. These books have a different style and present mathematics on different levels. Several of them were published by *Prószyński i S-ka* editors in a special series *Na ścieżkach nauki* (*In the ways of science*). Nevertheless, it is not easy to find popular books on mathematics in bookshops—even if booksellers order them, they do so generally in small quantities and usually put them on almost "invisible" shelves. However, a few years ago in a large bookshop in Kraków, a famous book by Ian Stewart *Does God Pay Dice?* was placed in the religious books' department and was selling very well from there.

2. Another very important activity is speaking about mathematics, especially to young people. Many such talks are organized, in different styles. Sometimes longer meetings take place, where a few lectures are presented (not just mathematical). The meetings are organized in different cities in different ways. Sometimes they are a part of a bigger event, like "Festival of Science" or "Night of Scientists". Sometimes they are special mathematical events. An interesting example is "A Day in the Department of Mathematics and Computer Science of the Jagiellonian University" (organized once a year) where more than one thousand secondary school students come for one day to the department and take part in lectures and workshops intended especially for them. However, "single" individual lectures are much more frequent. Some of them are organized by several institutions connected with mathematics (like the Polish Mathematical Society or mathematical departments of universities). On the other hand, it is quite common that the authorities in schools, mainly secondary schools (heads or teachers) invite mathematicians from different universities to deliver lectures for their students. Several special lectures for schoolteachers are organized as well.

The Polish Centre for Mathematical Culture was created in 1987 by a group of mathematicians very actively involved in popularizing mathematics. The main activities of the Centre are the Schools of Popular Mathematics. Two schools are organized each year (a winter school and a summer school), each of them lasting five days. The schools are intended for mathematicians and students, but the idea is also to present different areas of mathematics for people specializing in other areas of mathematics. Nevertheless, teachers and students participated in those schools, too. Schools are not devoted to a particular branch of mathematics, but are on topics with some common features or connections with titles such as: "Mathematical picklocks", "Space", "It happened in the 20th century", "The most important terms in mathematics" and "Mathematical pearls".

The Centre has also organized special talks for teachers for many years. The special journal of the Centre *Matematyka, Społeczeństwo, Nauczanie* (*Mathematics, Society, Teaching*) is published twice a year.

3.   We have been involved in very many of the activities mentioned above. We often meet the opinion that popularizing mathematics for school students should concentrate on solving problems. Indeed, this is very important. Solving problems and precise proofs are crucial in mathematics, which when popularizing mathematics, should be emphasized. The importance of proofs and reasoning must be pointed out. Nevertheless, interesting problems are not the only mathematical topic that may be presented to young people. For many years, besides such problems, we spoke about higher mathematics, its terms and ideas, including modern results. This task is not easy, especially when there is only one hour to speak about mathematics; however, it is possible. Of course, it is impossible to present all the details of modern mathematical achievements and this is the main reason that some mathematicians are sceptical about such methods of popularizing. In our opinion, this way of popularizing is worthwhile. It may really enlarge the mathematical knowledge of outsiders, many of whom will not encounter mathematics in the future.

What is important in such a presentation? First of all, it must not be oversimplified and it must not be too advanced. The basic ideas have to be shown but not trivialized. Of course, it must not contain sophisticated terminology or complicated mathematical expressions. Overly technical discussions are not suitable. On the other hand, it is very useful to give several examples. Sometimes it is not easy to give examples for people without a suitable background, so then it is useful to present some analogies. Some historical remarks and comments about mathematicians connected with the theory and results help to diversify the talk. If possible and if they are relevant to the main topic, some anecdotes or mathematical jokes are useful. Nevertheless, the most important thing is perhaps not to make the talk too difficult. Of course, this depends on the audience and the speaker has to know that sometimes the plans have to change and much less said than intended at the beginning.

Fig. 1 A strangely looking planet

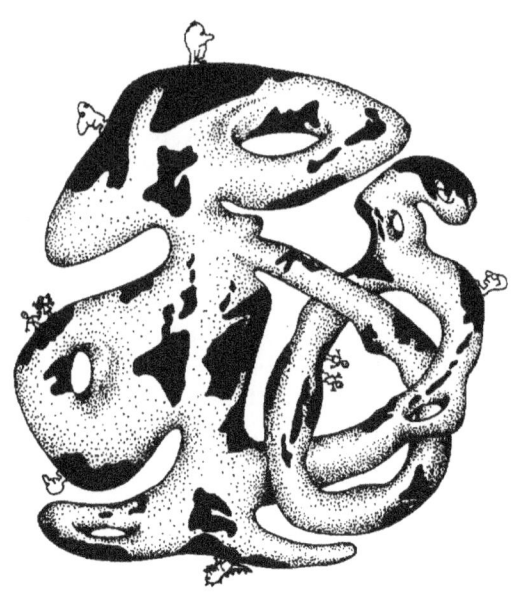

We have talked about mathematics to young people and teachers very many times, on many topics. One of the topics is manifolds. Below, we give some examples of how we explained manifolds to non-mathematicians, especially secondary school students. Several ideas and pictures are taken from our books *Bezmiar matematycznej wyobraźni* (*Boundlessness of mathematical imagination*, three editions so far) and *Diamenty matematyki* (*Mathematical diamonds*). These two books were regarded by the Polish Foundation for Science Advancement as the best books popularizing science written by a Polish author between 1995 and 1998 and were awarded the Hugo Steinhaus Prize.

4.   When one speaks about mathematics, it is always useful to answer the question: "Why are we interested in this particular topic?" Speaking about manifolds, there is a very good historical background. What is the shape of the Earth? Everybody knows. However, in ancient times, people used to think that it was flat. Why did they think so, as they were not idiots? The answer is simple: the Earth looks locally like a piece of a plane. And, in fact, if we investigate the properties of a particular object, in many cases it is enough to do so locally, in the neighborhood of a given point. So, regardless of the global shape of the object we only need to analyze it locally. However, the whole object could look very strange … (see Fig. 1).

And, in fact, for many practical purposes the shape of the Earth is really not important. Do people in Luxembourg care about what's going on in Patagonia? Moreover, the plane is very convenient to research, as it has a simple analytic description. This gives the basis for the introduction of the term "manifold" as a natural generalization of a local structure.

The definition should always be accompanied by several examples. In two-dimensions there are several nice examples: the torus (see Fig. 2), a torus with more

Fig. 2 A planet in the shape of a torus as a manifold

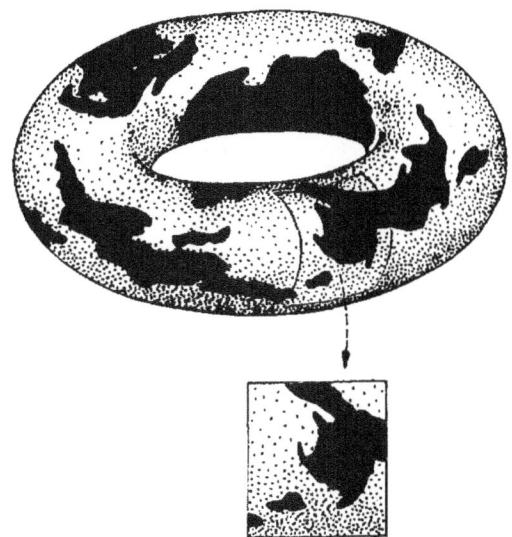

holes and the Klein bottle. Nevertheless, one cannot restrict investigations to the dimension 2: higher dimensions must be considered.

What about the third dimension? It is of great importance. Now we know how the Earth looks from space, so we have no problem with the question about its shape. But our ancestors could not see such pictures. They saw the Earth locally. We cannot see the whole Universe, we see it only locally. What about our space? Where do we live? It is a good question for young people. What we expect is that we are somewhere inside a three-dimensional manifold ...

It is difficult to imagine non-trivial examples of three-dimensional manifolds that are different from subsets of the space. No wonder. Our intuition is three-dimensional. We present below how we usually describe two "classical" manifolds—the three-dimensional sphere and the three-dimensional torus.

First, we ask for the help of the friendly creatures invented by Edwin Abbott a long time ago, in 1884, the Flatlanders. School students (and others as well) like very much stories about figures who live in a two-dimensional space and do not know the third dimension. What is possible for three-dimensional people but for Flatlanders would look at least mysterious? For example, to leave a locked room. To empty a closed bottle. To rob a guarded bank. See Figs. 3 and 4.

Fig. 3 Emptying a Flatlander's closed bottle in a mysterious way

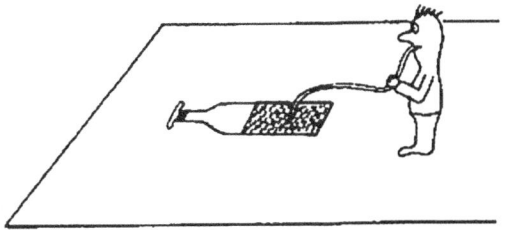

Fig. 4 Robbing a
Flatlander's bank not being
noticed

So, how could a Flatlander imagine a two-dimensional sphere? He can use time as the third dimension. We can speak about a two-dimensional sphere as a collection of the intersections of this sphere with parallel planes. In other words, we have circles "moving" from the bottom to the top. First we have a point, which changes to a circle and the radius gets larger and larger. At a certain moment we get the largest circle, then the radius becomes smaller and smaller, until at last we see a point and then we do not have anything. For us it is easy. However, when a clever Flatlander tries to imagine such a collection of figures, a good method for him is to consider time as the third dimension. At 6.00, say, a point appears suddenly; it becomes a circle that grows and grows, at 7.00 we have the largest circle, then the circle becomes smaller, at 8.00 we have a point and after 8.00 the point disappears, see Fig. 5.

We see the two-dimensional sphere and we know how it is visualized by Flatlanders. We cannot be stupider than the Flatlanders! So, we can think about the three-dimensional sphere in an analogous way. At 6.00 we see a mysterious point, then we see a two-dimensional sphere, which grows up etc., see Fig. 6. This analogy helps to explain how "geometrically" we can think about a three-dimensional sphere.

What about a torus? Again, the Flatlanders will help us. How can we tell the Flatlanders about a two-dimensional torus, when they are two-dimensional? In this case, we do not need to use time as the third dimension but we can apply a well-known characterization of the torus as a suitably described subset of the plane. The

Fig. 5 A sphere imagined by
Flatlanders by means of "time
wandering"

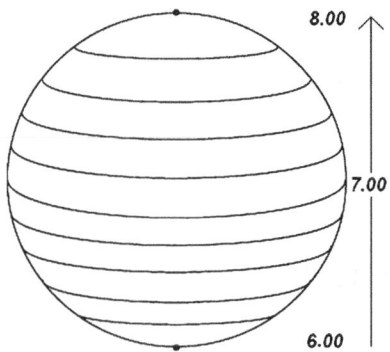

Fig. 6 Three-dimensional
sphere imagined by means of
"time wandering"

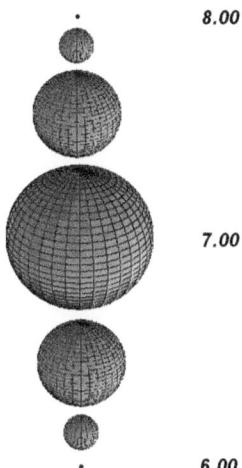

torus is equivalent to $[0, 1] \times [0, 1]$ with an identification of the sides. We can show this with a picture, but it is also very helpful to take a rectangular piece of paper or elastic band and glue one pair of opposite sides together, and then do the same with the other pair, see Figs. 7 and 8. This enables us to think of the torus as a planar set and avoid the use of the third dimension.

Consider $\mathbb{R}^2$ divided by a suitable equivalence relation. It is good to mention that a man is not "broken" when moving from one side of the rectangle to the other. Think of the rectangle as a TV screen so that a moving point that disappears from the right side of it reappears on the left side. Now, imagine a large bank of TV sets, arranged in a grid, as seen sometimes in shop exhibitions all showing the same picture. We see each particular point many times, see Fig. 9.

So, a Flatlander may imagine himself walking inside a rectangle. When he leaves it from the right side, he immediately appears on the left, entering the opposite side. Leaving up, he enters at the bottom.

This is an excellent way to speak to an audience about a three-dimensional torus. Imagine that we are sitting in the lecture room where the talk takes place. Think of

Fig. 7 Thinking of a torus as
a planar set

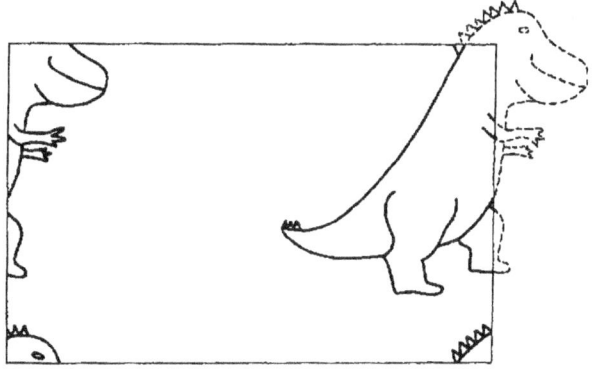

Fig. 8 A piece of paper after
a suitable gluing

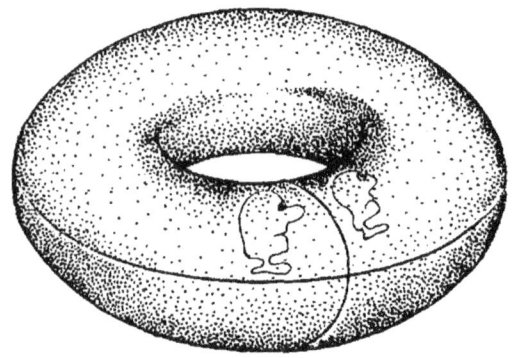

this room as a three-dimensional torus and identify the matching points of opposite faces. We are in a magic room—it is impossible to leave it! You penetrate the front wall and poke your nose back into the room, through the back wall (see Fig. 10). Sometimes in a room there are two doors on opposite walls, or—more frequently— a door and a window on opposite walls. Suppose we have a guest we did not invite. It is impossible to get him out. If we throw him out by the door, he immediately comes back in by the window. Whichever way we choose, he re-enters the room from the opposite side. We can also speak about dangerous events. If somebody has a revolver, it could be extremely risky to shoot: you will be killed by your own bullet. And what about spitting on the floor?

The perfect identification of opposite faces of the room results in a strange effect: there are no walls any more, no floor, no ceiling. You look up and see your feet; look down and see your scalp. Stretch out your arm and catch your other hand, provided the room is sufficiently narrow (see Fig. 11). Suppose it is a ball-room; dancing pairs are viewed simultaneously from many sides.

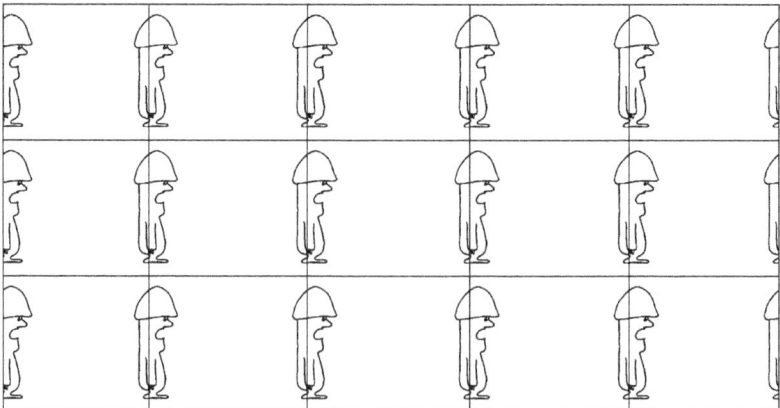

Fig. 9 Thinking of a torus as a plane divided by an equivalence relation

Fig. 10 Thinking of a
three-dimensional torus as a
subset of space

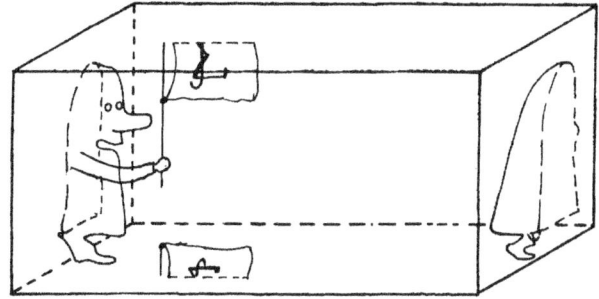

Listening to such stories, the audience may be stretched by a comparison to football. Do you think that football matches are not interesting enough? Do you think that our national team plays badly? So, we know what to do to make the competition more interesting. Put the playing field inside a three-dimensional torus. Then, when a player misses the goal, even very close to the post or to the bar, it immediately causes a dangerous situation for his own team. There is an interesting way to pass the ball to a teammate close to a touchline; it is sufficient to kick the ball out. What will happen if one kicks the ball so that it goes very high? By the way, a playing field embedded in a torus is a very good example for Poland, as Euro 2012 will be held here (so we need matches that will excite the audience) and the Polish national team is rather poor. Do we want to win? No problem, we are going to prepare playing fields this way and not tell the others. Other teams will be very surprised and our team will be well prepared … Probably this is the only way for the Polish team to get a good result.

Of course, we do not need to identify opposite sides. Consider again the two-dimensional case. With a different identification of the sides of the rectangle, we can get a Klein bottle or a projective plane. This is also a good method of speaking about those manifolds, as they can be perfectly visualized only in four-dimensional space. For a Klein bottle, "time wandering" can be used here as well. When we

Fig. 11 A narrow room as a three-dimensional torus

Fig. 12 A Klein bottle imagined by means of "time wandering"

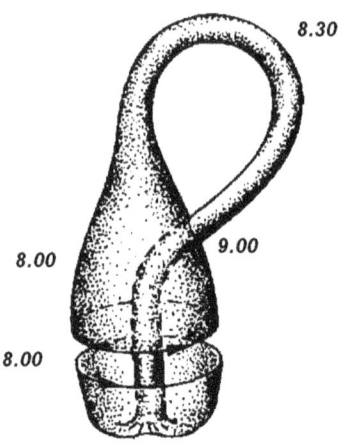

8.30

8.00   *9.00*

8.00

made a Klein bottle, we took an "elastic" bottle without a base and stretched its neck. Pulling on the neck is like traveling in time. We had the main part of the bottle at 8, as we pulled the neck, we simultaneously moved in time and reached 9 o'clock. Then, we moved "inside" the bottle. This was at 9 but the boundary of the bottle existed at 8, so we did not meet it, which explains why the Klein bottle is a manifold and indeed each point has a neighborhood looking like a piece of the plane. Then, we finished making the bottle by gluing the two ends, simultaneously coming back in time to 8, see Fig. 12.

With a suitable gluing of the sides of a rectangle, we can get not only a two-dimensional torus, but also various other two-dimensional manifolds. So, besides a three-dimensional torus, much more different and strange manifolds can be obtained by a box with its faces identified in another fashion. We can imagine that "left" and "right" interchange when the front was attached to the back. Having disappeared through the front you reappear through the back, but now as your own mirror image. An observer who stayed in the room and did not move will see that now you have your heart on the right side of your chest. By means of various identifications we can obtain several other objects. We do need to consider only a box, we can use other solids.

We can imagine certain three-dimensional manifolds in another way. Let us call the totality of positions of a moving point (or of a system of moving points) a *configuration space*. Such spaces are excellent examples of manifolds of different dimensions. For example, let us attach to the end of a pendulum, which moves in a circle in a plane, another pendulum that moves independently in space. We may conclude that the configuration space of these two pendulums is a three-dimensional manifold, since in a neighborhood of each point its ability to move is determined by three independent directions.

Such stories have another advantage. Generally, in the mathematics programmes in schools only very old results are given, much older than mathematics from the 20th century. Modern mathematical problems are not known by ordinary people and it seems impossible to speak about them. Here we have an excellent counterexam-

Fig. 13 Saddle-shaped
Flatland

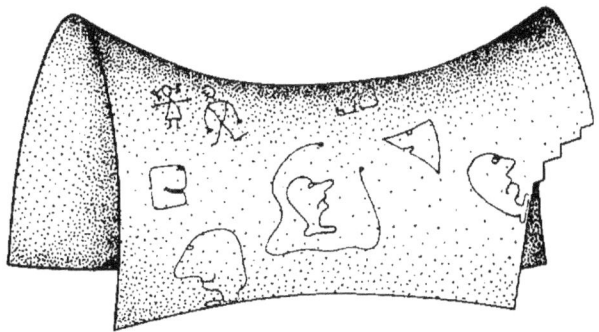

ple! One of the greatest challenges for modern mathematicians was the classification of three-dimensional manifolds. Outstanding scientists recently obtained magnificent results, which ended with Perelman's extremely advanced work. Without explaining the subject precisely or "semi-precisely", one can say that a "pleasant" manifold can be split into "elementary bricks" equipped with some basic geometry. When we know how to describe those bricks and how to split a manifold, we get a satisfying description.

This is very important! One could say that this is in some sense the classification of possible models of the Universe, if we think about the Universe as a three-dimensional manifold, at least approximately. Perhaps the Flatlanders live in a strange world, which is bent in the third dimension, without them knowing about it (see Fig. 13).

This topic is excellent not only because it shows the audience some results of recent years. It is a great chance to say something on another related topic. Have the audience heard about the Fields Medal? What about some recipients of this prize? It is a good chance to mention the Poincaré conjecture and Poincaré himself.

By the way, Poincaré could have been mentioned earlier: he was the first to consider the identification of the faces of solids.

If we speak about the Poincaré conjecture, we may want to formulate it, though for some audiences the exact formulation may be too difficult. However, even if we do not formulate the conjecture precisely, we can talk about some important types of mathematical theorems. Very frequently we meet theorems like: "An object fulfils particular properties." But, of great importance for mathematics there are theorems of another type: "If an object satisfies properties A, B and C, it must be object X", i.e. the given properties definitely determine the object. As we know, the Poincaré conjecture is one of these. We can make an analogy with real life. Imagine that we have a secret message for Mr. Spy, who is supposed to be here, but we do not know him. What we know is that he has a beard, glasses and wears a pink cap with a green bobble. If there is only one such person in the audience, we have no problem. But if there are several, we do not know to whom we should pass the secret message.

It might be useful to mention higher dimensional manifolds. This is not easy, but why not? It is enough to consider a configuration space created by two pendulums such that both move in the space. We get a four-dimensional manifold. We may think about more pendulums suitably attached to each other, moving in the plane or in the space ...

For a very different but practical example, imagine a spaceship moving in space. Six parameters describe it independently (three of position and three of movement), so we get a six-dimensional manifold. Now, imagine two spaceships moving independently. The parameters describing their motion yield a 12-dimensional manifold!

5.   Finally, we would like to give ten rules we try to follow when we speak about mathematics.

– If you want to explain a piece of mathematics to others, you have to know it yourself; in fact, you have to be knowledgeable about the subject.
– Do not try to say everything you know about a particular topic; in many cases if you say more, listeners will remember less.
– Try to explain things in a simple way. Do not try to give all the details. However, remember about the so-called Einstein's rule: "explain it in as simple a way as possible, but not simpler!"
– Even if you tell "stories", always try to include something connected with mathematical reasoning and logical thinking. On the other hand, when you present problems (even very elementary) and their solutions, try to include in your talk something about higher mathematics, even if in an extremely informal way.
– If possible, include information about real people in your lecture. Let listeners know something about the mathematicians who worked on the subject.
– Remember to speak to the audience, not to the blackboard or to the computer screen. You will benefit when you see faces interested in your talk!
– Observe the audience. If you see that your lecture seems to be too difficult, make changes during the talk.
– Do not be too technical! During a short lecture, keep calculations to a necessary minimum.
– Try to include some analogies from very elementary mathematics or real life.
– Mathematics is fun. It is fun for you. It will be wonderful, if you can share your joy with the listeners.

Mathematics and Interdisciplinarity: Outreach Activities at the University of Coimbra

João Fernandes, Carlos Fiolhais, and Carlota Simões

Abstract It is well known that informal learning may and should supplement formal learning. They are both known to help each other. Based on our experience of outreach activities at the University of Coimbra, we discuss the deep connection of mathematics with astronomy, physics and art. We show with some examples how these connections may facilitate the bridge between mathematics and the public, in particular young people, raising the public awareness of mathematics.

Astronomy and Mathematics

Astronomy and mathematics have a long tradition of contact to mutual benefit. This is a truly symbiotic relation, which may be illustrated by several cornerstones in the history of science: in the third century BC, Eratosthenes' estimation of Earth's radius; in the sixteenth and seventeenth centuries, the discussion about the true world system (heliocentric versus geocentric) by astronomers such as Copernicus, Clavius and Galileo; in the early seventeenth century, Kepler's laws; later in that same century the subsequent formulation of the law of universal gravitation by Newton; and, in the nineteenth century, the prediction of the existence of Neptune by Le Verrier and Adams, which required complicated calculations based on Newtonian mechanics (the new planet was discovered with the pencil before being discovered with the telescope!). This fruitful relation between astronomy and mathematics made fur-

J. Fernandes (✉)
CFC—Centro de Física Computacional, Departamento de Matemática, Observatório
Astronómico da Universidade de Coimbra, 3000 Coimbra, Portugal
e-mail: jmfernan@mat.uc.pt

C. Fiolhais · C. Simões
CFC—Centro de Física Computacional, Departamento de Física da Universidade de Coimbra,
3000 Coimbra, Portugal

C. Fiolhais
e-mail: tcarlos@teor.fis.uc.pt

C. Simões
e-mail: carlota@mat.uc.pt

E. Behrends et al. (eds.), *Raising Public Awareness of Mathematics*,
DOI 10.1007/978-3-642-25710-0_18, © Springer-Verlag Berlin Heidelberg 2012

ther progress in the twentieth century, with the development of astrophysics. An illustrative example is Einstein's perfect description, based on his general theory of relativity, of the perihelion advance of Mercury, which was known from careful observations, and the confirmation of that theory, which required differential geometry and non-Euclidean spaces, achieved with Eddington's observation of the deflection of light by the Sun's gravitational field, at the island of Príncipe—by then a Portuguese colony—in 1919.

There are recent examples of that relation such as the use of powerful N-body simulations to describe the interaction and even merging of galaxies and the use of complex integration to solve the Navier–Stokes equations of the three-dimensional grids used to describe the highly turbulent plasma in some parts of the interior of a star. We see no end in sight for the synergy between mathematics and astronomy.

Clearly, given their mathematical complexity, some of these applications are not suitable examples to use to raise the public awareness of mathematics. So which are, then, good astronomical examples for that purpose?

We propose to answer this question based on three conditions that must be fulfilled by astronomical examples:

- *The mathematics used must be adapted to the secondary school level (or lower).* The reason is quite simple: we may only hope to raise the public awareness of mathematics using astronomical examples if the mathematical language is understood.
- *The examples should use real astronomical data.* For the sake of realism, whenever possible, the astronomical observations should be obtained directly by the public being addressed. If this is not possible, there is a large amount of astronomical observations in several public databases. The most paradigmatic is the International Virtual Observatory,[1] which assembles and makes available to everyone astronomical observations in digital form. The use of real data shows that measuring Nature implies the admission of errors, which should be properly treated.
- *The examples should allow to talk about astronomy in particular and science in general, and not used only as a mathematics laboratory.* This condition is particularly suitable for truly interdisciplinary presentations.

Keeping these conditions in mind we have chosen, from our experience, the following examples to illustrate how astronomy can raise the public awareness of mathematics.

Earth's Radius Estimations by Measuring the Solar Altitude in Two Places

This example is an adaptation of Eratosthenes' experience, namely choosing two places with the same (or approximately the same) longitude and separated by about

---

[1] http://ivoa.net/.

Fig. 1 Scheme of
Eratosthenes' measurement
of Earth's radius

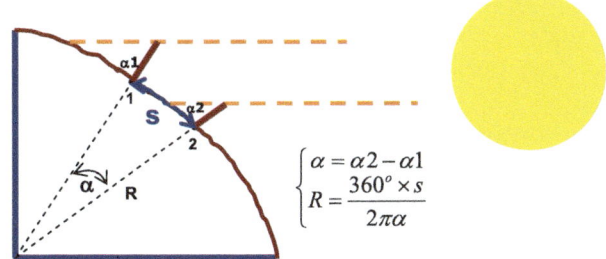

$$\begin{cases} \alpha = \alpha2 - \alpha1 \\ R = \dfrac{360^\circ \times s}{2\pi\alpha} \end{cases}$$

200 km. The idea is to measure simultaneously the astronomical altitude at solar midday in both places ($\alpha1$ and $\alpha2$, in Fig. 1).

During the International Year of Astronomy 2009, a project named *La medida del radio de la Tierra*[2] was carried out by 639 schools in Spain, under the supervision of Pere Closas (Agrupación Astronómica ASTER) which obtained an estimation of the Earth's radius with an error of only 3 %.

This example, which is relatively easy to implement in practice (teachers and students did the experiment in Portugal for sites along the same meridian but separated by more than 500 km), can be used to discuss mathematical concepts such as proportionality, arc circle length, basic trigonometric relations and linear regression. On the other hand, it is a very good opportunity to explore astronomical concepts as astronomical altitude, solar midday (versus civil midday), the apparent motion of the Sun and the Earth's rotation.

Sunspots and the Solar Cycle

The internal activity of the Sun, caused by the solar magnetic field, manifests itself in the solar atmosphere in several ways, such as sunspots, protuberances, flares or eruptions. One of the most interesting characteristics of the solar activity is its 11-year cycle: the number of sunspots observed on the solar surface has a variation with that period (Fig. 2).

The Astronomical Observatory of the University of Coimbra has been developing, since 2007, a project named *Sun4All* (Fig. 3),[3] which has made available to the public more than 30,000 images of the solar surface obtained since 1926 (one year after the inauguration in Coimbra of a spectroheliograph similar to the one in Meudon, Paris), allowing simple analysis of the solar cycle. Several schools in Portugal and in Europe have been using that data in the program *Hands-On Universe—Europe.*[4]

---

[2]http://www.astronomia2009.es/Proyectos_de_ambito_nacional/La_medida_del_Radio_de_la_ Tierra.html.

[3]http://www.mat.uc.pt/~sun4all.

[4]http://www.euhou.net/index.php?option=com_content&task=view&id=205&Itemid=13.

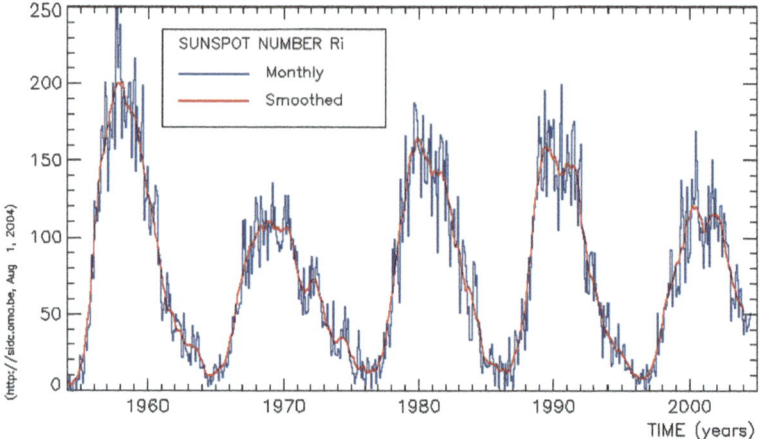

Fig. 2 Number of sunspots as a function of time, showing the solar cycle, which will have a new peak in 2012

Fig. 3 Cover of the guidebook for Sun4All

This activity motivates the discussion of mathematical topics such as histograms, periodical functions and period estimation. In astronomy side, we may discuss sunspots, the solar cycle and solar activity. It is also an excellent opportunity to discuss the Sun–Earth relation and space weather.

Hubble's Law and the Age of the Universe

Edwin Hubble at the Mount Palomar Observatory, USA, in the 1920s, found that the galaxies are, in general, moving away from us and that the recession velocity is larger for the most distant galaxies (Fig. 4). Hubble's law, shown by the regression line in the figure, was pivotal to the establishment of the Big Bang theory.

The slope of the graph, called Hubble's constant, has the unit of an inverse of time. It is straightforward to extract from it the age of the Universe, which is 14

Fig. 4 Hubble's law describing the recession of the galaxies (Original image from "Imagine the Universe", NASA)

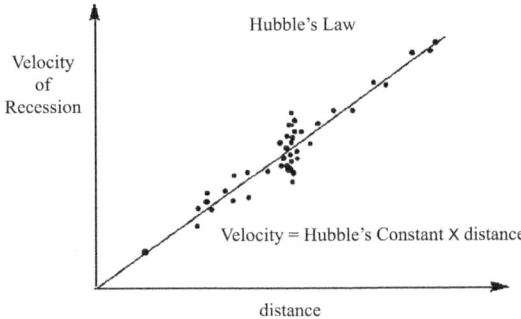

thousand million years. This example has a great advantage: by using basic mathematics, we can derive a very strong astrophysical result, the age of our Universe.[5] We have proven in several outreach sessions that astronomical topics such as galaxies, redshifts, the Big Bang and the error of the age of the Universe may be fruitfully discussed.

Physics and Mathematics

Galileo and the Free Fall Law

Galileo is generally recognized as the "father of Physics". He was indeed the first to infer a physical law, with a mathematical expression, from careful experimentation and observation. Despite the legend, he did not actually do any free fall experiments with objects of different mass from the top of Pisa's tower. Instead he used an inclined plane to create conditions where gravity has less effect. Relying on crude water clocks (some say that he used his own pulse!) to measure time, he measured distances along the inclined plane to arrive at the conclusion that the distance covered by a rolling ball is proportional to the square of the elapsed time, regardless of the plane's inclination (the distances go like $1, 2, 4, 9, \ldots$, with units of time, so that the distance differences in equal time intervals go as the odd numbers: $1, 3, 5, \ldots$). Simple mathematics can describe a natural phenomenon! He wrote in his book *The Assayer* (1623):

> Philosophy is written in that great book which ever lies before our eyes—I mean the universe—but we cannot understand it if we do not first learn the language and grasp the symbols, in which it is written. This book is written in the mathematical language, and the symbols are triangles, circles and other geometrical figures, without whose help it is impossible to comprehend a single word of it; without which one wanders in vain through a dark labyrinth.

---

[5]For an application see http://ngala.as.arizona.edu/dennis/instruct/ay14/hubble.html.

Fig. 5 Instrument to
demonstrate free fall (along a
parabolic trajectory) from the
Museum of Science of the
University of Coimbra
Physics Collection (Cabinet
for Experimental Physics,
created in 1773)

This is today totally true, so that the example of this historical experiment is
perfectly adequate to express the intimate connection between physics and mathematics as it was in the late eighteenth century when a *Cabinet for Experimental Physics* was established in Coimbra (Fig. 5). We have on many occasions told
Galileo's story in outreach lectures and performed the real experiment or shown
videos (nowadays it may be easily videotaped with a very good time measurement)
or computer simulations to convey the message that physics and mathematics are
intertwined. Roentgen, professor of Physics and Mathematics at the University of
Würzburg, Germany, and the first Nobel laureate in Physics due to his discovery of
X-rays, remarked once:

> The physicist when preparing for his work needs three things: mathematics, mathematics
> and mathematics.

Interestingly, these initially mysterious rays were named after the most common
mathematical variable ...

Newton and the Law of Universal Gravitation

Newton is known to have said: "If I have seen further it is only by standing on the
shoulders of giants." Among these giants were Galileo, who found laws of motion
that hold at the Earth's surface, and Kepler, his contemporary, who discovered mathematical regularities in the motion of the planets. But Newton managed to bring
together the physics on Earth and the physics of the Heavens in a unified framework. His point may be simply conveyed by the story of the famous apple (see
Fig. 6), whose truth cannot be guaranteed: the apple and the Moon are subject to
the same force, the gravitational force, which is therefore universal. The methods of
infinitesimal calculus, invented by Newton himself, allow one, given the mathematical expression for this force (inversely proportional to the square of the distance

Fig. 6 Cartoon by José
Bandeira on the cover of the
popular science book by
Carlos Fiolhais, Física
Divertida [1]

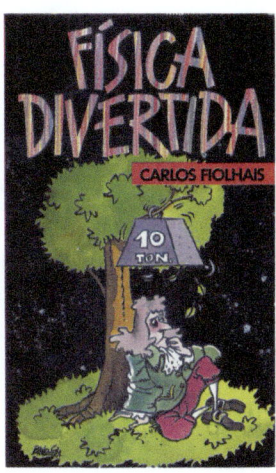

between two bodies) and the values of the initial conditions (position and velocity)
to predict the motion of the Moon at all times. In popular lectures we used to say,
qualitatively, that an apple may be transformed into a Moon if it is thrown with an
appropriate velocity or, instead, that the Moon may be seen as a very big apple if
we imagine an initial zero velocity and a straight fall to the Earth. To illustrate the
point we can use simple numerical simulations with a graphical output. In this way
geometric concepts such as the transition from a straight line to a circle and an el-
lipse, with a parabola in the middle, can be illustrated. But this reasoning, which
brings together the Earth and the Heavens, is more quantitative if we calculate, as
Newton did, the gravitational force that the Earth exerts on the Moon using only the
Moon's period (27 days) and the Moon-Earth distance (60 Earth radii), and compare
it to the gravitational force that the Earth exerts on the apple: the inverse square law
of universal gravitation can be easily checked using simple algebra, the conclusion
being that the gravity felt by the Moon is $1/3600 = 1/60^2$ of the gravity felt by the
apple.

Einstein and the Most Famous Equation of Physics

Einstein (Fig. 7) managed in a genial way to climb to the top of Galileo, Kepler
and Newton. His special theory of relativity modified our notions of space and time
(they became linked in the unified concept of space-time, for which a nice mathematical framework was proposed by Minkowski, one of Einstein's mathematics
teachers) and modified our understanding of mass and energy. These two concepts
were unrelated: mass having to do with inertia (resistance to motion) and energy
being associated with motion (kinetic energy is proportional to the square of the
velocity). In a radical move, Einstein arrived at the conclusion that mass and energy
are synonymous: energy is proportional to mass, the constant of proportionality uses

M.E.A. campanha de dinamização cultural

muito prazer em conhecer votência

Fig. 7 Cartoon by João Abel Manta showing Einstein being introduced to the Portuguese peasants after the 25th April 1974 revolution. Descartes may also be seen in the row, among many other savants

the speed of light: $E = mc^2$. To arrive at this he had to change the mathematical expression for mass: instead of being a constant it became a variable depending on velocity, so that, for small velocities, the old expression of the kinetic energy was recovered. Hawking, the author of the best-selling popular science book *A Brief History of Time*, included Einstein's famous equation as the only one in his book, but not without noting:

> Someone told me that each equation I included in the book would halve the sales. I therefore resolved not to have any equations at all. In the end, however, I did put in one equation, Einstein's famous equation, $E = mc^2$. I hope that this will not scare off half of my potential readers.

Not satisfied with associating space and time and energy and mass, Einstein also managed to unify mass-energy, which expressed the gravitational force, with the geometry of space-time. Differential geometry and non-Euclidean spaces, which were already available in the mathematical toolbox, are necessary to read the Book of Nature. As with Newton mechanics, in general relativity differential equations proved to be essential tools in physics. Einstein once commented with some irony:

> Thus the partial differential equation entered theoretical physics as a handmaid, but has gradually become mistress.

Art as a Source of Motivation for Mathematics

During the last decade, many recreational activities about mathematics, with an interdisciplinary nature, have been developed for young children in many Portuguese primary schools led by scientists working at the University of Coimbra. These mathematical activities intersect with areas as diverse as architecture, modern art or literature. Many of the proposed activities start with a monument, an art object, a concept or a myth, which leads to a mathematical problem, motivating explorations in

Fig. 8 Experiments developed under the project have been published regularly, since 1999 [2]

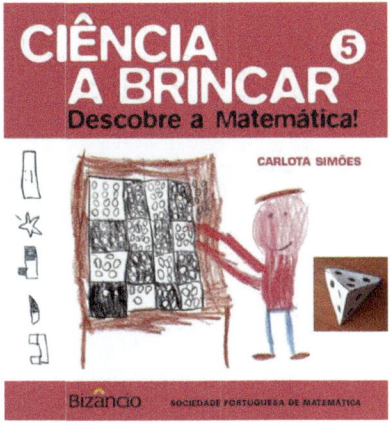

mathematics. During the last ten years, these experiments have been regularly published in the collection *Ciência a Brincar* (Fig. 8). Here we present some of these activities, revealing in particular how our historical heritage can be used to motivate young pupils towards basic mathematics.

Mathematics and Tiles

The walls of St Michael's Chapel at the University of Coimbra are all covered by tiles (in Portuguese, *azulejos*) dating from the seventeenth century, some of which are shown in Fig. 9.

The panel is constructed by repeating a tile pattern formed by four tiles (Fig. 10), which are placed so that it looks like a blue and a golden ribbon moves alternately over and under the white frame. Children were invited to identify repetitions, patterns and the smallest set of tiles that would reproduce the whole panel.

We can easily teach children to create several beautiful patterns from a set of tiles, but for a real panel one must be careful because, once built, it is not easy to change the position of any of the tiles. That is the case at St Michael's Chapel. Children are asked to discover, inside the Chapel, the places where the panel does not follow the pattern (Fig. 11).

A visit to the Chapel, identifying the pattern, and searching for the wrong tiles was a strong motivation for the study of patterns, symmetries and tiling of the plane [3].

Mathematics and Modern Art: The Cologne Cathedral Stained Glass Window

In 2007, Cologne Cathedral, in Germany, received a totally new and modern stained glass window. With 113 m^2 of glass, the window was created by the German artist

Fig. 9 Detail of the wall tiles of St Michael's Chapel at the University of Coimbra

Fig. 10 The unit pattern, with four tiles

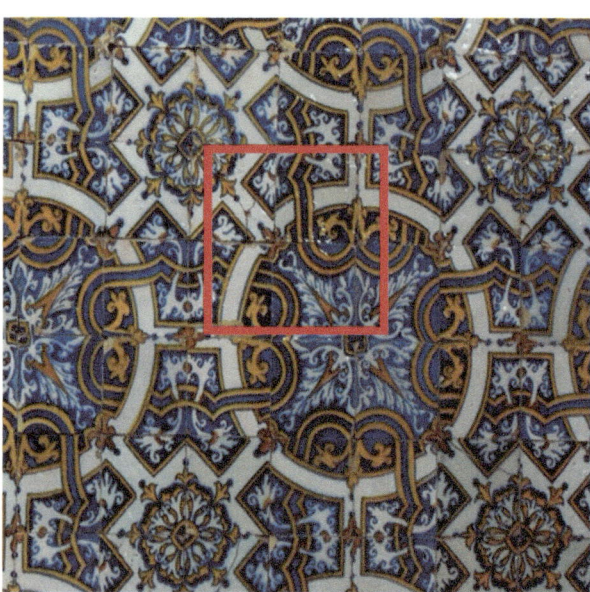

Gerhard Richter (Fig. 12). It is composed of 11,500 identically sized pieces of 72 different colors, the glass resembling pixels, randomly arranged by computer.

Despite the fact that Richter's stained glass window is randomly produced, we used it as a motivation for organized mathematics in the classroom. A picture of the stained glass window was distributed among the children. They were told to

Fig. 11 Detail of St Michael's Chapel at the University of Coimbra with several misplaced tiles

Fig. 12 Cologne Cathedral stained glass window, by Gerhard Richter

7 colours and 11 columns   12 colours and 11 columns   15 colours and 11 columns

Fig. 13 Some results obtained by children

Fig. 14 Several "stained glasses windows" obtained by the children

make their own versions of it, and were given a number of different colors to use, in sequence and without repetitions (Fig. 13).

The "stained glass windows" obtained by using different numbers of colors turned out to have several forms such as vertical lines, diagonal lines, tartan or even the movement of the horse in chess (Fig. 14). The observation of different patterns was a motivation for concepts such as multipliers, divisors and the remainder in division.

Mathematics and the Classics—The Three Graeae

> So under the guidance of Hermes and Athena he [Perseus] made his way to the daughters of Phorcyus [...]. The three had but one eye and one tooth, and these they passed to each other in turn. Perseus got possession of the eye and the tooth, and when they asked them back, he said he would give them up if they would show him the way to the nymphs. Apollodorus (ii. 4. § 2)

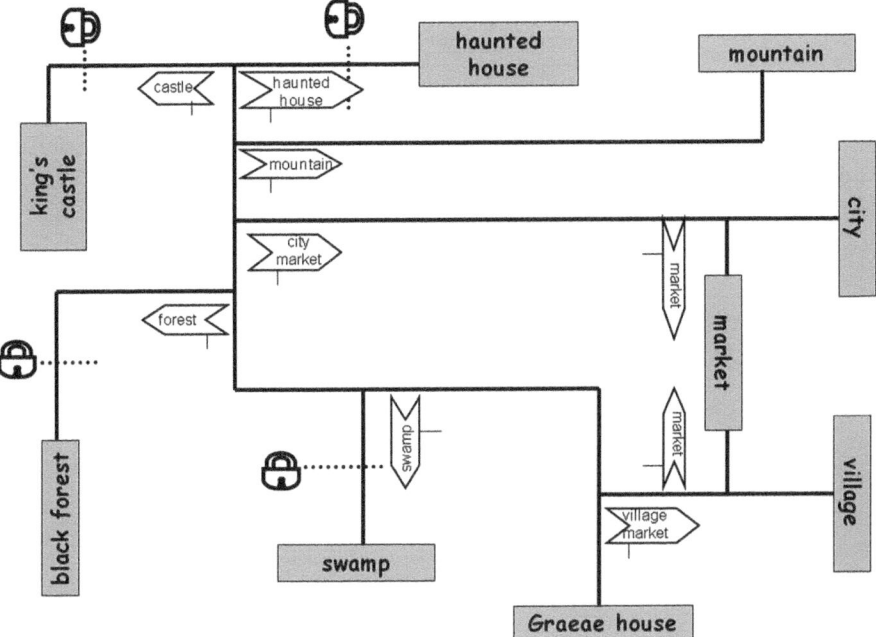

Fig. 15 The voyage of the three Graeae

The myth of the three old Graeae [4] was inspiring for several activities involving maps. Since the three sisters have only one eye to share, we can imagine several situations leading to a mathematical problem. For that in Fig. 15, we supposed that the three sisters have complementary skills: only the first one can drive; only the second can read and only the third can make calculations. The three sisters have a car, which is driven by the first while using their eye, but when there are directions to read the second must have the eye, and when there are calculations to make the third must have the eye. On this map, the arrows indicate directions (to read) and the locks indicate codes (which can be broken by a calculation). How many times must the three Graeae change their eye from one to another in order to drive from their home to, for example, the haunted house?

It was a complete surprise for us to realize that seven-year-old children already knew the myth of the Graeae. They were very enthusiastic, describing the myth of Perseus and the gory details of the eye of the three Graeae. We realized later that the three Graeae are present in the children's universe and are very popular. Just to name some examples, in the video game *Titan Quest: Immortal Throne*, one has to kill the Graeae in order to give their single eye to Medea; in the game *King's Quest IV: The Perils of Rosella*, the player has to steal the eye from the three Graeae; in the Disney film *Hercules*, the Graeae are shown sharing their eye. Does this mean that future adults will be experts in classic literature? Probably not, but we should pay more attention to in video games children's films . . .

Conclusion

Based on our examples taken from astronomy, physics and art (historical heritage) we have shown how interdisciplinary approaches may help to raise the public awareness of Mathematics. This discipline is still considered by many people as difficult and dry. This description of the links with astronomy, physics and art clearly are contrary to this common view and may help to dissipate it.

References

1. Fiolhais, C.: In: Física Divertida. Gradiva, Lisbon (1991)
2. Simões, C.: Descobre a Matemática. Bizâncio, Lisbon (2006)
3. Providência, C., Fiolhais, C.: Descobre o Património! Bizâncio, Lisbon (2008)
4. Grimal, P.: The Dictionary of Classical Mythology. Wiley-Blackwell, Oxford (1996)

From PA(X) to RPAM(X)

Steen Markvorsen

Abstract How can we use the well-established Public Awareness of some phenomenon X, i.e. PA(X), to Raise the Public Awareness of the Mathematics of—or
within—this X, i.e. RPAM(X)? There are several examples illustrating particular
assets for mathematics in this way within such phenomena X. Here we will discus
only one phenomenon, which, however, contains a particularly dramatic momentum
for arousing awareness among all of us, namely $X = $ wildfires. The mathematics
of this and of similar phenomena range—among several other topics—from elementary K–12 studies of ellipses to deep research questions in Finsler geometry.
Moreover, in this context RPAM(X) may even save lives!

Keywords Raising public awareness of mathematics by example · Wild fires ·
Finsler distance geometry · Geodesic fire tracks · Bear hugs

Mathematics Subject Classification 00 · 53 · 58 · 35 · 92

What Is Public Awareness?

Public awareness in the large (or in the small) is usually (but not always!) concerned
with and centered around *concrete* phenomena which for a variety of reasons attract
the attention of many people (or just a few) for a shorter or a longer period of
time—mostly shorter. This is definitely not a definition: it is only a rough attempt to
apply the two descriptors *concrete* and *time limited* to what might be called public
awareness. What we might call mathematics, on the other hand, is mostly *abstract*
and mostly *timeless*. This apparent dichotomy is, however, just apparent. In fact
there is ample room for mergers, as witnessed by the following quotations:

> In the teaching of mathematics, and when explaining the essence of mathematics to the
> public, it is important to get the abstract structures in mathematics linked to concrete man-
> ifestations of mathematical relations in the outside world. Maybe the impression can then

Dedicated to Professor Vagn Lundsgaard Hansen on the occasion of his 70th birthday.

S. Markvorsen (✉)
Department of Mathematics, Technical University of Denmark, Copenhagen, Denmark
e-mail: S.Markvorsen@mat.dtu.dk

E. Behrends et al. (eds.), *Raising Public Awareness of Mathematics*,
DOI 10.1007/978-3-642-25710-0_19, © Springer-Verlag Berlin Heidelberg 2012

Fig. 1 Forest firefighting

be avoided that abstraction in mathematics is falsely identified with pure mathematics, and concretization in mathematics just as falsely with applied mathematics. *Vagn Lundsgaard Hansen* [1]

I find it difficult to convince students—who are often attracted into mathematics for the same abstract beauty that brought me here—of the value of the messy, concrete, and specific point of view of possibility and example. In my opinion, more mathematicians stifle for lack of breadth than are mortally stabbed by the opposing sword of rigor. *Karen Uhlenbeck* [2]

Guided by the spirit of these parallel quotations we will thence concentrate and focus upon only one single very specific, very concrete, and very messy example from real life, a phenomenon which, nevertheless, is known to have moved everybody into alert mode since—and probably even long before—the Mesolithic era (see Fig. 1).

The Phenomenon of Wildfires

A wildfire, also known as a forest fire, vegetation fire, grass fire, brush fire, or bush fire (in Australia), is an uncontrolled fire often occurring in wild land areas, but which can also consume houses or agricultural resources. Common causes include lightning, human carelessness and arson. One main component of Carboniferous north hemisphere coal is charcoal left over by forest fires. The earliest known evidence of a wildfire dates back to Late Devonian period (about 365 million years ago). [3]

Wildfires are frequent and extremely threatening phenomena that we must try to understand from first principles. How do we most effectively prevent them from happening? How do they evolve once they have started, how do we most effectively escape from them, or conversely—the firefighters' quest—how do we most effectively fight them?

A working knowledge of the effects of wind and other weather elements on fire behavior, supported by accurate fire intelligence, is vital for good suppression planning. Without good fire behavior information firefighters are unable to:

- determine the number of firefighters and level of equipment necessary;
- identify the location of suitable areas for backburning; and
- ensure that the general public is informed about the precise fire situation. [4]

Fig. 2 Mathematical
Horizons, 2009. Freely
available via [5]

The existence of a well-informed public as well as professional up-to-date aware-ness of the general behavior of wildfires is a necessary prerequisite for tackling these tasks.

Mathematical Horizons

Clearly we have PA(X) en masse for this wildfire X. How do we then generate momentum into RPAM(X), which is our main concern here?

How do we engage the public, how do we engage students, teachers, and re-searchers through K–12, high school, and university, to look for, to appreciate, and to apply known mathematics and search for new mathematics and thereby contribute to the solution of such tasks?

The following is but a brief account of the paper [5, pp. 50–61], which aims to show by example that this can be done. This example has recently appeared in a selection of similar examples in a book that is freely distributed on the web; it has also been distributed as an ordinary hardback book to the schools in Denmark, see Fig. 2 and [5]. In the book you can find exciting mathematical unfoldings (with ex-ercises) of similar phenomena under headlines such as: "Mathematics through the millennia"; "Mathematics and evolution"; "Fire!"; "How a vending machine actu-ally works"; "Wavelets"; "Secret codes made public"; "Math in medicine"; "Tour de France mathematics"; "Women and mathematics"; "Error correcting codes"; "Beer and flat screens"; "Mathematics in the computer and vice versa"; "The science of the better"; "Artificial intelligence"; "Mathematical modeling of climate and energy"; "The Mars mission"; "The mathematics of shape."

Some Details from the Wildfire Case

The simplest possible two-dimensional model of a fire front propagating in a per-fectly homogeneous domain and in a wind that is directed along the y-axis is usually

Fig. 3 Elliptic fire zone

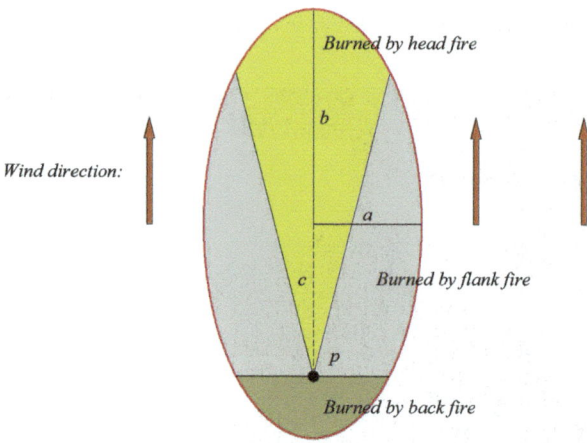

modeled as a parametrized time foliation through ellipses:

$$x(t, \phi) = at \cos(\phi),$$
$$y(t, \phi) = bt \sin(\phi) + ct, \tag{1}$$

where a, $b \geq a$, and $c \leq b$ are constants depending on the fuel material and the speed of the wind. When $b = a$ and $c = 0$ this model gives elementary circular propagation with constant radial speed a from the initial center point.

In the general case of elliptic propagation we may define three values for the fire-front propagation speed, see Figs. 3 and 4:

$$v = b + c \quad \text{(Downwind front speed)},$$
$$u = a \qquad \text{(Flank front speed)}, \tag{2}$$
$$w = b - c \quad \text{(Upwind front speed)}.$$

Note that the fire will also propagate upwind when $b - c > 0$.

This elliptic model is in actual use in Canada:

> The Canadian Forest Fire Behaviour Prediction System (CFFBPS) assumes elliptical growth and has documented values of u, v, and w for a very large set of constant parameters affecting a fire. It has also been observed that, within certain limits, the ratio a/b is a function of wind speed only; this is also an assumption of the CFFBPS. [6]

There are so-called pocket cards for firefighters, which recommend that the fire front should be attacked on the flanks. However, when the fuel density is not homogeneous, or when the topography is not perfectly flat, or when the basic model cannot be assumed elliptic but is some other oval-shaped generator as in Figs. 6 or 7, then the flank attack strategy may not be optimal.

The more advanced mathematics needed to see and understand this is concerned with geodesic sprays in Riemannian geometries (and Finslerian geometries when the wind is blowing). This will be discussed and exemplified in some detail in the next section.

Fig. 4 Elliptic foliation of a homogeneous fire zone with a constant wind from the south-west

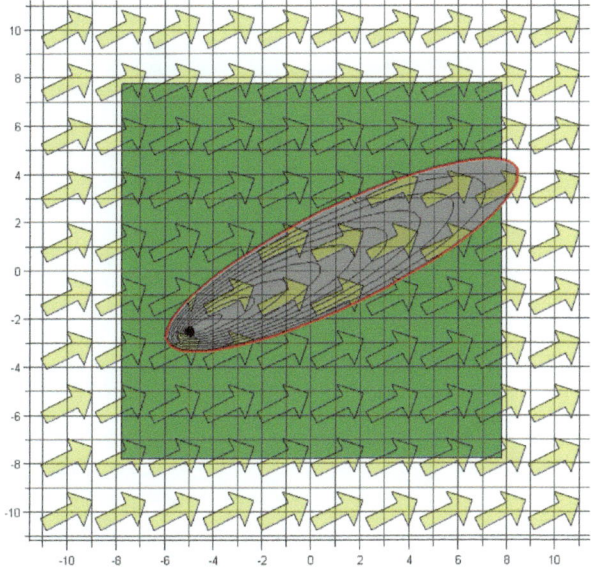

In Fig. 5 we indicate how a fire front (without wind) may attack the fuel domain—and also the firefighters—from more than one side when propagating— like a pincer movement. This occurs precisely when the geodesic spray from the initial point of ignition creates so-called cut points in the domain. The set of these points are indicated by yellow dots in Fig. 5. The blue color in Fig. 5 indicates the Moorish area through which the fire does not burn easily and where it therefore progresses only slowly.

The Cramer's Creek accident is but one such dramatic case featuring the formation of dangerous cut points. In [7] it is described in detail how fire unexpectedly approached from both sides of a ridge between two valleys and eventually killed two firefighters who were trapped by such a pincer movement.

Finsler Geometric Analysis and Modeling

In this and the following sections we discuss some of the tools and concepts from Finsler geometry, which have only been alluded to above.

The possibilities of studying and applying asymmetric length functionals had been suggested by Riemann in his famous and foundational *Probevorlesung* in 1854 [8], but they were first developed in detail by Finsler in his *Inauguraldissertation* in 1918 [9]. Within the last 10 years the methods and results of global Finsler geometric analysis have experienced a renaissance—not least inspired by the seminal works of e.g. Chern and his collaborators and students. See for example the survey paper by Chern [10] and the works [11–13].

Like every Tour de France racing cyclist we all know that it is much harder to cycle uphill or against the wind than it is to freewheel downhill or with the wind

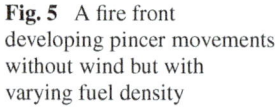

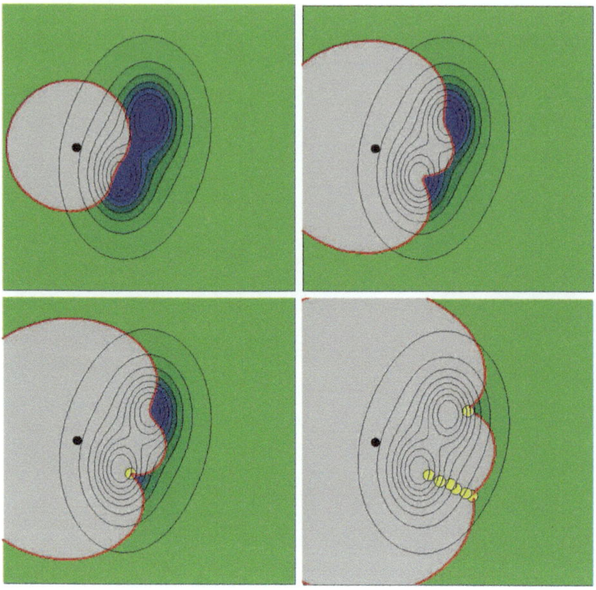

Fig. 5 A fire front developing pincer movements without wind but with varying fuel density

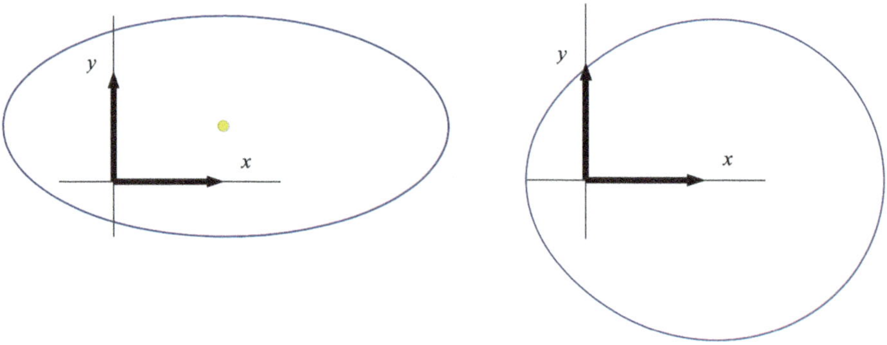

Fig. 6 Two indicatrices consisting of (the endpoints of) F-unit vectors. Every vector from the origin to the oval is in each case of F-length 1

pushing comfortably on your back—although the (classical Euclidean) *length* of the road of course is the same, whether we measure it in one direction or the other.

In a similar way (but note the important up-down reversal) a fire front will move much faster uphill (!) or with the wind than it moves downhill or against the wind.

A Finsler geometric model of a forest fire has this asymmetry built directly into the so-called indicatrix field $\mathcal{I}_q$ of unit vectors at each point q. Two such possible indicatrices are shown in Fig. 6, and indeed the oval shown in Fig. 7 is yet another possible indicatrix. Note that in Fig. 7 the origin is very close to the indicatrix at the left. This indicates that in this model the upwind speed of the fire front is very small. See also [3] and Figs. 6 and 8.

Fig. 7 An alternative oval
propagation generator
(indicatrix) model, see [3]

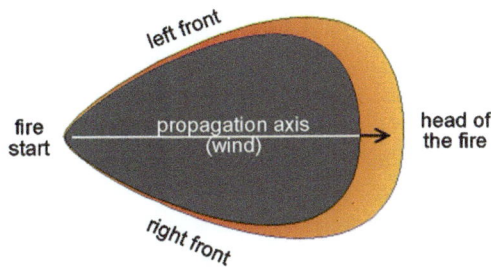

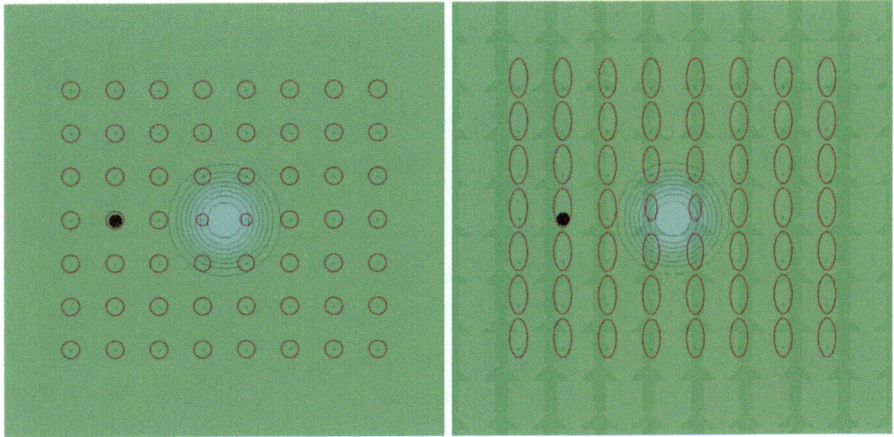

Fig. 8 Two simple fields of indicatrices, one without and one with wind

In all cases the value of the Finsler length F_q is defined to be 1 for all the vectors connecting the origin to the points on the indicatrix oval:

$$\mathcal{I}_q = \{u \mid F_q(u) = 1\}. \tag{3}$$

Note that in some directions the F_q-unit length is much larger (by ordinary Euclidean standards) than in the opposite direction. So if we measure the cost of transport—or the cost of propagation—by F_q-standards it may be much cheaper to go a long (Euclidean) distance in one direction than the same (Euclidean distance) in the opposite direction.

As illustrated in Figs. 6, 7, and 8 the origin (of the local vectors) need not be at the center of the respective indicatrix ovals (in fact the ovals need not even have well-defined centers). The shape and size of each oval as well as the position of the center inside it can be chosen to depend on the wind, on the topographical slope, and on the quality of the fuel at the point in question.

The Finsler Length Functional

Euclidean geometry is obtained in the special case where all indicatrices are identical circles with their centers at the origin. Riemannian geometry is obtained when all indicatrices are circles of possibly varying sizes but again with their centers at the origin. Riemannian geometry thus corresponds to the no-wind and no-slope situations because the winds and topographical slopes essentially shift the positions of the indicatrix centers and break the otherwise centered elliptic symmetry of the indicatrices. The possibility of varying the *radii* of the circles, however, corresponds to varying fuel conditions in the area.

In Fig. 8 small circles model the wet moorland shown in blue and the larger circles model the more homogeneous forest-like area, shown in green. To the left in Fig. 8: The field is Riemannian in the sense that all indicatrices are circles. Note, however, that the circles are smaller in the blue area, so that it takes effort (long time for the fire) to go straight through the moorland. To the right in Fig. 8 is shown a genuine Finslerian field of indicatrices, which consists of wind-shifted ellipses. Note that the shifted ellipses are again smaller in the blue moorland area.

In general, when we choose an indicatrix $\mathcal{I}_q$ of F_q-unit vectors at every point q, then the F_q-length of any other vector at q is simply defined by homogeneous scaling:

Definition 1 Suppose we know the F_q-unit vectors at every point q, i.e. we assume that we have chosen the indicatrix field $\mathcal{I}_q$ already—as exemplified in Fig. 8. Then, since every (other) vector y is a factor λ times some unit vector u, $y = \lambda \cdot u$, we simply define the F_q-length of y to be that factor:

$$F_q(y) = F_q(\lambda \cdot u) = \lambda \cdot F_q(u) = \lambda, \quad u \in \mathcal{I}_q. \tag{4}$$

The F-length of a *curve* is then (as usual) the integral of the F-length of its tangent vectors:

Definition 2 Suppose $c(t) = (c^1(t), c^2(t))$, $t \in [0, T]$, denotes a regular smooth curve in the plane. Then the *F-length* of c is given by:

$$\mathcal{L}(c) = \int_0^T F_{c(t)}(\dot{c}(t)) \, dt, \tag{5}$$

where $\dot{c}(t)$ denotes the tangent vector of the curve c at the point $c(t)$.

Note that the length of $c(t)$, $t \in [0, T]$, is not necessarily the same as the length of the reversed curve $\hat{c}(t) = c(T - t)$, $t \in [0, T]$. This is because $F_{c(t)}(\dot{c}(t))$ is not necessarily the same as $F_{c(t)}(-\dot{c}(t))$. And this is exactly what we want! The length functional $\mathcal{L}$ measures and takes into account that it is easy to go one way along the curve ($\mathcal{L}(c)$ is small) but possibly difficult to go the other way ($\mathcal{L}(\hat{c})$ is large).

Fig. 9 Without any wind the fire front will penetrate through and around the symmetric blue Moorish area

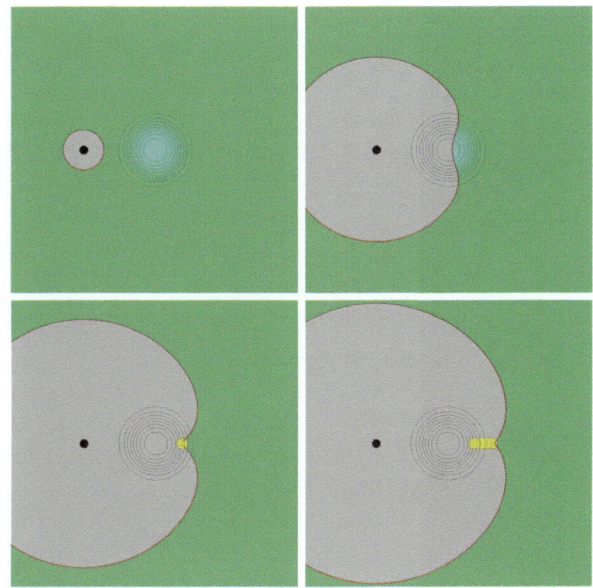

The Geodesic Paradigm for the Fire Front Propagation

One key ingredient in the forest-fire model presented here is that the fire front is formed by the union of all F-shortest curves—of the same F-length—issuing from a given ignition point. Such F-shortest curves are called F-geodesics. Due to the asymmetric measure of the F-length they extend further in the direction of the far ends of the local indicatrices than in the opposite directions. Therefore the endpoints of these geodesics (the geodesic "circle," the fire front) initially (for small radii) look like the indicatrix at the ignition point, but as the geodesic circles and fire front extend to further radii they may take on very different shapes as illustrated by the simple examples in Figs. 9 and 10. In both figures the fire front will progress through and around the symmetric blue moorland area and create pincer cut points along a curve east of the moorland. When the wind is blowing from the south then the pincer curve is clearly shifted towards the north of the moorland.

The F-geodesics satisfy a system of nonlinear ordinary differential equations:

Proposition 3 *A given curve $c(t) = (c^1(t), c^2(t))$ is an F-geodesic, i.e. a locally F-shortest curve between any pair of its points, if it satisfies the differential equations:*

$$\ddot{c}^i(t) + \gamma^i_{jk}\big(c(t), \dot{c}(t)\big) \cdot \dot{c}^j(t) \cdot \dot{c}^k(t) = 0, \quad i = 1, 2, \tag{6}$$

where the so-called connection (or Christoffel) functions $\gamma^i_{jk}(c(t), \dot{c}(t))$ are given by a (somewhat complicated) mixture of suitable derivatives of the Finsler function F at the point $c(t)$, see e.g. [14].

Fig. 10 With a wind from
the south (shown by the
arrows) the fire front will be
elliptical and the first part of
the cut locus will be shifted

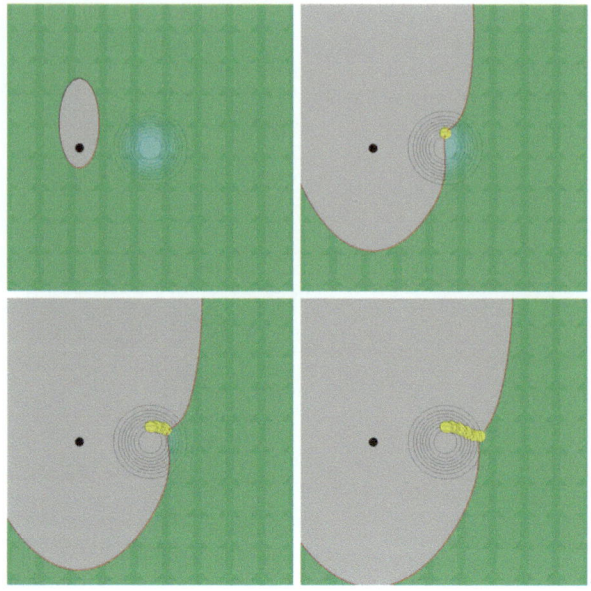

Constant Indicatrix Fields

In particular, if F has a constant indicatrix field, i.e. if the background fuel and
topography is completely homogeneous, then the Christoffel functions all vanish,
and the differential equations for the geodesics reduce to

$$\ddot{c}^i(t) = 0, \quad i = 1, 2, \tag{7}$$

so that every geodesic issuing from the ignition point $p = c(0)$ is a straight line; the
solutions are linear functions in t:

$$
\begin{aligned}
c^1(t) &= \alpha_1 t + \beta_1, \\
c^2(t) &= \alpha_2 t + \beta_2.
\end{aligned}
\tag{8}
$$

This is in precise accordance with the model equation (1) if we let

$$
\begin{aligned}
\alpha_1 &= a\cos(\phi), \\
\alpha_2 &= b\sin(\phi) + c, \\
\beta_1 &= 0, \\
\beta_2 &= 0.
\end{aligned}
\tag{9}
$$

This linearity of geodesics is displayed in Figs. 11, 12, and 13. Initially the
geodesics, the fire particle tracks, from the point p of ignition are directed straight
away from p because the region in front of the "moor" is essentially homogeneous

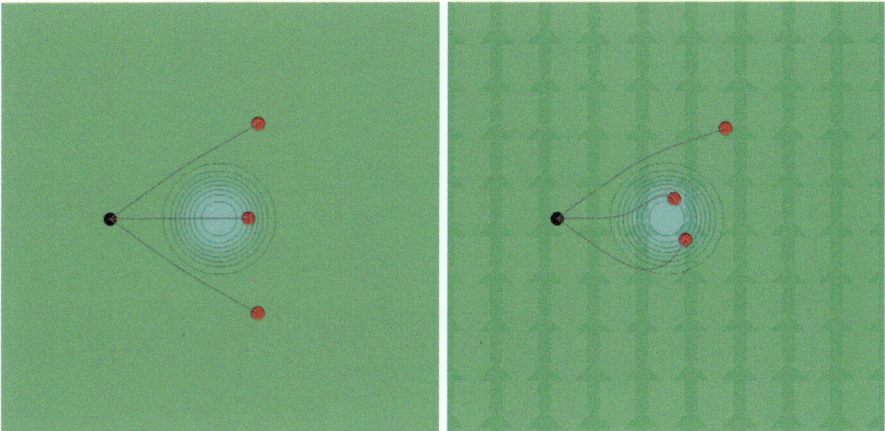

Fig. 11 Three geodesics of the *same* F-length without wind and with wind (shown by the *arrows*), respectively

Fig. 12 The individual geodesics issuing from the ignition point bend around the moorland to form the fire front movement in Fig. 9

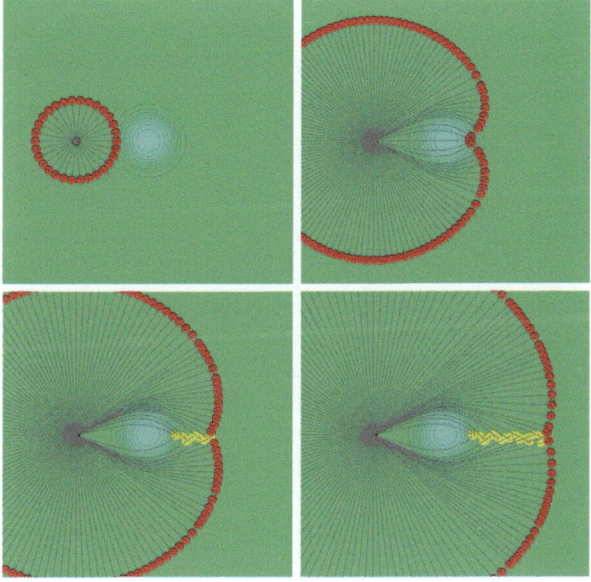

and flat. Note (in Fig. 11) that when you approach the moorland area you get further to the right side of the moor when you go around it. When the wind is blowing it clearly also matters which way you choose to go around it!

In Figs. 12 and 13 we show *all* the geodesics issuing from a common ignition point and up to a common propagated F-length. The different cases with or without wind are also indicated. The ensuing endpoints of the geodesic fire tracks (which

Fig. 13 The individual fire particles that create the fire front movement in Fig. 10. The influence of the wind is evident

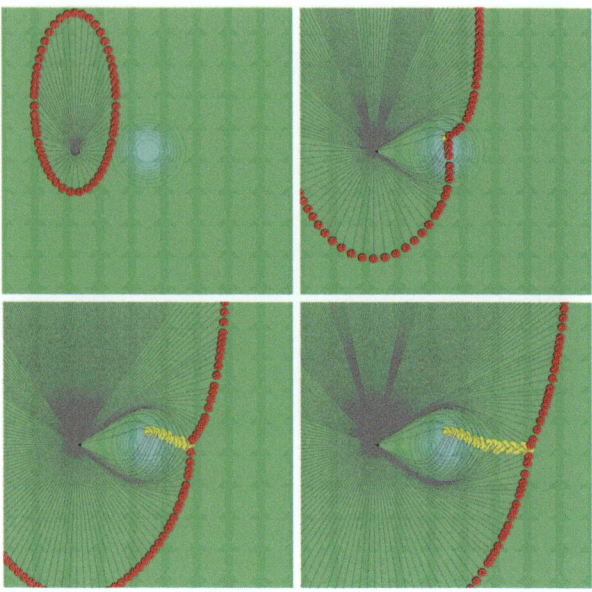

together form the geodesic "circle" fire front) are marked together with the burnt-out area in the accompanying Figs. 9 and 10.

Bear Hugs

A fire front initially forms an oval, which is very similar to the F_p indicatrix $\mathcal{I}_p$ at the ignition point, but as it develops further away from the point of ignition the front will bend around lakes or any other Moorish domains that cannot be easily penetrated by the fire, and the front may thus create self-intersections, pincer movements (also known as "bear hugs") behind these obstacles—as already mentioned and observed in Fig. 5. As before we mark such "pincer points" by yellow (warning) dots on the figures. They are the positions where firefighters may be in danger of attack from multiple (at least two) sides by the fire front. Technically these points of self-intersection form a continuous set of points Cut(p), which is called the *cut locus* of p.

Conclusion

Cut loci are of current research interest for several other reasons and for many other applications than the particular one addressed here, but in the present setting, where we have been concerned with understanding wildfires, they are obviously of particular direct impact and importance. With a suitable geometric analysis and an aroused and raised public awareness, i.e. RPAM(X) to hand, their formation may even be predicted—and catastrophes thus prevented.

References

1. Hansen, V.L.: Popularizing mathematics: from eight to infinity. In: Proceedings of the International Congress of Mathematicians, Beijing, ICM 2002, vol. III, pp. 885–895 (2002). Free access: http://www.mathunion.org/ICM/ICM2002.3/Main/icm2002.3.0885.0896.ocr.pdf
2. Atiyah, M., et al.: Responses to: A. Jaffe and F. Quinn, "Theoretical mathematics: toward a cultural synthesis of mathematics and theoretical physics". Bull. Am. Math. Soc. (N.S.) **29**(1), 1–13 (1993). Bull. Am. Math. Soc. (N.S.) **30**(2), 178–207 (1994). Free access: http://www.ams.org/publications/journals/journalsframework/bull
3. Wikipedia, Wildfire. http://en.wikipedia.org/wiki/Wildfire
4. Fogarty, L.G., Alexander, M.E.: A field guide for predicting grassland fire potential: derivation and use. Forest and Rural Fire Research, Fire Technology Transfer Note, No. 20 July 1999. Free access via: http://nofc.cfs.nrcan.gc.ca/publications?id=18643
5. Hansen, et al.: In: Hansen, C.B., Hansen, P.C., Hansen, V.L., Andersen, M.M. (eds.) Matematiske Horisonter (in Danish). DTU 2009. ISBN: 978-87-643-0453. A free download of the complete collection is available via: http://www.imm.dtu.dk/Om_IMM/Informationsmateriale/MatematiskeHorisonter.aspx
6. Richards, G.D.: An elliptical growth model of forest fire fronts and its numerical solution. Int. J. Numer. Methods Eng. **30**, 1163–1179 (1990)
7. Close, K.R.: Fire behavior vs. human behavior: why the lessons from Cramer matter. In: Eighth International Wildland Fire Safety Summit, April 26–28, 2005 Missoula, MT. Free Access: http://www.wildfirelessons.net/documents/Close.pdf
8. Riemann, B.: Über die Hypothesen, Welche der Geometrie zu Grunde Liegen. In: Neu Herausgegeben und Erläutert Von H. Weyl. Zweite Auflage. Springer, Berlin (1921)
9. Finsler, P.: Über Kurven und Flächen in Allgemeinen Räumen. Birkhäuser, Basel (1951)
10. Chern, S.-S.: Finsler geometry is just Riemannian geometry without the quadratic restriction. Not. Am. Math. Soc. **43**, 959–963 (1996). Free Access: http://www.ams.org/notices/199609/chern.pdf
11. Bao, D., Chern, S.-S., Shen, Z.: An Introduction to Riemann-Finsler Geometry. Graduate Texts in Mathematics, vol. 200. Springer, New York (2000)
12. Bao, D., Robles, C., Shen, Z.: Zermelo navigation on Riemannian manifolds. J. Differ. Geom. **66**, 377–435 (2004)
13. Shen, Z.: Lectures on Finsler Geometry. World Scientific, Singapore (2001)
14. Wikipedia: Finsler geometry. http://en.wikipedia.org/wiki/Finsler_manifold

Promoting the Public Awareness of Mathematics in Developing Countries: A Responsibility and an Opportunity

Yasser Omar

Abstract In this article we present a case for the importance of promoting the public awareness of mathematics in developing countries. We start by discussing the factors that attract the public to mathematics and argue they are universally valid, including in developing countries, with less educated populations. Based on the experience of the NGO SiW—Scientists in the World, which has developed projects in Africa and South-East Asia, we present several concrete examples of activities raising the public awareness of mathematics in those regions, as well as ideas for possible future projects and recommendations on how to implement them.

Promoting the public awareness of mathematics is important in any inhabited region of the world. Mathematics is an intellectual heritage of humanity, which should be accessible to all. It not only offers numerical literacy, it also contributes in a unique way to develop the capacities of abstract, rigorous and logical reasoning. Furthermore, mathematics plays a substantial role in facilitating scientific literacy. And all this important knowledge represents power, a power which should be as homogeneously and as democratically spread as possible within society and around the world. Nowadays, with the advent of universal basic education, access to this knowledge has reached unprecedented levels in the history of humankind. Yet, the scenario is still unsatisfactory, both at the quantitative and qualitative levels. In many developing countries, the levels of mathematical literacy are appallingly low. Even in the richest and most developed countries, the effectiveness of mathematics education for the average student is arguable. And yet—or maybe precisely because of that— these students will essentially be the ones constituting the *public* in RPAM (Raising the Public Awareness of Mathematics) activities in such countries. Of course, there are naturally talented students who take good advantage of their education in mathematics, but they tend to be the minority. Most people, unfortunately, associate mathematics with a frustrating part of their education, something that was hard, unnatural, that they never truly understood, sometimes even an obstacle to graduation.

Y. Omar (✉)
SiW–Scientists in the World & CEMAPRE, Department of Mathematics, ISEG, Technical University of Lisbon, Lisbon, Portugal
e-mail: yasser.omar@iseg.utl.pt

E. Behrends et al. (eds.), *Raising Public Awareness of Mathematics*,
DOI 10.1007/978-3-642-25710-0_20, © Springer-Verlag Berlin Heidelberg 2012

Is this what motivates RPAM, a reconciliation with mathematics after (or before!) a possibly traumatic experience in school? And if so, what can RPAM teach us about education? And how about countries where education is practically inexistent: does this change the motivation and the strategies for RPAM there?

There are distinct and complementary motivations for RPAM, defining different targets and approaches. The most important motivation is for the wide dissemination of scientific knowledge and intellectual skills, which should not be the privilege of a minority. This is subdivided into basic mathematical knowledge (distilled and refined over time) and recent discoveries, with a novelty value. Other motivations are the promotion of the role of mathematicians in society, attracting young people to mathematical study programmes and careers, and justifying their respective funding by tax payers. Yet, this does not mean that the responsibility and initiative for RPAM lies solely on mathematicians, nor that it is their exclusive domain, despite their technical authority. Raising the public awareness of mathematics is a communication challenge. A range of expertise and talent may be used to tackle it as from science communicators, reporters, educators, mathematicians, scientists in general or even artists. The channels are equally diverse and include the media, press releases, blogs, books, software, exhibitions, museums, etc. But what can be done in countries where most of these are missing (starting with mathematicians!) or at least very distant from mathematics?

Let us first discuss what attracts the public to mathematics. There are certainly several factors, depending on the public, but they invariably include:

A. Applications, i.e. the role of mathematics in familiar things, typically in technology and technological challenges, both current and historical (for instance, navigation, weather forecasting, telecommunications), but also in societal aspects (such as the economy or the spread of pandemics) and in artistic aspects (e.g. musical notes and rhythms)
B. Identity, e.g. the national pride in a Mathematical Olympiad medal or Fields Medal, the heroization of a local mathematician, the contributions of a particular city or country to the development of mathematics, etc.
C. Historic discoveries and the anecdotal stories around them, e.g. Grigori Perelman's proof of the Poincaré conjecture and his refusal to accept the Fields Medal and the Clay Millennium Prize
D. Creativity, when exploring mathematical concepts and tools one does not necessarily need to understand to create aesthetically pleasing visual art or music
E. Intellectual curiosity and intellectual challenges, i.e. understanding new things (e.g. the mysteries and discoveries of mathematics) or solving a numerical or logical puzzle, sometimes hidden in ludic activities
F. Simple ignorant fascination about things people assume they cannot understand

All these, presented in the arguable order of decreasing interest to the public, have one thing in common: typically they are absent from mathematics education (the latter justifiably so). Yet, many RPAM activities are targeted at children and their educators. This is a clear sign that RPAM is competing with education, even if involuntarily. And it does so quite successfully, not only because it typically subverts the school space and time, but specially because good RPAM caters to the public's interests, curiosity and creativity. And although the public for RPAM is not just students, RPAM is by nature a project of (continued) education at large. So, should the unsuccessful mathematical school education absorb RPAM activities, approaches and methods? In theory, yes. Unfortunately, in developed countries, and in particular in Europe, education has been increasingly dominated by pedagogy theorists as well as policies aiming at good-looking statistical results, strongly limiting the influence of scientific experts and expertise.[1] Their (our) intervention is reduced to RPAM-like activities, complementing and subverting the school education, instead of improving it in a decisive and sustainable way and freeing RPAM for society's post-school continued education.

In poorer countries, namely in Africa and Asia, the educational and social scenarios are quite different. And yet the motivations to promote the public awareness of mathematics there are exactly the same, if not even stronger. Developing countries are typically rich in natural resources, but poor in human qualifications, depending on foreign powers for technological and scientific expertise. Education is thus a strategic bet for the independence of such countries, a bet which should go beyond linguistic literacy into basic scientific and mathematical literacy. In developing countries the education system is commonly dysfunctional: the curricula are often non-existent, very old fashioned or simply copies of the ones from their former European colonial power (suffering from the above-mentioned problems); the teachers are not paid regularly and are frequently absent, as are the students who go to work or to pick up wood for their families; there is no electricity in many places; transportation and infrastructure are usually very poor; and science news, exhibitions or museums are extremely rare. This situation is both a tragedy and an opportunity. It is possible to act effectively for education, given the level of disorganization. To do so in a durable and sustainable way, that is the challenge, although it does not necessarily require large resources. Thus, this represents an opportunity to expand RPAM's typical geographical horizons. But it is not only an opportunity, it is a responsibility! Outside the school community there is also a curious and interested society. RPAM should target both. The question is then: what can be done, in practice?

SiW—Scientists in the World[2] has been promoting and developing projects along both these directions and has examples to answer the above question. SiW is an international non-profit NGO founded in 2007 with the goal to promote science and technology within the framework of international cooperation and development. In its projects in East Timor, Angola, Cape Verde and São Tomé Príncipe (see figures) SiW has experienced on the ground how the promotion of scientific and mathematical literacy is both important and welcomed. This is in fact the main lesson that can be drawn from our projects: it makes sense to develop RPAM activities in Africa, and in developing countries in general. Not only is it possible, even with scarce

---

[1] See, for instance, *La débâcle de l'école: Une tragédie incomprise*, Laurent Lafforgue & Liliane Lurçat (editors), F-X. de Guibert, 2007.

[2] See www.siw.org/en site, or www.facebook.com/SiW.Scientists.in.the.World.

Fig. 1 Okutiuka orphan's shelter, where SiW developed its project in Huambo, Angola, 2010

resources, it is furthermore appreciated and can have an impact locally. This is apparently a non-trivial result, since many had doubted it a priori. Yet, it should not be that surprising. The factors that attract the general public to mathematics, previously described (see above), are universal, as is human curiosity. Here are some examples and proposals on how SiW has been exploring these factors, in a context where the usual RPAM driving agents, resources and channels are absent:

Mathematics Through Musical Instruments In the summer of 2010 SiW developed a project in Huambo, Angola, together with Okutiuka, a local NGO that hosts war orphans (see Fig. 1). The children were between 12 and 18 years old and had limited schooling, mainly because of the dysfunctional local education system. The project consisted of the construction of string, wind and percussion musical instruments with common objects and cheap materials (see Figs. 2 and 3) as a means to teach wave mechanics and the inherent discrete mathematics in musical harmonies and notes: multiples, dividers, sequences, exponentials, etc. These concepts appeared naturally in the construction and use of the instruments; they were never introduced formally, although by the end of the two-week project the students were, in our assessment, ready and motivated for more formal mathematical education. Unfortunately we could not find reliable local teachers to pursue this, as we intended. But we were very satisfied with the results obtained, a true and successful increase in the awareness of mathematics in those children. What attracted them

Fig. 2 SiW project in Huambo using home-made musical instruments, Angola, 2010

Fig. 3 SiW project in Huambo using home-made musical instruments, Angola, 2010

were clearly the factors A and D identified above. SiW is planning to repeat this project in 2012.

Local History of Mathematics   Finding bridges to the local identity and history (factor B) is an extraordinarily powerful approach for RPAM. In Africa, during the colonial period, mathematics was taught by glorifying the supposed intellectual superiority of European men: any difficulties the locals had in understanding mathematics were evidence of this superiority. Furthermore, there is this underlying feeling that truly important scientific discoveries always happen very far away, in other continents. Yet, there are counter examples. For instance, in 1919, Arthur Eddington, an astronomer from Cambridge, performed eclipse observations on the island of Príncipe, off the west coast of Africa, which were the first experimental evidence of Einstein's theory of General Relativity, a key moment in the history of science. In 2009, SiW used the 90th anniversary of these experiments to develop a couple of science awareness projects in São Tomé and Príncipe (see Figs. 4, 5, 6 and 7) promoting

Fig. 4 Exhibition about
Einstein's theory of General
Relativity and Eddington's
experiments, SiW project in
São Tomé and Príncipe, 2009

Fig. 5 Understanding curved
geometry, SiW project in São
Tomé and Príncipe, 2009

the role of the territory and local population in this important discovery, a source of
pride that further motivated students, educators, parents and local authorities to develop science education. Following in the same spirit and concerns, SiW has other
projects in preparation. One involves activities exploring the differences and similarities between the mathematics developed by a local culture or civilization (namely
their counting methods, computation algorithms, geometry, odds estimation, logic,
coding, . . .) with modern mathematics. Some of these native methods still survive in
sub-Saharan Africa, in Andean regions or in South-East Asia, amongst other places,
but even where they have disappeared one can do these comparative studies and activities from a historical perspective. Another project is simply to promote popular
mathematics activities by referring to work by local mathematicians, working locally or abroad: these are identity references, which, via talks, articles, exhibitions
or even simple interviews, can attract their fellow citizens' attention and curiosity to
mathematics, as well as that of local policymakers.

Fig. 6 Interacting with local students, SiW project in São Tomé and Príncipe, 2009

Fig. 7 Local volunteers get involved, SiW project in São Tomé and Príncipe, 2009

Popular Mathematics Talks   In many developing countries and regions, the concept of popular science is simply unknown. The idea of bringing scientific knowledge, whether recently discovered or well established, to the general population, outside school, is a complete novelty. Since its foundation, SiW always tries to annex popular science talks to its projects, aimed at the local general population. And every time this has proved a successful and fruitful experience. People are usually not sure what they are coming to, sometimes they think it is some sort of cultural event, but their curiosity is precisely a way to attract them to the venue, as is the presence of people from a different continent. And their interest and enthusiasm is always stronger if the subject is presented in a form they can relate to (typically factor A, but also B and C ...). To give a concrete example from mathematics: a popular talk on cryptography and information theory a few years back, in East Timor, attracted an audience of more than a hundred people who, to our surprise, found this scientific topic and its challenges very familiar. In fact, during the Indonesian occupation of the country, out of practical need the underground resistance and guerrilla movement developed their own methods of steganography and cryptography, including simply communicating in Portuguese, which was not known by the Indonesians! On the other hand, they were really surprised and interested in the mathematical formulation of such a familiar problem and the power of mathematical reasoning to solve it. That was pure RPAM! Popular mathematics talks can have an even greater impact in developing countries, given the novelty factor they usually carry there.

Popular Mathematics in the Media   Of course, there is no reason to limit popular mathematics to talks for a few dozen people and not to try to reach a wider audience. This can be done through the media, namely via articles in newspapers and magazines, or broadcast programmes, such as on the radio, which typically is the most pervasive medium in developing countries. The factors that will attract the audience to mathematics are always the same, therefore one should be sensitive to the local reality and to what people can relate to, aspects which can be quite different from our preconceived impressions. But the true novelty is essentially in bringing popular mathematics to a wide audience in places where the concept is simply unknown. SiW believes this is the right direction to follow and has been preparing a project to introduce popular science on a regular basis in the mass media in different African countries. This could also be extended and reproduced to include recreational mathematics and puzzles (factor E) as a complementary way of introducing popular mathematics.

Mentoring of Journalists   Further to the introduction of popular mathematics into the mass media, SiW also aims at raising the awareness of science in general, and mathematics in particular, amongst journalists and news editors. Science journalism is in crisis in developed countries, as is journalism in general, which can be seen from the growing number of media outlets reducing costs and personnel, or simply closing. Yet, this is not a reason why science journalism should not be promoted in countries where it barely exists, or does not exist at all. On the contrary. Science journalism is not the popularization of science in the media: it is a means to raise the awareness of the public of scientific discoveries and their implications for society, beyond the scientific perspectives and the scientists' interests. This is a job that (in theory) is reserved for journalists, given their professional responsibilities, ethics and regulations. Formal training in science journalism is not easy to find, even in developed countries, since it is a very specific topic and the current economic situation is putting pressure on journalists to become generalists rather than specialists. But there is good science journalism, and it should be protected and promoted beyond national borders. Together with the World Federation of Science Journalists, SiW has devised a programme for science journalism training in Portuguese-speaking African countries and East Timor. The project is based on mentoring by well-established science journalists who, via the Internet, advise and assist local journalists to produce scientific news stories, as well as helping them to conquer the space for those stories vis-à-vis their news editor. This SiW project has been on hold due to a lack of resources, but is expected to be launched soon. Science journalism is already quite a specific area, so it is difficult to further refine it to focus only on mathematics, but on the other hand mathematics should not be excluded from it. Scientific discoveries make the news, especially if they have an impact on society, and it is true this is more common for areas such as health sciences, environmental studies, technology, etc. Yet, mathematics also contributes: for instance, the discoveries regarding quantum algorithms may challenge the current security of telecommunications, and quantum cryptography can offer eavesdropper-proof communications, with serious consequences for society. Besides the applications (factor A), mathematical discoveries can also make it into the news given their historic importance—such as the proofs of Fermat's theorem and the Poincaré conjecture—or the consequent awards and stories (factor C), as well as when some mathematician originally from a particular region or nation is involved in the discovery (factor B). Science journalism can be a very significant channel for RPAM, with unique responsibilities and the potential to reach the widest audiences.

The Human Compiler Robot This was an idea first proposed by Jeffrey Warren (MIT). Our project is to design a game where one person, the *compiler robot*, acts out the logical instructions and actions written by others, as a means for the group to acquire and practice logical reasoning. For instance, a very simple programme could be:

```
1. Face North.
2. FOR i=1 to 5 DO:
   2.1 Take one step ahead.
   2.2 IF hit wall THEN turn 180°.
   2.3 Turn 90°.
   2.4 IF touch table THEN:
       2.4.1 FOR j=1 to 3 DO put cookie(j) in bag.
       2.4.2 Turn 180°.
3. IF bag ≠ empty THEN FOR i=1 to 3 DO eat cookie(j).
```

The programmer(s) can then execute and debug the programme, a very instructive procedure. As is, of course, the programming itself. This is a simple but powerful game whose only limit is the players' imagination. It can also be used as an education tool, especially where computers are not available. But above all it is a way to experience the logical and structured reasoning offered by mathematics, as well as concepts such as algorithm complexity and recurrence.

These are just some examples of projects SiW—Scientists in the World has developed or will develop in Africa and South-East Asia. Over the last few years SiW has accumulated experience on how to raise mathematics awareness (and science awareness in general) in developing countries and will gladly share these details with anyone interested in promoting similar initiatives. But some of our RPAM activities have an ulterior agenda: projects like *Mathematics through musical instruments* or *The human compiler robot* are used also to motivate and introduce mathematical education projects. Rather than using RPAM to complement (or in some case correct) an unsatisfactory education in mathematics, as is the case in many developed countries, in these poorer countries, given the degree of dysfunctionality, is it actually possible to act effectively on education. And complementing or even helping to shape a modern education in mathematics would be a major contribution for such countries. One very important message to transmit is that it is possible to develop good educational projects with little material resources. Of course, if computers are available (and they may well be: see the *One Laptop Per Child* project[3]), we have a terrific educational tool to explore. This is certainly an idea worth investing in. But otherwise it is perfectly possible to create mathematical educational activities using only everyday objects.

The examples of the projects presented above can illustrate some key aspects for the success of RPAM projects in developing countries. First, it is crucial to be on the ground and demonstrate the ideas. Second, one needs to work with local partners, who should contribute with their local expertise and culture to the design and development of the project, and should be convinced and trained to reproduce and continue the activities once we leave. Third, the students and educators should be empowered by building and participating in the experiments themselves. Finally, it pays to find and explore any bridges to the local identity within the proposed activities.

People often ask why one should promote mathematical awareness and literacy in developing countries. The answer is simple: it is because they need it, because they know they need it, because they want it, and because they like it and appreciate it! There is a lot of intelligence, curiosity and creativity waiting in these countries to be sparked. It is our responsibility to respond and our privilege to learn new things by doing so.

---

[3]one.laptop.org.

Mathematical Pictures: Visualization, Art and Outreach

John M. Sullivan

Abstract Mathematicians have used pictures for thousands of years, to aid their own research and to communicate their results to others. We examine the different types of pictures used in mathematics, their relation to mathematical art and their use in outreach activities. (This article is based on a talk on 28 September 2010 at the conference *Raising Public Awareness of Mathematics* in Óbidos.)

Mathematical visualization can be defined as the use of pictures to convey mathematical ideas. Diagrams have of course been used for thousands of years, especially in geometry and related areas of mathematics. In recent decades, computers have become increasingly able to easily create high-quality, mathematically accurate graphics; this has led to an increasing number of well-illustrated research papers.

Furthermore, many researchers have started to use numerical simulations (computer experiments) as a research tool, for instance to test and refine conjectures. The interplay between these experiments and rigorous proofs is what allows progress on both fronts. Especially in geometry, the results of a simulation are usually understood most easily and quickly through pictures. Of course the same pictures that have helped a researcher understand her own experiments (and thus to formulate and prove better theorems) can often be adapted to convey the results to colleagues, students, and the general public.

Our focus here is on pictures of mathematics. Of course there is a separate but related story about the mathematics of pictures, which first came into focus with the mathematical study of perspective projection during the Renaissance. All computer graphics algorithms are based on this sort of mathematical background. Current research in geometry processing and other related areas of graphics increasingly uses mathematical techniques like those of *discrete differential geometry* [2]. Here classical notions from differential geometry (like curvatures of smooth surfaces) are adapted to the discrete (triangulated) surfaces typically used as computer models.

J.M. Sullivan (✉)
Institut für Mathematik, Technische Universität Berlin, Berlin, Germany
e-mail: sullivan@math.tu-berlin.de

E. Behrends et al. (eds.), *Raising Public Awareness of Mathematics*,
DOI 10.1007/978-3-642-25710-0_21, © Springer-Verlag Berlin Heidelberg 2012

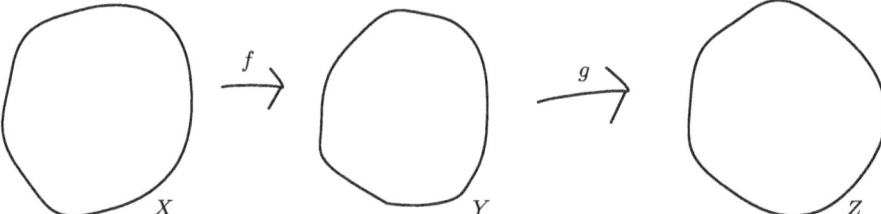

Fig. 1 Symbolic sketch: the composition $g \circ f$ of two maps

Fig. 2 Examples of
topological sketches include a
Venn diagram (*left*) and a
knot diagram (*right*), as well
as more general plane graphs

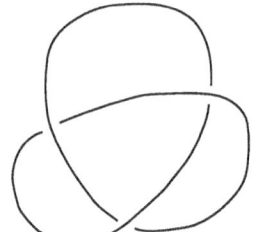

Types of Mathematical Pictures

Some mathematical pictures are very useful even though they convey no precise
geometric content. For example, it seems almost impossible to discuss in a lecture
the composition of functions $f: X \to Y$ and $g: Y \to Z$ without drawing the *symbolic
sketch* shown in Fig. 1: three blobs with two arrows between them.

The next class of pictures could be called *topological sketches*. This would include knot diagrams, drawings of planar graphs, or Venn diagrams showing the
various intersections of several sets, as in Fig. 2. Here, the topology of the sketch—
for instance which curves cross which others in which order—is very important, but
the specific geometry is not. This freedom often makes it easy to draw topological sketches by hand—there is no need to worry about accuracy—but can make it
surprisingly difficult to draw them with computer graphics.

Some diagrams seem to convey their content so well that no further explanation
is needed. Such *proofs without words* (like that in Fig. 3) have been for instance
regular features in the journals *Mathematics Magazine* and *College Mathematics
Journal* [1]. Of course they illustrate just one case of the result to be proved, but in
a way very suggestive of the general case.

Of course many proofs without words—like the various dissection proofs of the
Pythagorean theorem—are much more geometric and are really examples of our
next class of mathematical pictures: accurate *two-dimensional geometric diagrams*.
These have been a mainstay of elementary geometric reasoning since the time of
Euclid (despite all the acknowledged dangers they entail, like hidden assumptions
about the existence or relative order of points).

Two-dimensional (2D) geometric diagrams are of course among the easiest to
draw accurately by computer. The geometry to be displayed translates directly into

Fig. 3
$1 + 2 + \cdots + n = n(n+1)/2$

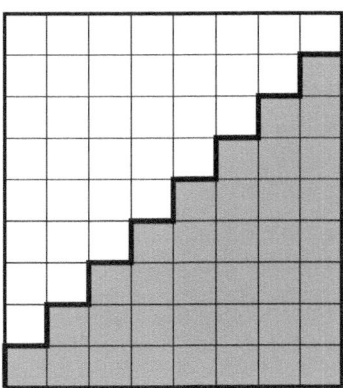

the picture shown on the screen or on the printed page (neglecting for a minute the important aesthetic decisions about colors, line widths, and so on). In recent years, several marvelous software packages[1] have become available that make this kind of diagram fully interactive. Elementary constructions—showing, say, that the medians of a triangle intersect—can immediately be tested for many different possible triangles.

The mainstay of computer graphics is rendering 2D pictures of 3D objects and scenes. The computer of course has no difficulty accurately projecting the 3D scene onto the 2D canvas. Trickier is the choice of an appropriate rendering style. Much work in computer graphics has been directed towards making renderings as photo-realistic as possible, for instance for Hollywood feature films. But a mathematical object isn't a real physical object; there's no mathematical meaning in choosing to depict a torus as if made out of, say, wood, metal, plastic or glass.

Sometimes a skilled artist can use just a few brush strokes to create an image that helps the viewer focus on the essential features of a scene. The typical topologist has not learned these techniques in general, but has learned to draw a torus with three strokes: the silhouette edges where the surface tangent plane is parallel to the line of sight, as in Fig. 4 (left). Topologists who wish to learn to draw better by hand might like to read the *Topological Picturebook* [3] by George Francis.

Increasingly, research in computer graphics has been directed towards obtaining such "nonphotorealistic renderings" automatically [5, 8, 13], for instance with special shaders for emphasizing edges in technical drawings, or for cartoon effects as in Fig. 4 (right).

A pair or renderings of the same 3D scene—from slightly different viewpoints—can be combined for a stereoscopic effect. The two different images are presented to the viewer's two eyes. This is best achieved with special devices, often involving lenses or polarizing filters. But it can also work with a pair of pictures side-by-side

---

[1]These include Cabri <cabri.com>, Cinderella <cinderella.de> and Geometer's Sketchpad <dynamicgeometry.com>.

Fig. 4 The typical mathematician's sketch of a torus (*left*) gives just the silhouette edges, three black lines. Standard 3D computer graphics instead aims for photorealism, often much fancier than the default plastic material (*middle*, rendered in Blender, <www.blender.org>). Nonphotorealistic rendering aims to recreate various styles of hand-drawing, for instance cartoon cells (*right*, rendered by Sami Hamlaoui's cell-shader <nehe.gamedev.net>); the thin black lines here are an accurate version of the silhouette lines mimicked in the hand drawing

Fig. 5 A stereo view of the (presumably) tight configuration of the Turk's head knot 8_{18}, rendered by Charles Gunn. The *middle image* is for the right eye: to see the stereo effect, view the left pair wall-eyed or the right pair cross-eyed

on a printed page, as in Fig. 5, for those people who have taught themselves to view such a pair either cross-eyed or wall-eyed.

As a final type of picture, we should mention three-dimensional models or sculptures. Although it may seem strange to call these "pictures", they are often the best imaginable visualizations of three-dimensional geometric objects. The California-based sculptor Bathsheba Grossman—in her *Math Models* series—has turned some well-known mathematical models into truly artistic metal sculptures, including those in Fig. 6.

Complementary to this list of types of mathematical pictures is the choice of how they can or should be made. Any of these types can for instance be made by hand or with a computer. (A camera can also be a useful tool for "converting" a 3D model into a 2D picture—or for illustrating real-world connections in mathematics.) Certain tasks (like proper 3D perspective) are trivial for the computer and harder by hand, while others (like rough topological sketches) are easiest by hand. Modern 3D printers allow a computer to directly create a complex physical model (out of plastic, metal or other materials)—indeed this is how Grossman's sculptures are produced.

Finally, one can add a time dimension to any of these mathematical visualizations, creating an animation. For instance, a changing sequence of 2D pictures gives a movie; a flexible sculpture is usually called a mobile. Here I would suggest an im-

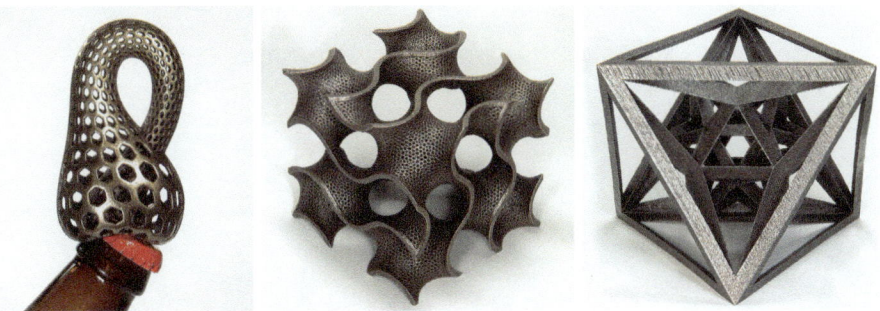

Fig. 6 Three *Math Models* by Bathsheba Grossman: the Klein bottle (*left*) as a functional bottle opener, the gyroid minimal surface (*center*) and the 24-cell (*right*), a four-dimensional regular polytope in cell-first projection. Photos by Grossman, used by permission

portant distinction between two types of animation. By a *narrative animation* I mean a fixed time sequence telling a story, often with voice narration for further explanation. Often a narrative animation is created by choosing a good fixed path through the (high-dimensional) parameter space. This kind of animation can be saved as a video (say on a DVD) and is well-suited for presentation to a large group.

By contrast, an *interactive animation* lets a (single) user navigate through the high-dimensional parameter space. Such an animation is most often implemented as an interactive computer program, and is good for individual learning: the user can easily choose to spend more time exploring the parts or configurations of most interest.

It is, however, important for the author of the animation to give the user guidance on how to explore the possibilities. More freedom doesn't necessarily help the user here—instead, in order to demonstrate a particular mathematical phenomenon, the user should be guided there (say through restrictions on the parameters). I find this analogous to the way some artists say artistic constraints help their creativity. For instance most classical poetry used fixed forms like the sonnet. Species counterpoint in Renaissance music (partly developed as a pedagogical tool) had fairly rigid rules that almost automatically led to interesting and pleasing harmonies. In more modern music, Arvo Pärt is an example of a composer who sets very rigid rules for himself: his pieces since 1976 are written with the technique he calls "tintinnabuli"; different voices in a given piece are linked through set rules.

Of course movable 3D models can also have varying degrees of freedom. The "jitterbug" is Buckminster Fuller's name for a motion that folds a cuboctahedron down through an icosahedron to an octahedron, preserving pyritohedral $(3*2)$ symmetry. If one simply attaches eight equilateral triangles together at their vertices, this motion is possible but hard to find, since there are too many degrees of freedom— the model is floppy. If, however, a hinge is used at each of the twelve attachment points to fix the dihedral angle between the two triangles, then only one degree of freedom (that of the jitterbug) remains. Such models have been implemented and popularized by Dennis Dreher and Stuart Quimby, among others. Figure 7 shows an *Octabug* and a *Vector Sphere II* purchased at the wonderful AHA store in Zürich

Fig. 7 The *Octabug* (*left*) has the one degree of freedom needed to accomplish the jitterbug motion, folding a cuboctahedron down to an octahedron. The *Vector Sphere II* (*right*) allows this while also having a second degree of freedom

<www.aha-zurich.ch>. Jochen and Conrad Valett <www.valett.de> have also produced some interesting stainless-steel models, for instance of knots and Möbius bands, that have usefully limited freedom of motion.

Similarly, one can view different geometric modeling kits as enforcing different constraints. For instance several kits (like Polydron and Geofix) have rigid polygons that click together to form polyhedra. Zometool on the other hand has sticks of given lengths that fit into balls in one of 31 fixed directions—the system is perfectly tuned for icosahedral symmetry. Either system makes it easy to build the kinds of objects it is designed for (as in Fig. 8), but it is completely impossible to build most objects from the other system.

Visualization Challenges

There are several special challenges in mathematical visualization. Mathematicians might want to view objects in curved spaces (like spherical or hyperbolic space) or higher-dimensional spaces; they are often interested in periodic or symmetric structures. Finally, the objects of interest often have complicated internal structure. (Of course, these issues are not unique to mathematics: cosmologists also deal with curved spaces and crystallographers with periodic lattices. Similarly, anatomy books and car repair manuals have to show internal structures.)

To illustrate some of these challenges, let me turn to my own research in geometric optimization problems. Here we are given some topological object and seek to find an optimal geometric shape for it, typically by minimizing a geometric energy. A classic example is the soap bubble, which—as a round sphere—minimizes its surface area for the given enclosed volume of air.

Fig. 8 A Zometool model (*left*) with icosahedral symmetry and a Polydron model (*right*) of a deltahedron. Neither model could be built with the other system

Fig. 9 The 120-cell is the four-dimensional analog of the dodecahedron, and can also be viewed as the universal cover of the famous Poincaré homology sphere from topology. Under stereographic projection it has exactly the geometry of this bubble cluster

There are many open mathematical problems related to clusters of several bubbles or to foams of infinitely many bubbles filling space [11]. Here of course we suddenly have lots of interesting internal structure, namely the various sheets of soap film separating the different bubbles. Some bubble clusters (like the one in Fig. 9) have particular symmetry and interesting connections to other parts of geometry. The Kelvin problem is to find the foam that divides space into unit-volume cells while using the least total interface area. The Irish physicists Weaire and Phe-

Fig. 10 The Weaire–Phelan foam (*left*, with two shapes of cells) does better than the Kelvin foam (*right*, with congruent cells) for the problem of optimally dividing space into equal-volume cells

lan discovered in 1994 that Kelvin's proposed solution from 1887 is not optimal. Instead their less symmetric foam (Fig. 10), with two shapes of cells in the A15 crystal lattice, does better.

A *sphere eversion* is the process of turning a sphere inside out under certain mathematical rules. We don't allow the surface to have holes or creases at any time (otherwise the eversion would be too easy), but do allow it to pass through itself (otherwise the eversion would be impossible). Steve Smale proved in 1959 that an eversion is possible in principle, but it took many years to find an explicit eversion. (One of the first was by Tony Phillips in 1966.) Most eversions have been designed by hand. If we work from a so-called *half-way model* (showing the two sides of the sphere equally), it suffices to simplify this to a round sphere. Our minimax eversion [4], illustrated in the video *The Optiverse* [10], was computed automatically by minimizing the Willmore bending energy to get from the half-way model down to the round sphere. In particular, our half-way model (as suggested by Rob Kusner) was one of the Willmore-critical spheres classified earlier by Robert Bryant. In a sphere eversion, as in a bubble cluster, there is lots of important internal structure.

The visualization challenge here arises from the fact that our eyes are used to seeing mainly outer surfaces, not inner structure. Most of the transparent objects we encounter in everyday life—like glass windows—are meant to be looked *through*, not looked *at*. But artists have developed several useful ways to depict internal structure (see also [3]). For instance, in different scenes of *The Optiverse* we used transparent

Fig. 11 Three scenes from *The Optiverse* with different rendering styles: transparent, solid, and with triangles shrunk to leave large gaps

Fig. 12 In another scene from *The Optiverse*, a fly-through of an intermediate stage of the eversion, a window is opened within each triangle. This still image has attracted much publicity

renderings (like soap films), solid rendering (to show the outer shape), and renderings with gaps or windows (to focus on the self-intersections). See the examples in Figs. 11 and 12.

Sometimes a completely transparent rendering is hard for the eye to read. In the picture of the nonstandard double bubble shown in Fig. 13, the soap film has been artificially darkened to help the viewer see that one bubble is a toroidal belt around the other (which itself extends from the very top through the middle to the very bottom of the picture).

Although a 3D model or sculpture is usually the best way to depict a three-dimensional mathematical shape, internal structure is hard to show even here. At the International Snow Sculpture Championship 2004 in Boulder, Colorado, I joined a team led by Stan Wagon to carve the half-way model from a sphere eversion by Bernard Morin. Competing against 11 other teams from around the world, we had four days (using only hand tools) to carve a 20-ton, $3 \times 3 \times 4$-meter block of

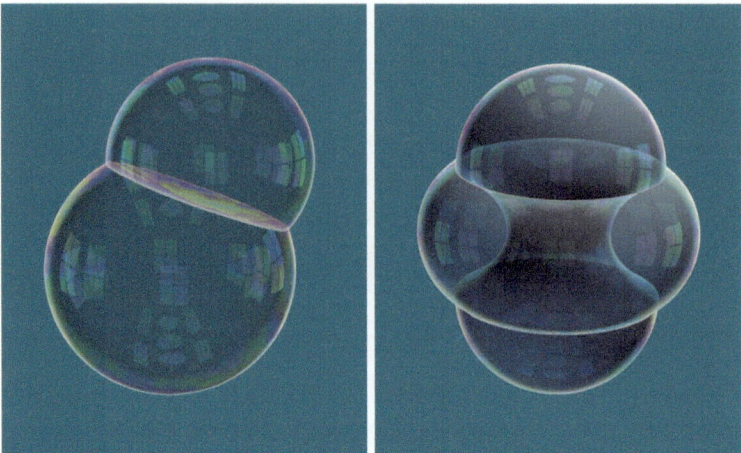

Fig. 13 The standard double bubble (*left*) is formed whenever two soap bubbles attach; it encloses and separates the two volumes using the least surface area. It was hard to prove rigorously that this is best; the difficulties lay in ruling out various nonstandard double bubbles (*right*) composed of surfaces of revolution, where one bubble might have a toroidal shape or even be disconnected

Fig. 14 From the 3D-printer model (*left*, photo by Séquin) we sculpted a Morin surface four meters tall in snow (*right*)

snow into the Morin surface (Fig. 14). To show the inside and outside without color, we used a framework vs. solid depiction, allowing a view of the interior structure through the gaps. We worked from a spline-based design by Carlo Séquin, who brought a small 3D-printer model as a guide.

Fig. 15 Bathsheba Grossman's sculpture *Alterknot* (*left*) has tetrahedral (233) symmetry, while her *Soliton* (*center*) has 222 symmetry. Brent Collins' *Atomic Flower II* (*right*) has 223 symmetry

Fig. 16 John Robinson's sculpture *Genesis* (*left*) has pyritohedral (3 ∗ 2) symmetry. Charles Perry's *Eclipse* (*center*) has icosahedral (235) symmetry, as do many of George Hart's sculptures, including *Eights* (*right*, photo by Hart, used by permission)

Mathematical Art

While there is a very long history of art inspired by mathematics, in recent decades interest in mathematical art has been growing. This is represented, for example, in the annual Bridges conferences <bridgesmathart.org> on "mathematical connections in art, music and science", attracting hundreds of active participants, and in the peer-reviewed *Journal of Mathematics and the Arts* founded in 2007.

There are certain themes which often recur in mathematical art. For instance, there are various types of pictures that are well-suited to mathematical analysis, including (Celtic) knotwork, tilings of the plane, and other objects with interesting symmetries. Mathematicians tend to enjoy looking at this type of art. M.C. Escher is famous for his prints of various tilings usually using animal figures; he corresponded with H.S.M. Coxeter to learn about hyperbolic geometry in order to make his *Circle Limit* prints. Other artists make beautifully symmetric artwork with little explicit knowledge of the mathematics. Sculptors who are known for highly symmetric work include Brent Collins, Bathsheba Grossman, George Hart, John Robinson and Charles Perry; some of their work is shown in Figs. 15 and 16.

Some artists borrow themes from mathematics without any deep mathematical connection (even if there is some supposed mystical connection). Famously, Dalí used an unfolding of the hypercube as a cross in his painting *Crucifixion (Corpus Hypercubus)*. Many artists have been inspired by the golden ratio, or have used patterns of prime numbers almost algorithmically to help create artworks.

In some areas, computer-generated mathematical visualizations have such immediate apparent beauty that the programs can immediately be adopted by artists, who then mainly concern themselves with artistic decisions like color schemes. Here the classic example is fractals: Much fractal art has been made simply by finding visually interesting parts of the Mandelbrot and Julia sets, which have also been intensively studied by mathematicians. (Recently, for example in the search for 3D analogs of the Mandelbrot set, there has been a move towards using iteration formulas that have little mathematical meaning, but simply produce interesting pictures.) The traveling exhibition *Imaginary* <imaginary-exhibition.com>, organized by the Mathematisches Forschungsinstitut Oberwolfach, started in 2008 and mainly presents renderings of algebraic surfaces (mostly with interesting singularities). Visitors to the exhibition can use interactive software to design new pictures, which may get included in future exhibitions. Minimal surfaces form another example of well-studied mathematics easily adapted to make nice mathematical art (like the gyroid in Fig. 6).

Of course no short survey like this can attempt to capture all the multiple facets of mathematical art. Perhaps I will end just by mentioning the sculptor Helaman Ferguson, a trained Ph.D. mathematician who carves large-scale stone sculptures, largely by hand. His work is often inspired by advanced mathematics, though it is not meant as a precise visualization.

To turn a mathematical visualization into art, one must make decisions about the colors and materials to be used: as mentioned above, mathematical objects have no intrinsic colors. Sometimes, aesthetic judgments even lead to modifying the mathematics to be visualized. For instance, my *Minimal Flower* sculptures shown in Fig. 17—an homage to Brent Collins, whose work inspired me—were planned as minimal surfaces spanning certain knotted boundary curves. For a better aesthetic effect, they are actually minimal surfaces in a conformal ball model of hyperbolic space, which enhances the U-shaped cross-section of the outer arcs. (See [12] for more details on the creation of these pieces.)

Mathematical Outreach

Mathematicians often speak of the beauty of mathematics, referring to the abstract elegance of results or their proofs. Indeed the search for this beauty can be said to drive progress in pure mathematics. Its abstractness, however, makes it hard to share with the general public. Mathematicians are thrilled by the "Aha!" moment when the "proof from the book" suddenly becomes apparent—but how can we describe this to nonmathematicians?

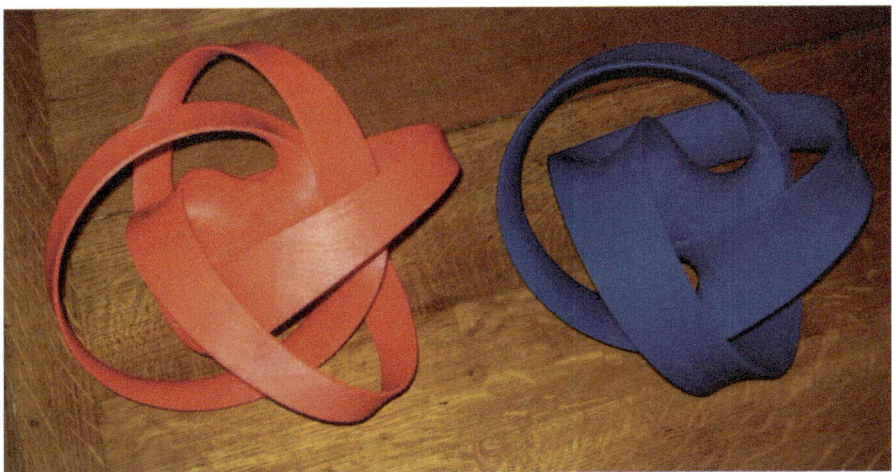

Fig. 17 My sculptures *Minimal Flower 4* (*left*) and *Minimal Flower 3* (*right*), as printed on a 3D printer, are close to minimal surfaces, but with certain aesthetically motivated modifications

One approach to mathematical outreach is to emphasize applications. Coming after a historical period where many pure mathematicians seemed to feel it was beneath them to even consider possible application of their work, this is certainly a welcome trend. Modern mathematicians realize that almost any mathematical development is potentially applicable, perhaps to quite different areas of mathematics, to natural science, or even to industry. But I feel it is a mistake to completely hide the new mathematics behind these applications. When at an outreach event mathematicians speak only about applications, the audience may simply end up with the feeling that people trained in math can then go on to do good work in physics, engineering and other fields. What is important for the mathematical community, however, is to convey the fact that *new* mathematics must be developed (for its own sake as well as for applications) and that there are many interesting open mathematical problems.

We should also attempt to make the abstract beauty and elegance of mathematics visible to the public—and perhaps even to give a taste of the "Aha!" moment by explaining a simple proof. But, on the other hand, we don't have to hide the fact that mathematics is difficult.

One source of difficulty in getting publicity for mathematical results may be surprising to those who have not experienced it. Newspaper reports, for instance, have to be keyed to a newsworthy event. In many sciences, the results of a paper are kept secret until publication; journals like *Nature* and *Science* have official embargo policies, giving journalists advance notice while ensuring that the news coverage happens on the day of publication. The most newsworthy event in a mathematical result is probably the "Aha!" moment behind the proof. But this is usually followed by a long period of working out the details. Mathematicians give talks about work that they are just starting to write up, and they often distribute preprints well before journal submission. The refereeing process is much slower than in most other fields

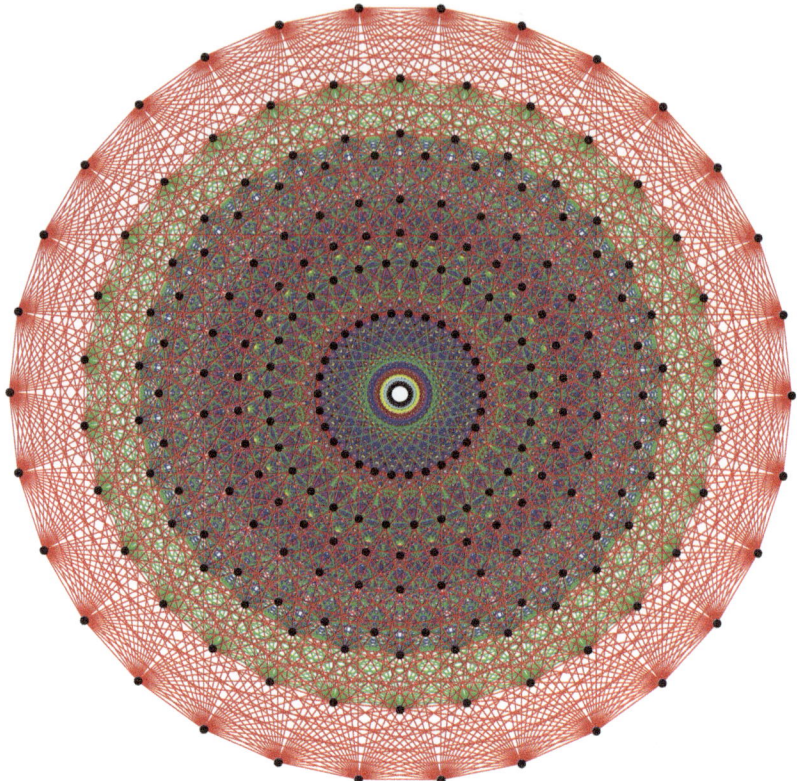

Fig. 18 The eight-dimensional Gosset polytope, also called 4_{21}, is the vertex figure of the E_8 lattice. This particularly symmetric orthogonal projection to the plane is slightly modified from the version rendered by John Stembridge (used by permission)

(perhaps because it aims at a higher standard of rigor). All of this unfortunately means that there is no particular point in time at which a mathematical story is most newsworthy.

Of course, one good way to increase journalistic interest in a piece of mathematics is to have nice pictures. Good visualizations help attract publicity. I first experienced this when I made pictures (Fig. 13) to illustrate the double-bubble conjecture; although I was not involved in its proof, my images were printed at least two dozen times in newspapers and magazines across the US and Europe. (Perhaps the publicity was also greater because the conjecture was first proved for equal volumes [7] and only years later for the general case [6]. Unusually for mathematics, each paper was also preceded by an electronically published research announcement.)

My own work that has received the most publicity is the minimax eversion shown in *The Optiverse*—despite the fact that it is just a numerical simulation with (as yet) no theorem behind it to say the Willmore-gradient flow really gives a sphere eversion. The image shown in Fig. 12 was used for the cover of *Science News* and in a dozen other print reports on three continents; the animations were used on television news reports.

Another case in which there was intriguingly much publicity for a mathematical result—presumably helped by John Stembridge's illustration (Fig. 18)—was the "media frenzy" in March 2007 about the E_8 Lie group. Tony Phillips gave a nice discussion of this in his "Math in the Media" column online [9]. Rather than a new mathematical discovery, the event here was the completion of a large computer calculation in representation theory, meaning that the American Institute of Mathematics (AIM) could easily time their press-release <aimath.org/E8/>. It was prepared with the help of a public-relations firm, and invoked purported applications to theoretical physics as well as comparisons to the Human Genome Project. Phillips concludes that "A press release with some good 'hooks' and a good list of contacts really works."

In the end, the mathematicians involved seem to have been somewhat embarrassed by the success of the AIM campaign. Their webpage <liegroups.org/AIM_E8/technicaldetails.html> starts out saying "people have been asking what the fuss is about" and ends with a comment on "the attention this story has gotten in the press": The E_8 calculation was just "an important test", "a small step", and "not remotely as important as the original work" of others; however, AIM decided it would be "an excellent opportunity to educate the public about research in pure mathematics". The authors conclude that they are "happy to have been successful in raising awareness of mathematics research worldwide", and perhaps the rest of us can learn something from their example.

References

1. Alsina, C., Nelsen, R.B.: An invitation to proofs without words. Eur. J. Pure Appl. Math. **3**, 118–127 (2010)
2. Bobenko, A.I., Schröder, P., Sullivan, J.M., Ziegler, G.M. (eds.): Discrete Differential Geometry. Oberwolfach Seminars, vol. 38. Birkhäuser, Basel (2008)
3. Francis, G.: A Topological Picturebook. Springer, Berlin (1987)
4. Francis, G., Sullivan, J.M., Kusner, R.B., Brakke, K.A., Hartman, C., Chappell, G.: The minimax sphere eversion. In: Hege, H.-C., Polthier, K. (eds.) Visualization and Mathematics, pp. 3–20. Springer, Berlin (1997)
5. Gooch, B., Gooch, A.: Non-photorealistic Rendering. AK Peters, Wellesley (2001)
6. Hutchings, M., Morgan, F., Ritoré, M., Ros, A.: Proof of the double bubble conjecture. Ann. Math. **155**, 459–489 (2002). doi:10.2307/3062123
7. Hass, J., Schlafly, R.: Double bubbles minimize. Ann. Math. **151**, 459–515 (2000). doi:10.2307/121042
8. Hertzmann, A., Zorin, D.: Illustrating smooth surfaces. In: SIGGRAPH, pp. 517–526 (2000)
9. Phillips, T.: Math in the Media (April 2007). ams.org/news/math-in-the-media/mmarc-04-2007-media
10. Sullivan, J.M., Francis, G., Levy, S.: The Optiverse. In: Hege, Polthier (eds.) VideoMath Festival at ICM'98. Springer, Berlin (1998). Narrated videotape (7 min)
11. Sullivan, J.M., Morgan, F.: Open problems in soap bubble geometry. Int. J. Math. **7**(6), 833–842 (1996)
12. Sullivan, J.M.: Minimal flowers. In: Bridges Proceedings (Pécs), pp. 395–398 (2010)
13. Winkenbach, G., Salesin, D.: Computer-generated pen-and-ink illustration. In: SIGGRAPH, pp. 91–100 (1994)

Part IV
Popularisation—Why and How

Image, Influence and Importance of Mathematics as Directives for Public Awareness

F. Thomas Bruss

Abstract The image of mathematics in society has many different sides. If we interview people then we get different answers depending on whom we ask. The level of education has a substantial role, though it is not clear whether it is the dominant role. The variety of answers is large if we ask people to give their opinion about the influence of mathematics and mathematicians in society. We cannot expect the public to have a clearer view of mathematics than mathematicians themselves, and here things become interesting. Mathematicians have a rather coherent perception of the image and influence of mathematics but their perception of what is important in mathematics, and of the importance of mathematics as such, may be very divergent. Do different perceptions of importance possibly harm our field? Where do we see the greatest influence of mathematics on society? And then, where should we feel responsible for the reputation of mathematics? The objective of this article is to analyse these questions and to give examples showing that strategies for raising the public awareness of mathematics should take these questions and partial answers into account.

Mathematics Subject Classification 00-XX · 00A30

Introduction

What role does mathematics play in society?

None at all, some people may answer, nobody I know needs mathematics.

No, others may object, it plays a role, but a sad one. I never understood mathematics, and for me mathematics is a permanent reminder that there are things in life I simply cannot do.

Again others may differentiate a bit more and admit some importance by arguing that, after all, people should be able to calculate.

F.T. Bruss (✉)
Département de Mathématique, Université Libre de Bruxelles, Campus Plaine, CP 210, 1050 Bruxelles, Belgium
e-mail: tbruss@ulb.ac.be

E. Behrends et al. (eds.), *Raising Public Awareness of Mathematics*,
DOI 10.1007/978-3-642-25710-0_22, © Springer-Verlag Berlin Heidelberg 2012

Answers of this kind to the question about the role of mathematics in society are, fortunately, only one half of the story. Nevertheless they are not flattering and must be taken into account when we think about the public awareness of mathematics. Moreover, such answers are in striking contrast to a more official view of the importance of mathematics as advertised by a committee of the British government on strategies for education (see http://nationalstrategies.standards.dcsf.gov.uk/node/ 16073). This committee addresses the importance of mathematics in society, and since importance is a likely candidate of anything which plays a role in society, we cite:

> Mathematical thinking is important for all members of a modern society as a habit of mind for its use in the workplace, business and finance, and for personal decision-making. Mathematics is fundamental to national prosperity in providing tools for understanding science, engineering, technology and economics. It is essential in public decision-making and for participation in the knowledge economy.

The same source deals a few lines later with other aspects of mathematics, namely, creativity, internationality and prestige:

> Mathematics is a creative discipline. The language of mathematics is international. The subject transcends cultural boundaries and its importance is universally recognised. Mathematics has developed over time as a means of solving problems and also for its own sake.

As much as mathematicians would agree with this description of mathematics, many would nevertheless doubt that this view could be seen as a generally accepted point of view in public opinion. And, to be fair, the UK-government source does not say this in the text. The quotations may therefore be interpreted as a well-chosen mission statement for mathematics as seen by UK scientists as well as, it seems, by the UK government.

It is comforting for mathematicians to know that several great philosophers would not only support this statement, but go further, such as for instance Plato (427–347). We know that Plato, seen by many as one of the greatest philosophers of all times, was a mathematician himself. It may be less known how far his appreciation of mathematics really went and how rigorously he defended the overall importance of mathematics. After having studied with the Pythagoreans Plato founded the Academy of Athens where he lectured for the rest of his life, and where mathematics played a central role. He focused on the foundations of mathematics. In particular the seventh book of *The Republic* [12] makes it clear that we should consider him as a second-to-none advocate of mathematics in every aspect. He stresses the necessity of rigour and precision and underlines the importance of encouraging and inspiring people to study mathematics in a remarkably strong way.

One may wonder why Plato was such a fervent advocate of mathematics. The answer is in his seventh book of The Republic [Sect. 4]. The ultimate goal of philosophy should be *the good*. In his eyes there is no alternative to mathematics for approaching the goal of truth, a necessary condition to get access to *the good*. Plato writes:

> In the world of knowledge the idea of good appears last of all, and is seen only with an effort; and, when seen, is also inferred to be the universal author of all things beautiful and

right … and (for) attaining the idea of good, one has to study arithmetic and geometry. Arithmetic has a very great and elevating effect, compelling the soul to reason about abstract number and repelling against the introduction of visible or tangible objects into the argument. Those who have a natural talent for calculation are generally quick at every other kind of knowledge. [12, The Republic, Sect. 4, pp. 179–188]

With a somewhat surprising optimism Plato continues:

Even the dull, if they have had an arithmetical training, …, always become much quicker than they would otherwise have been; anyone who has studied geometry is infinitely quicker of apprehension than one who has not. [Sect. 4, pp. 188–189]

Let us try to balance the point of view. In Sect. 2 we add to Plato's flattering statements and to the UK mission statement by looking at some indicators of the importance of our discipline.

Independent Indicators of the Importance of Mathematics

We have started with a few somewhat discouraging answers. Nevertheless, we know of course much more. The higher the degree of education, the clearer becomes the perception of the role of mathematics. Here we meet in particular those people who regret not having been good at mathematics. We sometimes hear reactions like "I would have loved to study astronomy (engineering, physics) but I was simply not good enough at mathematics. And there was nothing I could do about this".

It is true, mathematics is not a subject, it is a discipline. It needs not only knowledge but also a lot of training and practise. A minimum of mathematical talent is needed to succeed in doing something however small in mathematics, even simple problems. Moreover, seen as a discipline, mathematics is a highly distinguished one.

What this distinction means is probably nowhere in Europe more evident than in France. The top grandes écoles of engineering or science give mathematics traditionally the greatest weight during entrance exams. Distinction means distinction by competition, in a *concours*, as the French say. Ability in mathematics and solid specific training provide the best to the difference between competitors and to bring the bright candidates to the top of the list. In order to get entrance to the most prestigious schools where everybody wants to go, getting to the top of the list for admission is a necessary condition. Seen that way it is no real surprise that France, normalised by the number of inhabitants, is the world leader for Fields Medal awards as again confirmed by the recent announcement at the International Congress of Mathematicians in Hyderabad. In Great Britain, Germany, Italy and I think in most other European countries, good marks in mathematics are also important for the top universities, although in general this is more confined to admissions in mathematics or physics.

Now, there are of course prestigious schools everywhere in the world, which pay less importance to mathematics. Some of them do not even test the mathematical skills of candidates. Nevertheless, it is safe to say that good marks in mathematics, whenever they appear in the results of entrance exams, or are visible otherwise, are almost nowhere ignored and usually taken seriously.

Other well-noticed distinctions for the discipline of mathematics should be mentioned. The national as well as the international mathematics Olympics for instance are very informative. People who look, year after year, at data from students who have done very well in mathematics, can see what is going on. These competitions provide large sets of data, and the statistical inference from these data is clear: there is no better predictor of brightness measured in terms of high marks in all subjects than high marks in mathematics. Recall that this is very coherent with what Plato said some 2400 years ago.

Consequently, the story of the boy or girl who is brilliant in mathematics in school but "that's all" is in general not true at all. Nevertheless, the prejudice of seeing mathematics as a single-minded talent seems to survive in most countries. We note that a prejudice is like a saga: often enough we neither know where it comes from nor whether it contains much truth. Also, it is easy to find explanations why certain prejudices seem to survive forever. We should say, they are not really surviving, they are rather re-born all the time. And why are stories about a specific subject or a specific talent re-born all the time? Psychologists refer to similar cases, which usually have nothing to do with mathematics. Typically they would advise us to keep for such cases jealousy in mind. Whether this is true or not for mathematics, we cannot say, although we cannot exclude it. However, knowing this may be helpful as a directive for raising public awareness (RPA) activities, we may ask good statisticians to try to find out.

Apart from distinction, and being a convenient measure for admission competitions, mathematics has many other faces. Some of them are a long way from giving rise to any kind of jealousy. They must be called weak points, and we should discuss the important ones.

In certain aspects, mathematics really has to struggle, and the more one is interested in raising the public awareness of mathematics, the more likely one feels this. One of these is when the question of how to "sell" mathematics arises, because politicians and deans increasingly expect us to do so. Seemingly it is not easy to sell mathematics. Somewhat loosely we may say: Those who are good at mathematics need not buy it. Those who are weak at mathematics have little incentive to buy it. People of the second group often reason that time is money, hence they do not waste time on things they do not understand. And then they ask: What is it good for? Consequently, between the good ones and the weak ones at mathematics there is little instigation left for communication, and less for discussion.

Another difficulty for our discipline is a problem which we should call, more than anything else, a language problem. Mathematics is a language with words, semantics and grammar. Those who have not learned the language cannot understand. This implies a considerable bias against mathematics. A mathematician can read in English, say, a book about international law. He/she may not be familiar with all the specific terminology but may guess their meaning in the context, just as an economist, social scientist or engineer would guess from the context. Moreover, he or she will understand the main arguments on which the conclusions are based. In particular, the mathematician will then be able to discuss the arguments with a specialist in international law. The other way round, things are usually very different. If the specialist in international law does not have sufficient education in the language of mathematics, he or she would be lost with most books written in English about mathematics. This reminds me with a smile of the complaint I once heard about the inaccessibility of mathematics: "I know only two sorts of mathematics books. Those where I am lost after the first page, and those where I am lost after the first line."

We often hear from non-mathematicians that it is easier to remember from school a language like French, say, than the language of mathematics, and this seems true. This shows again that mathematics is a discipline rather than a subject. The asymmetry is dangerous for us, because people have a tendency to mistrust what they do not understand or cannot manage. Mathematicians have to live with this regrettable situation. Popular science publications in mathematics certainly do help, but we know that they can only solve a part of the problem.

Before discussing further weak points of our subject which will turn up naturally in later discussions, let us now discuss other important and more enjoyable aspects of mathematics. It is what we may call the joy of mathematics but what we usually call enjoying mathematics.

The Enjoyment of Mathematics

Why do mathematicians or others with mathematical ability usually enjoy doing mathematics? If we try to analyse the image of mathematics or mathematicians we cannot do so without trying to understand what people think is the reason we do mathematics. If we could say *it is for the money* this would be easy because people usually understand that one is doing something for money. The problem is that this is not the full truth. If we turned our ability for mathematics into a degree in engineering or economics, say, many of us would probably earn more, and educated people know this. So we run a risk of being considered as somewhat weird if we do not clarify that we enjoy mathematics.

If I were asked to give summarising keywords for the reason why we enjoy mathematics, I would hint at some sort of hunting fervour, to the enjoyment of seeing the essence of a problem by reduction, the satisfaction of finding the solution and to the satisfaction of achieving completeness. It makes no sense to try to give a complete list. Several colleagues would want to add items, and others would use different names for the same thing.

We find some answers in good books about mathematics or mathematicians. For books by famous mathematicians I think in particular of the lovely *Littlewood's miscellany* [9], or of *A mathematician's apology* by George Hardy [8], although the latter is not a story with a happy ending. Another well-known book about enjoying mathematics and being a mathematician is Stan Ulam's *Adventures of a mathematician* [14]. It contains implicit statements of his enjoyment experienced in his adventures. However, as far as I remember, Ulam never tried to make things explicit, that is, to make the enjoyment really comprehensible and seizable. Unlike Ulam, Fields medallist Timothy Gowers (Cambridge) puts real effort into this. In his article on the importance of mathematics [7], the examples are tailored to make both the enjoyment and excitement for mathematics seizable. A direct way to sense the joy and enjoyment of mathematics is of course a hands-on approach, that is reading a good book concentrating on ideas in mathematics, as for instance the book by Courant and Robbins [6], or alternatively, a collection of interesting science communication articles in mathematics, as e.g. in Aigner and Behrends [1].

From my own experience I would like to add a few items, which I have learned from or experienced with a few admired colleagues.

Jean Mawhin (Université Catholique de Louvain) does not speak of hunting fervour but makes another interesting comparison [10]. He argues that the work of a mathematician is very much like the work of a gold-digger. Hope brings him to a river where he starts rinsing (looking for structure). After some time little grains may show up, and, at the same time, excitement. Now he becomes eager to examine the contents very carefully, finding finally perhaps a gold nugget (interesting result). Looking backwards, he realises that each step of the search was important and equally enjoyable.

Seeing the essence through translation is an aspect of the enjoyment of mathematics which may not hit the eyes, but if one perceives it, it is impressive. The celebrated J.W. von Goethe is reputed to have said "Mathematicians are a sort of Frenchmen. Once you talk to them they translate everything in their language and then it becomes something completely different." If so, Goethe may have found no better proof than Freddy Delbaen (ETH Zürich) with whom I have the honour to collaborate. His quick translation ends up so *pure* that it surprises even mathematicians. He has the ability to make astonishing use of this translation by working with surprising speed in assembler language rather than "soft macros". This is great talent, of course. The speed of translation itself, however, can only be explained by the excitement felt in the act of translating. It must be the joy of seeing the power of the right language in terms of rigour.

A third aspect, probably more common, is the joy of completeness, i.e. the joy of getting complete results in a field or an interesting sub-field. Thomas Ferguson (UCLA) once told me: "When a problem is a game or can be seen as a game, I feel I *must* think about it." And Ferguson does this with much success, solving many hard problems in games and optimal stopping.

Staying faithful to particularly hard problems is also a symptom of the search for completeness. If it does not work out, or has not worked out yet, it can become, as many of us know (including Ferguson and myself), a time-consuming obsession. There is no need to cite here famous problems like the Riemann hypothesis or the Goldbach conjecture. Most of us have our own experience with problems that are, from the individual point of view, equally important.

Responsibility of Mathematicians

This section consists of a few arguments taken from my recent article on the self-assertion of mathematicians written for the European Mathematical society [3] because they should figure here.

My former student Yvik Swan once asked me to explain the difference between a introverted mathematician and a self-assured mathematician. I was not sure what to answer but I certainly enjoyed Yvik's description: "the latter would look on *your* shoes when he talks to you." We know that this is in general not true. But isn't it most interesting? Even though we know this is not true we mathematicians still enjoy stereotype descriptions of our breed by others. Is it only because we know better?

Having said this, many people think mathematicians should be more open to communicate what they are doing. Communication of what a mathematician is doing is sometimes difficult as we well know. Nevertheless, some of us feel, we should be more willing to advertise what we *think* whenever we believe there is a very good reason to do so.

When a medical doctor we meet for the first time tells us right away that he or she was always very bad in mathematics, I don't think we should say that we have heard this from others. Still worse would be adding that we know others who were very bad in mathematics and still succeeded in their lives. I think a better answer is: "Oh, that is very sad!" Indeed we should point out the truth without compromise, that is, being very bad in mathematics is very likely to impair his or her medical decisions. Elementary logic is a must for medical doctors, and failing in elementary logic is one of the problems with doctors who say that they were a zero in mathematics. Typically, the immediate reaction of the surprised doctor is to say that they are of course very "logical" and that they simply wanted to say they could not do the other stuff with the x's or $f(x)$'s and so on. Realising that we doubt their common sense logic, annoyance will show up on the doctor's face. Still, I think we should not compromise. Polite and merciless we should insist on our sadness by telling the further truth: unfortunately, not knowing how to handle x's or $f(x)$'s at high school is highly correlated with failures in elementary logic throughout.

A physician who cannot negate correctly a statement containing one or two implications or one or two *and-or* attributes could do harm to a patient. The same holds for the problem of using correctly our beloved equivalence of *A implies B* and *not-B implies not-A*. Mathematicians use the latter mostly in the more convenient form *A and not-B implies a contradiction* for indirect proofs, and both forms are important for everyday decisions.

Another typical and sometimes dangerous failure is with elementary probabilistic or statistical reasoning, in particular when involving conditional probabilities. The example of 75 % completely wrong answers in a test of Bayes' formula at Harvard medical school, arguably one of the most famous medical schools in the world, shows how serious the problem can be (see e.g. [5] (1978)). This seems long ago, but we should not be mistaken. Many colleagues confirm that things have not improved.

I dare to say that no doctor who proudly advertises his mathematical inabilities should be trusted in important medical decisions, as for instance analysing a multi-target blood test and then proposing a treatment, or speaking about threshold values but having no idea about their variance. Errors are likely, and we should point this out. If we leave the doctor's surgery politely on the basis of such a conversation, he or she will remember. This may be beneficial for society, and certainly for the good of the due recognition of our subject.

It would be unfair to exemplify medical doctors without mentioning other examples which come easily to mind, and I shall give a second one. We had a round-table discussion with a member of the European Commission who spoke, shortly before the introduction of the euro, about the laws and recommended rules that would make the European Central Bank an institution at least as reliable as the Deutsche Bundesbank and hence the euro at least as strong a currency as the Deutsche Mark. His arguments were: "if you increase the size of an institution (central bank) by a factor c you increase control by the same factor c to maintain the same level of safety" and assured us that the European Union has (of course) the means to do so. I was not sure how an increase of control was supposed to be measured, quite apart from the fact that in the European Central Bank member states with national interests would all have their own views. So I asked why he could assume such a simple linear relationship between control efforts and safety. My question was hardly helpful for the continuing discussion but I am still glad that I had tried to draw attention to conclusions which seemed naive.

Mathematicians should, I think, react in similar cases. It is our responsibility to do so, both towards society and for the defence of mathematics. If we are interested in raising the public awareness of mathematics then the feeling of responsibility, and the required courage to stand up, and this in situations which may be sometimes delicate, is, I think, a must.

As much as we feel that this opinion may be shared by a considerable number of other mathematicians we can be less sure about our feelings when we pass on to the big issue of research.

Interest and Importance of Research

There is a lot of divergence among our peers in saying whether a research problem in mathematics is interesting and important. Moreover the divergence seems to be enhanced if the issue of possible applications intervenes. To see this we cite Gowers [7]:

> Most mathematicians, including me, lie somewhere in the middle of the spectrum, when it comes to our attitude to applications. We would be delighted if we proved a theorem that was found to be useful outside mathematics—but we do not actively seek to do so. Given the choice between an interesting but purely mathematical problem and an uninteresting problem of potential benefit to engineers, computer scientists or physicists, we will opt for the former, though we would certainly feel awkward if nobody worked on practical problems.

This statement raises questions. How can a problem of potential benefit to engineers, computer scientists or physicists be uninteresting? Mathematically uninteresting? Clearly Gowers is not thinking of trifling computations. But then, if the problem is (mathematically) difficult and useful to engineers, computer scientists or physicists how can it be called an uninteresting problem? Is this inconsistent?

Gowers continues by shedding some more light:

> Actually, this attitude is held even by many of those who work in more practical-seeming areas. If you press such a person, asking for a specific example of an application in business, industry or science of their own work as opposed to an application of a result in their general area, you will often, though not invariably, witness an uncomfortable reaction. It turns out ... this phenomenon is a natural and desirable consequence of what it means to view the world mathematically.

And now things become clear. There is no misunderstanding and no enigma of inconsistency either. Existence statements on a set may become wrong on subsets, of course. Gowers sees the classification regarding the true interest of a problem confined to the *scope* of mathematics. If a problem is of great significance for applications, if the need for mathematics is apparent, if it demands moreover effort to bring it into a mathematically well-defined form, and even if it may require distinguished skills to make it mathematically tractable, all this is a priori not enough to call it *mathematically interesting.* We conclude that Gowers sees no entrance ticket for getting into the world of mathematical interest. The interest should be apparent beforehand in the world of mathematics right in its own. Gowers seems to leave little room for a-posteriori justifications of mathematical interest.

The point I want to make here is that the perception of the importance in research can be very different, even among mathematicians. Personally I admit that Gowers' statement is not an easy one to accept. For the few of my own results of which I am proud I must doubt they would pass Gowers' test of importance in mathematics. Still, I think that results that are not void of mathematical elegance and which, with respect to applications go further than originally expected (e.g. [13] and [15]) are arguably important.

Having said this, the divergence of perception in a specific domain (research) does not imply divergent perceptions for the whole discipline. Viewed mathematically, different perceptions are just different definitions. Indeed, I found no other instance of different perception in Gowers' thoughtful article. This supports the feeling that all mathematicians at all levels are bound to have much in common. Concerning our efforts in raising public awareness this means that mathematicians usually have a common basis for strategies, but we still have to think who should target whom to win people over for mathematics.

Back to the Selling Argument

In the introduction we argued that mathematics is difficult to sell to the outside world. But now, is this not true for any specialised subject?

It may be true to some extent for all subjects, but the phenomenon seems pronounced and clear-cut for mathematics. The most important reason I see is that most people have few chances to see the importance of mathematics in things or ideas they encounter every day. We all live with so many things around us: objects, devices and machines we need, tools we use and things we truly like. And then there are things we use and admire at the same time like computers, airplanes, and many others. Now, who would think about differential equations and simulation techniques when flying comfortably from one continent to another? Similarly, when a doctor looks up from the screen and says with a smile "It's a boy", the young mother may marvel the news and greatly praise the incredible progress of medicine. We excuse the fact that she seems to have forgotten physics and engineering. But which of us would dare to hope that she would exclaim "How wonderful are these Fourier transforms and deconvolution techniques!"

No great exception should be made here for us as mathematicians. How often would we think of fast algorithms when we Google Lebesgue's theorem, colleagues, movie stars or butterflies, how often would we think of routing optimisation and coding-decoding when we call somewhere in the world? The point is, we can all see and appreciate technology and slick design, we may guess at the use of radio waves, magnetism or other physical facts and tools, but we are essentially blind for everything that is one step removed. Only a few physicists and engineers still see the mathematics underneath, and know that without mathematics the technology would hardly work. We say a few, not all, because usually only those who create can see and understand. Not those who produce. For them it is too late, just like for us who simply consume. In the chain of creating a product or anything else of direct use, mathematics is at least one step too far away for almost everybody, and this step is a big jump because mathematics is hard for almost everybody. This is why mathematics has a problem.

Behrends [2] put it as concisely as one possibly can: "The public understanding of mathematics is in marked contrast to the importance of this science in society." Mathematicians understand well why this is so. Do we have to live with this? Some of us are afraid that it may be so forever. People who are active in RPA can flatter themselves that they are trying to do something about it.

Changing the Perception of Mathematics

If nothing changes, it is indeed likely that the marked contrast will survive forever, much to the distress of people who try their best to communicate mathematics.

However, things may change indeed. Probably not so because the interest in mathematics within society may change through a reversal of the attitude of mathematicians. More likely, it will be the necessity for society to change the focus of its objectives which will make the difference. What we mean by focus is the type of progress which society will have to recognise. More than ever before we have to worry about the environment, about nuclear energy, about burning fossils and climate change, about renewable energy, about water scarcity and new methods of growing food, about poisoning, and, last but not least, about a drastic overpopulation of the earth. Solutions which, in our industrialised world, were able to mature to a state of near perfection, sometimes over decades or even centuries, are suddenly seen in a different light.

We have to worry about alternative solutions and thus have to look for means to lay out a targeted search, that is, to formulate search strategies for alternatives. Becoming aware stands at the beginning. Almost every day we are confronted with instances of becoming aware. We see a beautiful car, a powerful plane, and we wonder what all this is doing to the environment. And then we hear of flight offers that seem much too cheap, or see golf-course lawns that look much too green, and again we have second thoughts.

The notion of a solution becomes shaky once we realise we have to look for alternative solutions. By definition, an alternative solution is a solution found from a reduced set of possibilities. The more restrictions we have, typically the harder the new optimisation problem. It is therefore likely that sophisticated optimisation techniques will become increasingly important. But we can easily see more. Alternatives need a feasibility study, which comes with a multitude of problems for planning. New mathematical and statistical models of all kind are needed, simply because so many new questions will naturally pop up everywhere and lead, stochastically, to other new questions. And almost everything new will have to be tested or simulated. We conclude that mathematical modelling, deterministic for some problems, stochastic modelling for others, may be among the frontrunners in the future of mathematics, possibly with numerical methods and simulation techniques as runners-up.

Being needed or being useful does not mean being in demand. Hence the problem of selling mathematics to those who can benefit from our subject may remain more or less the same, and the danger that mathematicians will miss the bus is not negligible. Leaving this point aside one can be confident that mathematics will be able to deliver, exactly as other sciences will deliver. For mathematics we should cautiously say "deliver at the end" because we all know that the time when a mathematical tool will have a real impact on applications is often enough unpredictable.

This brings us to another point. One may get the impression that the arguments given so far are somewhat confined to applied mathematics. They are not. It is true that we argued in terms of possible applications of mathematics rather than of pure mathematics since the word application fits better with the idea of selling. However, most mathematicians see no important difference between pure and applied, as many of us know. Indeed, if we insist on a difference between pure mathematics and applied mathematics then this is rather a difference of degree than a difference of nature. Mathematics is mathematics. With respect to applications the difference between pure and applied mathematics is just the time until a new theory (or a very theoretical result in an established field) is used, which seems in general longer for theoretical results than for tailored mathematical tools.

Still, any evaluation of the importance of a result for applications depends on the result being really used one day. And this is not evident at all, whether it is, on the one hand, a very theoretical result, or, on the other, an applied piece of mathematics that was tailored for immediate application. Only time will tell. Therefore people are allowed to have their opinions but should not normally judge how much interest there is in the mathematical research of others. An exception would be if they are specifically asked to give their opinion, such as referee reports for which we may have no real alternatives.

I remember something I once read as a student. A mathematically inclined biologist wrote: "It is my wish that mathematics will be able to achieve for biology what it has achieved for physics." I was somewhat impressed by this sentence taking it as a sign of deep insight, an insight which I did not have. The author seemed to predict the fundamental use of mathematical tools in his subject, so I gathered he must see new fundamental links between biology and mathematics. Today I think I understand better what he meant, and I dare to extrapolate: it should be our wish that mathematics will be able to achieve for the future challenge of alternative solutions what it has achieved so far on more established grounds of the task of finding *a* solution. There will be ample challenges.

Challenges often turn into opportunities, and opportunities give rise to optimism. Hence we can be optimistic for the role of mathematics. First of all, just by conservative extrapolation and without much further ado, we may reason that mathematics is an old discipline. It has always, or at least during very long periods, been considered to be important, otherwise it would not have survived as a major subject. Furthermore, it is currently considered to be important. Hence it is probable it will continue to play an important role in future. However, by our earlier arguments, we can hazard that its role will be more important than ever before. Again we can argue by the incredible number of new problems, which we will meet in the future, and which will, often enough, come up almost concurrently. Being a probabilist I dare to suggest that the number of new problems and hence mathematical problems will almost surely be incomparably higher than anything we have seen before. And the task of raising the public awareness of mathematics should become easier.

Having risked a prediction I can comfortably shield myself behind the wonderful wit of J. Michael Steele (Wharton School). Steele was speaking in Leipzig about interesting inequalities and announced about one of his preferred ones: "I pay you one hundred thousand dollars in cash if you can prove that you won't be able to use this inequality within the next hundred and fifty years!" Mathematicians are allowed to predict.

Mathematics for Kings and Ambassadors

It is a fair guess that few of us will ever prove one of these world-famous mathematical conjectures or, alternatively, develop a theory everybody will use forever. Having said this, each of us can contribute in an essential way. If we understand mathematics, we can communicate, that is, we can help others to understand and to use mathematics. We can become ambassadors of something, and for something, that really counts. Einstein reportedly said "Not everything which can be counted, counts, and not everything which counts, can be counted." What counts is that the statement is clearly true; whether it is really due to Einstein does not count. In the same vein we can argue: It would be great if we could prove in our life an important piece of mathematics! But nobody will count (after promotion), and nobody can really count what we do. It is important and honourable that we can admire major advances in our discipline and the people who achieved them; promoting these in one way or other (e.g. [4]) is an equally important task every mathematician may assume.

Great mathematicians should be seen as kings, glorious by definition but bleak without a court, and weak without good ambassadors who bring the message to others. No king can be powerful for long unless he has good ambassadors. This is what people who enjoy mathematics and want to serve mathematics should keep in mind.

What is a good ambassador? Ambassadors, like other professionals, come from different moulds. The worst are those who only work for their own interest. Then there are those who represent their country well but limit their activity to representing their king, meaning their country or the interest of the head of state or government. By kings we mean kings, dukes, counts, etc., or in the terminology of mathematicians, popes. But then there are good ambassadors who try their best in combining the interests of both the country they represent and the host country or countries, and some of these are truly excellent. They understand the interests of both sides better than anyone else and have honest intentions so that all sides benefit. Often enough, ambassadors speak or write better than the kings. Often enough, they are more convincing, and in many ways more successful. Still they know their place, and enjoy their place. Knowing their kings can do things they cannot do, these good ambassadors will serve them and their country and the people with great respect.

The kingdom we have in mind is the large kingdom of mathematics.

Conclusions

The first conclusion for RPA is that we may strengthen the enthusiasm of young people with an interest in mathematics by saying: firstly, if your mathematical skills get you to one of the top universities in your country, then this is very nice and probably very helpful for your career. But it is not the only or most important thing mathematics can offer.

Secondly, we can argue that mathematics is incredibly versatile, always open and adaptable to new problems of very different kinds.

Thirdly, mathematics is in many ways a subject of eternal youth, and I think RPA authors should point this out with great emphasis. Perhaps it is this eternal youth which explains also the following fact:

A larger statistical study published in the Wall Street Journal Europe (2009, see Needleman [11]) listed, among forty academic professions, mathematician as the profession with the top rate of satisfaction and fulfilment in life. This result came as a surprise to many, except, it seems, to mathematicians. Indeed, in the coffee room of our department, we were pleasantly surprised to read the study. When we discussed it we all agreed on its essence. Hence we were no longer really surprised. (By the way, in second place for professional satisfaction and life fulfilment were statisticians. From a Bayesian point of view this is the best proof that statisticians are mathematicians!)

Satisfaction and professional fulfilment in life are important, and a good reason why everybody with some talent in mathematics should be warmly encouraged to nourish his or her interest.

Final Comment

Undaunted by the fact that different perceptions may influence our attitudes, and thus our behaviour in defending the interests of mathematics, we should know our subject and defend it. Mathematics is worth being defended as a means to tell the difference between false and true, and much more. Therefore mathematicians must make their point as everybody should make his point in his field when seeing or hearing arguments which are shaky or even bluntly wrong.

Unfortunately, many colleagues do not do this. They rather prefer to confine themselves to telling wrong or naive arguments heard at events or discussions of importance just as a joke to another colleague. Why not react where it counts? Is it modesty? Is it politeness? Lack of courage? I do not know. But if some people summarise this behaviour as a lack of self-assurance of mathematicians it is hard to refute this reproach. Hence RPA ambassadors may suggest to teach how to react with adequate courage.

The author is aware that many people with interest in RPA activities for mathematics make wonderful contributions in very different directions for our common goal. They should be felicitated by the mathematical community for what they achieve. Some of these contributions are in important directions not discussed in this article, some others are in directions addressed here or at least related with our discussion.

Acknowledgements   I would like to thank Ehrhard Behrends (FU Berlin), Nuno Crato (U. Lisbon) and José Francisco Rodrigues (U. Lisbon) for their invitation to contribute to this book and for several interesting discussions. My thanks go also to Marcel Berger (corr. Acad. sc., Paris), M. Chaleyat-Maurel (U. Paris 5), Maria Dedo (U. Milan), Marc Yor (U. Paris 6), and Günter Ziegler (FU Berlin) for their comments on different sections.

References

1. Aigner, M., Behrends, E. (Her.geb.): Alles Mathematik—Von Pythagoras zum CD-Player. Vieweg, Wiesbaden (2008)
2. Behrends, E.: Private communication (2010)

3. Bruss, F.T.: Mathematicians' selfconfidence and responsibility. Newsl. - Eur. Math. Soc. **79**, 24–27 (2011)
4. Bruss, F.T.: Abel in Holland and a few reflections on the Abel Prize. Newsl. - Eur. Math. Soc. **80**, 20–21 (2011)
5. Casscells, W., Schoenberger, A., Graboys, T.B.: Interpretation by physicians of clinical laboratory results. N. Engl. J. Med. **299**(18), 999–1001 (1978)
6. Courant, R., Robbins, H.: What Is Mathematics? Oxford University Press, Oxford (1978)
7. Gowers, W.T.: The importance of mathematics. Millennium 2000 lecture, Collège de France. Available on www.dpmms.cam.ac.uk/~wtg10/importance
8. Hardy, G.H.: A Mathematician's Apology. Cambridge Univ. Press, Cambridge (1940)
9. Bollobás, B. (ed.): Littlewood's Miscellany. Cambridge University Press, Cambridge (1990)
10. Mawhin, J., Nirenberg, L., Schmidt, K.: The joy of differential equations. Electron. J. Differ. Equations, Conf. **15**, 221–228 (2007)
11. Needleman, S.E.: Doing the math to find the good jobs. Wall Street Journal, 9th January (2009)
12. Plato: The Republic. Dover, New York (2000). Translated by B. Jowett
13. Thomas, E., Levrat, E., Iung, B.: L'Algorithme de Bruss comme contribution à une maintenance préventive opportuniste. Sci. Technol. l'Automat. **4**(3), 13–18 (2007)
14. Ulam, S.: Adventures of a Mathematician. University of California Press Berkeley, Los Angeles (1991)
15. Bruss, F.T.: Sum the odds to one and stop. An. Probab. **28**, 1384–1391 (2000)

The Importance of Useful Mathematics: On Tools for Its Popularization, from Industry to Art

Jorge Buescu and José Francisco Rodrigues

Abstract Written popularization of mathematical sciences may have had its roots in Euler's "Lettres à une Princesse d'Allemagne", but have seen during the last decades an enormous increase of publications in newspapers, magazines and books. In this article we review some issues arising from our direct experience, on one hand, related to the connections of mathematics and art in the recreation of a virtual exhibition in Portugal during the WMY2000 and, on the other hand, in relation with a series of popularization articles for the magazine of the Portuguese engineering association. Concerning written popularization, we provide examples such as a natural logistic problem leading to optimization of functions without derivatives with counterintuitive results and to the use of check digit identification schemes in ID cards and in Euro banknotes, which correspond to mathematical problems with a "useful" nature or directly tangible meaning with a significant success of social communication. Illustrating the recent impact of elliptic curves in cryptography we describe the origin and the making of a recent short digital movie of mathematics and art, promoted by the "Centro Internacional de Matemática", in collaboration with "Imaginary".

Introduction

Although the role of mathematical sciences in civilization has been of central importance for centuries, the current trend to a global economy and a knowledge society has made information and innovation technologies increasingly dependent on scientific research, whose results and techniques are underpinned and driven by mathematics. This was recognized in a recent report on "Mathematics and Industry" by the

J. Buescu (✉)
FCUL (Faculdade de Ciências da Universidade de Lisboa), Departamento de Matemática, Lisbon, Portugal
e-mail: jbuescu@ptmat.fc.ul.pt

J.F. Rodrigues
FCUL/Centro de Matemática e Aplicações Fundamentais da Universidade de Lisboa, Lisbon, Portugal
e-mail: rodrigue@fc.ul.pt

E. Behrends et al. (eds.), *Raising Public Awareness of Mathematics*,
DOI 10.1007/978-3-642-25710-0_23, © Springer-Verlag Berlin Heidelberg 2012

Global Science Forum (GSF) of the Organization for Economic Co-operation and Development (OECD) [1]. Mathematics (or the mathematical sciences, including statistics and computing) is nowadays considered in its broadest sense, and industry is interpreted as "any activity of economic or social value, including the service industry, regardless of whether it is in the public or private sector".

The European Science Foundation, together with the European Mathematical Society, has promoted a Forward Look initiative on "Mathematics and Industry". The final document contains several observations and recommendations for policy makers and funding organizations, for mathematical societies and academic institutions. It refers, in particular, to the lack of recognition of mathematics by industry and the lack of recognition of industrial mathematics by the mathematical community: "It is a common interest of the entire mathematical community to outreach activities to make society and industry aware that mathematics is the common denominator of much that goes on in everyday life, activating the many sectors of society that can benefit from mathematics. Indeed, promoting such awareness will bring resources to all mathematicians" [2].

It has become clear that problems with a significant mathematical content above the more or less trivial must be supplied by mathematicians or scientists willing to invest the necessary time, work and effort. Mathematics changes. Technology changes. Innovation occurs, and mathematics is a crucial part of it. In fact, mathematics is a key and enabling technology not only for the other sciences but also for industry and society, as it "provides a logically coherent framework and a universal language for the analysis, optimization, and control of industrial processes". The use of mathematics in technology, whether in cell phones, DVDs or ATM machines, etc. changes constantly. It seems clear that the dissemination of emerging applications can only be made by those in the know. In hindsight, it is obvious that it is the mathematical community who should have the energy to make the outreach effort. Even if the task of raising public awareness in research and academic institutions will always be relatively small, it should be present as one of their essential responsibilities. Moreover those institutions should play an active role in promoting the change of attitude of mathematicians towards public awareness of their science, as well.

Popularization of Mathematics

Among current trends for the old and difficult process of the popularization of science, mathematics, in all its scope, is not only the one with the most recent activity but it also has special features, not only because of its role in the general education process but also because of its history and applications that are a natural link between the humanities and technology.

In contrast with the usual activity of mathematicians when they are creating new mathematics or when they are teaching it to students, a process that it is sometime referred as "doing mathematics", the popularization of mathematics is a very different

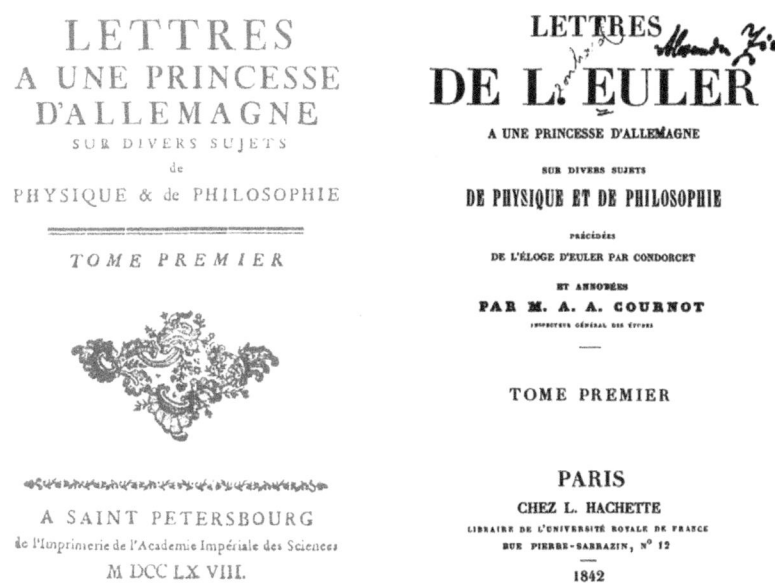

Fig. 1 Cover of the first French edition (1768) and one with the *Éloge* by Condorcet (1842) of Euler's *Lettres*

activity associated with talking, writing or just communicating "about mathematics".

One of the first successful examples was the letters by Euler on different subjects in natural philosophy addressed to a German princess [3] which were published in several editions in French and in English (see Fig. 1). The first letter, written from Berlin on 19th April 1760, concerns "Of Magnitude or extension" and is followed, three days later, by a second one with the aim "to unfold the idea of velocity, which is a particular species of extension, and susceptible of increase and of diminution". They may be considered a masterpiece of science popularization of the Enlightenment. As Condorcet wrote at the end of the 18th century, "they have deservedly been considered as a treasury of science" since "they convey accurate ideas respecting a variety of objects, highly interesting in themselves, or excite a laudable curiosity".

Another well-known example is the story of the Phoenician princess Dido and the foundation of the city of Carthage. This example of the maximization of the area that could be enclosed by an oxhide is quoted in most introductory courses on the calculus of variations as the oldest applied isoperimetric problem, solved by Dido in an empirical form. It is however less known that this story was popularized by the British mathematical-physicist William Thomson (Lord Kelvin) in a conference at the Royal Institution of London on 12th May 1893 [4].

In Portugal, the beginning of the popularization of science may also be traced in the Enlightenment with publications such as the *Jornal Enciclopedico Dedicado á Rainha*, a periodical "aimed to the general instruction with news about discoveries in all sciences and arts" that started publication in Lisbon in 1779 and had about

481 subscribers in 1790 [5]. In the 20th century the mathematician Bento de Jesus Caraça (1901–1948) was one of the pioneers of the modern "divulgação e difusão" of scientific knowledge in the Portuguese language to the general public. In 1933, in a conference on "The Whole Culture of the Individual, central problem of our times" [6], Caraça, recognizing the role and stating his confidence in the progress of science and its application to the well-being of mankind, advocated that as part of culture it cannot be the monopoly of the elites but should belong to the whole of society and contribute to the individual culture of its citizens. He himself contributed to the diffusion of mathematical ideas with an original exposition on the "Fundamental Concepts of Mathematics" [7], published in a collection of books on scientific popularization he had founded in Lisbon, the first of such created in Portugal.

Books, magazines and even newspapers, spreading the written and spoken word are the traditional media with a rich and long historical tradition of the popularization of science and, in particular, of mathematics. They are very powerful and may serve several purposes and different publics, but are far from being the only media for sharing mathematics with society at large. Another important means of popularizing mathematics is by exploring its relations and interactions with art.

Aesthetic and Mathematics

The relationship between mathematics and art has deep roots in many old and primitive cultures and has greatly influenced them. Since the time of the Hellenic civilization to the present, with particularly important moments during the Renaissance and at the dawn of the 21st century, this relationship cannot be neglected. In classical philosophy this relation was clearly established by Aristotle: "Those who assert that the mathematical sciences say nothing of the beautiful are in error. The chief forms of beauty are order, commensurability and precision"[8]. The very interesting example of the Golden Ratio in the Parthenon is well known, although it has been the subject of several controversies [9].

Mathematicians have used pictures to convey mathematical ideas for millennia and they have been instrumental in the creation of shapes, forms and relationships. Visualization is nowadays the subject of many studies and current research (see, for instance, [10] and [11]). The sculpture *Genesis* (Fig. 2), by the British sculptor John Robinson, was chosen as the logo of the project RPAMath, launched by the European Mathematical Society in the WMY2000. *Genesis* is composed of three rhombuses representing the Borromean rings, the hyperbolic knot that was adopted in 2006 as the International Mathematical Union logo. According to its creator, *Genesis* was inspired by Celtic culture and is a symbol of "The never ending renewal of life" [12]. Copies of this sculpture are at the University of Wales, Bangor, and at the front of the Isaac Newton Institute of Mathematical Sciences, Cambridge, UK.

On the other hand, the conscious or unconscious use of mathematical ideas to create art has been used by several artists. A remarkable contemporary example

Fig. 2 *Genesis*, by John Robinson (University of Wales, Bangor, UK, and the RPAMath logo in 2000)

is the *Umbilic Torus NC*, by Helaman Ferguson, which was displayed at the ACM SIGGRAPH show on computer art in 1989. Ferguson is an American mathematician and in 1977 was the co-author, with R. Forcade, of an integer relation detection algorithm, which generalizes the Euclidean algorithm and was included as one of the top ten algorithms of the 20th century by a 2000 survey of the American Institute of Physics and the IEEE Computer Society [13]. In his textured bronze sculpture *Umbilic Torus NC*, Ferguson has used two mathematical ideas: one for the body of the piece, using real binary cubic forms parameterized by four coefficients, a four-dimensional group action provided the umbilicus options; another for the surface texture, inspired by Hilbert's version of Peano's surface-filling curve. As the mathematician-sculptor himself explained, "I selectively previewed computer generated images of toroids of hypocycloid cross-sections, discarding and eliminating many possibilities to select the current. (. . .) My viewing computer graphics images as a sculptor brings a unique perspective." [14].

Another pioneer interaction between mathematics and art using computer graphics is the virtual gallery created by the mathematicians Thomas Banchoff and David Cervone in 1997, which was recreated for the WMY2000 as interactive itinerant exhibition *Beyond the Third Dimension* [3D]. In particular, the short movie *"In and Outside the Torus"* [15], which can be freely downloaded from cyberspace as an MPEG film or a GIF animation, may be considered as a masterpiece of digital art. Besides the quality of the visualization tools, which can be technologically improved, as was done in the 2000 recreation by *Arte Numérica* [16], the aesthetic value of this short film is in the animation of a three-dimensional projection of a rotating Clifford torus in four-space (Fig. 3). As the torus passes

Fig. 3 *In and Outside the Torus*, snapshots of the digital animations

through the projection point, its image in the three-space appears to extend to infinity, and it turns inside out: the region that was outside is now inside, and vice
versa [17].

The use of computer graphics in mathematics has increased, in particular, the number of mathematical videos and films, especially since the VideoMath Festival held at the International Congress of Mathematicians in Berlin, in 1998 [18]. A selection of sixteen short math films, the winners of the international competition held in Berlin ten years later during the German "Year of Mathematics", yields a contemporary survey of animated mathematical visualizations [19].

The concept of beauty in mathematics goes much beyond its direct relationship with the arts and is also a phenomenon described by mathematicians as "mathematical beauty" [20]. John von Neumann wrote about the discussion of the nature of their intellectual work: "The mathematician has a wide variety of fields to which he may turn, and he enjoys a very considerable freedom in what he does with them. To come to the decisive point: I think that it is correct to say that his criteria of selection, and also those of success, are mainly aesthetical. (...) These criteria are clearly those of any creative art, and the existence of some underlying empirical worldly motif in the background—often in a very remote background—overgrown by aestheticizing developments and followed into a multitude of labyrinthine variants—all this is much more akin to the atmosphere of art pure and simple than to that of the empirical sciences" [21].

Popularizing Mathematics with *Ingenium*

A long and direct personal experience of communicating mathematics in a popular way to the wider public, mainly through magazines, yielded a series of four books [22–25]. This experience began in earnest in 1995, with the invitation to write a regular column on the general subject of the popularization of science for the professional journal of the Portuguese engineering society, *Ingenium* ([26], see Fig. 4). These essays gradually morphed into short pieces, sometimes with a slight journalistic flavor, on mathematics proper.

Conceiving these essays was always a rewarding challenge to the first-named author. As opposed to strictly journalistic pieces, they had the enormous advantage of addressing a mathematically educated audience who were potentially very interested in learning more about mathematics, albeit in an informal setting. The questions addressed had a range, from Wiles's proof of Fermat's theorem to exotic n-spheres, from check digit schemes to the mathematics of sudoku, which would be impossible for a more generalist audience. The emphasis has been always to convey some meaningful mathematical content through "real-world" problems, which would appeal to an audience consisting mostly of engineers.

Reactions from the readers were extremely encouraging from the very beginning. Some columns were extremely popular; in one of them it was shown that the Portuguese ID card incorporates a check digit and its algorithm was explained, thus resolving wild urban myths then circulating about "the extra digit on the ID card". In another pair of columns, when the euro started circulating, the first-named author proposed finding the check digit scheme behind the Euro banknotes and enlisted

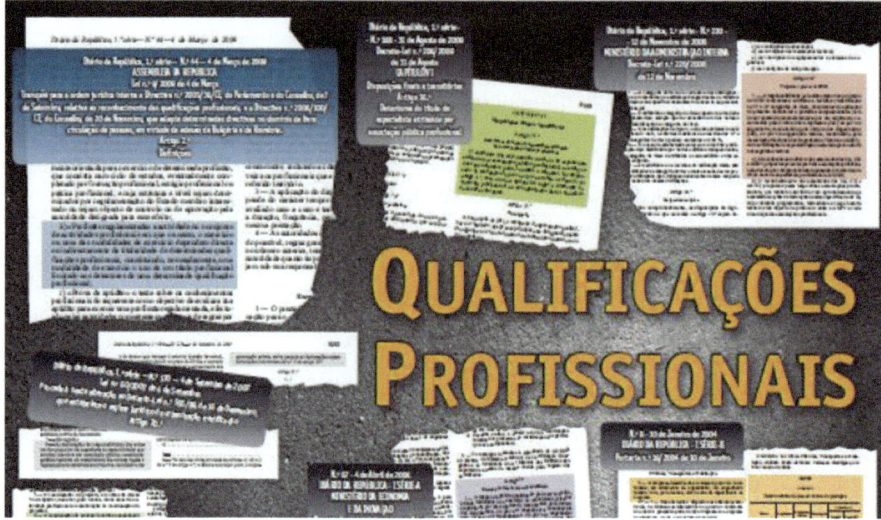

Fig. 4 Cover of a recent issue of the magazine *Ingenium*

readers to contribute euro banknote numbers. There were hundreds of replies. At some stage interest in these articles overflowed the intended audience. Clearly there was an interest in seeing mathematics in action in real world problems, in contexts where we would least expect to find it—whether in our ID cards, our euro banknotes, the sudoku puzzles in the paper, the graph of the Internet to coding theory and CD scratches. There were dozens of requests for authorization for classroom use by high school and college teachers. Many readers suggested that these essays should be published in book form.

So it was that in 2001 the first book in this series appeared [22]; its title was taken from the piece about the check digit of the ID card. It was a best-seller by Portuguese standards; there are currently 10 editions and over 10.000 copies have been sold. It was followed by three other books, in 2003 [23], 2007 [24] and 2011 [25]. The number of original papers in *Ingenium* is by now over one hundred. These books have had a large in the educational community, since they conveyed, in clear but rigorous language, examples of the use of mathematics to solve real-world problems. Examples of this effect in schools are many and, in some sense, unusual, as the following cases show:

1. Utilization of some of the problems described (ID card, Monty Hall problem, euro banknote . . .) in college exhibitions of mathematics (e.g. IPL Leiria, 2003).

2. Dozens of yearly and standing invitations to the first-named author to speak on mathematics in high schools, technical schools, colleges and universities all over the country.
3. Proliferation of Java applets in the web with the described algorithms, with reference.
4. Many high schools had students work on school assignments on some of the topics covered.

Also surprising is the fact that, due to the nature of the audience (engineers, mathematically knowledgeable but somehow retired from active mathematics, more interested in solving "practical" problems), some scientific (i.e. mathematical) collaborations have arisen. In this context it is worth mentioning the very rewarding interaction of the first-named author with Miguel Casquilho, an engineering professor at the Technical University of Lisbon, who is especially interested in the area of optimization. For some time, they both worked together in a problem of optimization, known as the *optimal facility location*, which in mathematical terms can be expressed as, given a set of m points in $\mathbf{R}^n$, finding a point which minimizes the sum of the simultaneous (Euclidean) distances. This is called the *space median*, which is sometimes (and wrongly) confused with the centroid. In fact, such a confusion led for instance to a wrong identification, by the census bureau of the USA, of the "population centre" between 1910 and 1930 [27].

A solution to the problem has obvious interest to an engineer thinking about logistics. Probing the literature, the problem was found to be extremely hard in the general case, and impossible to solve without numerical methods [27]. In the particular case of three points in the plane the problem admits an analytical solution, which is the Fermat-Weber point of the triangle with the points as vertices (as opposed to the centroid of the triangle). However, a few counterintuitive properties of this problem were found, and this resulted in a paper published in *Ingenium* (Fig. 5). A second example of this type of very fruitful interaction between a mathematician and an engineer was in a related optimization problem, which happened to be analytically solvable and gave rise to a joint mathematics paper [28].

There are two other especially significant episodes arising from these efforts for communicating mathematics through *Ingenium* that deserve a brief description.

The Case of the Portuguese ID Card and Check Digit Schemes

Sometime in the early 1990s government agencies decided to include an "extra" digit following the national ID card number, in an isolated box. Obviously, this was a check digit, in most likelihood a checksum digit. However, care was not taken to explain the reason for the new digit. As a consequence, by the late 1990s the wildest urban myths about this recently introduced extra digit were floating around. The most popular one was that it would represent the number of people with the same name as the card bearer. This urban myth was extremely widespread at the time (and

CRÓNICA

JORGE BUESCU
Professor na Faculdade de Ciências da Universidade de Lisboa

O mistério do armazém absorvido

Imagine o leitor que pretende construir um armazém para distribuição de bens por um certo número de cidades. Qual é a localização óptima do armazém, no sentido de minimizar os custos totais de distribuição?

O problema não tem (apenas) um interesse académico. Se o leitor trabalhar numa empresa de distribuição, quererá ter armazéns localizados de forma a minimizar os custos de transporte dos bens aos clientes. Se trabalhar em energia ou telecomunicações, quererá ter uma rede de cabos eléctricos ou fibra óptica com o menor comprimento total de rede, mas que sirva as necessidades de tráfego (que são diferentes para pontos diferentes: Lisboa não terá a mesma intensidade de tráfego do que, digamos, as Berlengas). Se for responsável pela construção de uma unidade fabril, quererá localizá-la de forma a minimizar a distância total aos clientes. Se quisermos localizar uma central de transportes (um aeroporto, por exemplo), um critério importante é minimizar a distância total aos centros populacionais servidos, ponderando cada um pela sua população.

Os exemplos multiplicam-se: saber resolver este problema pode, dependendo das circunstâncias, poupar milhões de euros. O seu interesse económico é claro.

Tentemos formalizar o problema. Temos um conjunto de n pontos no plano, $X_1,..., X_n$, com pesos $p_1,..., p_n$. Em geral, interessa permitir que os pesos associados de pontos distintos sejam diferentes: significa que o número de clientes em cada ponto de distribuição x_i é variável. Se eu estiver a construir um armazém para distribuição de leite na região de Lisboa e Vale do Tejo, a importância económica do ponto Lisboa é muito maior do que, digamos, do ponto Bombarral: vou fazer muito mais vezes a viagem para Lisboa e, portanto, interessa-me reflectir a cidade de Lisboa com um peso maior (talvez proporcional ao número de habitantes).

Se eu situar o armazém num dado ponto (a,b) do plano, o custo associado $C(a,b)$ será a soma, ponderada pelos pesos p_i, das distâncias dos pontos fixos $X_i = (x_i, y_i)$ ao ponto (a,b):

$$C(a,b) = \sum_{i=1}^{n} p_i \sqrt{(x_i - a)^2 + (y_i - b)^2}. \qquad (1)$$

O problema é, portanto, de optimização: o objectivo é escolher o ponto (a,b) de forma a minimizar o custo total $C(a,b)$ dado pela equação (1).

Em primeiro lugar, é fácil concluir que existe um ponto de mínimo: quando o ponto (a,b) se afasta do domínio convexo definido pelos X_i, a função custo C cresce (naturalmente!) sem limite. Por continuidade, tem de ter um mínimo nesse domínio convexo. Esse mínimo é, portanto, uma "média espacial" e, ingenuamente, poderíamos supor que essa média é um centro de massa.

Não é! De facto, esse ponto tem propriedades muito diferentes e contra-intuitivas. Para ilustrar a situação, suponhamos, como na figura 1, que temos cinco cidades diferentes, dispostas num pentágono regular. Comecemos com uma situação em que todas as cidades têm igual peso (20%); é óbvio, por simetria, que o ponto de máximo, representado a vermelho, está no cent do pentágono.

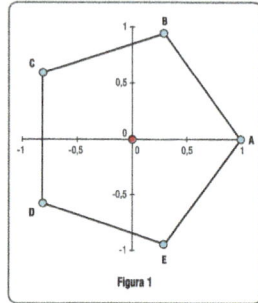

Figura 1

Façamos agora aumentar o peso relativo do ponto A, mantendo as outras quatro com peso igual entre si. Por simetria, o ponto óptimo estará sobre o eixo vertical. No entanto, desloca-se muito rapidamente à medida que o ponto A aumenta de peso. Na figura 2 está representada a sua localização quando o peso relativo de A é 43%: o ponto óptimo quase coincide com A!

Fig. 5 A numerical study of the optimal facility location problem

in fact still survives). But there were others, like the number of outstanding traffic tickets.

In the meantime Jorge Picado, a mathematician at the University of Coimbra, had gone to work on disproving this myth. He collected the ID numbers and corresponding check digits of a few dozen people and, under the assumption that the checksum algorithm was similar to the ISBN algorithm, programmed his computer to reverse engineer the problem and discover the checking algorithm. Some surprises were in store.

First of all, only 10 digits were used as checkdigits: 0 through 9. It is well-known (see e.g. [29]) that the additive algorithms require a prime number of check digits; thus for instance the ISBN algorithm uses the set of 11 digits $\{0, 1, \ldots, 9, X\}$, since 11 is the smallest prime number greater or equal to our usual numbering base 10. Now Picado's data included only a set of 10 check digits; there was no 11th digit. This might lead us to think that non-additive algorithms, such as non-commutative algebraic codings based on dihedral groups, like the Verhoeff scheme adopted by the Bundesbank (and conceived by A. Beutelspacher, a contributor to this volume) ([29, 30]) might be in use. Secondly, a very strange thing happened. When Picado ran the algorithm-detection code, the program did not converge, whatever the supplementary heuristic hypotheses applied. This seemed to indicate either an inconsistency in the data or a faulty error-detecting algorithm (e.g. one which would not preserve injectivity of the check digit).

In fact, the data revealed an obvious anomaly: the relative frequency of the digit 0 was *double* the frequency of the other digits. This suggested that something could be wrong with the check digit 0. In fact, simply eliminating all the 0s from the data and running the algorithm-detecting code ensured immediate convergence and allowed Picado to discover the identification system used by the Portuguese agency. This discovery was a double-edged sword. On the one hand, it was shown that the identification scheme is (almost) precisely the ISBN scheme. On the other hand, since there is no 11th check, the scheme cannot work efficiently. Indeed, what was observed is that the Portuguese ID card uses a version of the ISBN identification scheme *with a mathematical bug*: the non-existent 11th digit for checksums is replaced by a second (false) zero. Thus the identification system is not injective, and one half of the occurrences of the digit 0 as a checksum are false.

This is a rather embarrassing situation, since it implies that indeed the check digit cannot be relied upon for error-detecting purposes, therefore defeating the purpose of its own introduction and rendering it useless! In actual practice, no official agency using the ID card number ever bothers asking for the check digit—not even the agency which issues the cards themselves. Things get even more curious. Although the whole ID card was recently changed, with the issue of a *Citizen Card* incorporating the most modern biometrical and physical technologies, the *ID numbers did not change, neither did the check digits or the algorithm.* Thus the ID card bug propagated to the new identification scheme. Moreover, the exact same identification scheme is used in the Fiscal ID number—and thus the exact same mathematical bug occurs.

In May 2000 a column in *Ingenium* was published about this question, *O mistério do Bilhete de Identidade (The mystery of the ID card)*. In a short span of time there were reactions of awe, disbelief and the article was widely reproduced in the media and the Internet, where it can still be found today. The title of the paper was used the book [21], published the following year, which had a large impact and made the material available to a very different audience, the general public. The title of the book, the problem itself and the question of the bug in the identification system led to a large impact on the educational community: everybody has an ID card and number, and it is easy to explain and implement the mathematics and the algorithm behind the ISBN scheme.

The ramifications and implications of this work are quite surprising and show quite clearly how communicating mathematics to an engineering audience can have an "overspill effect", which quickly crosses the interface to the outside world and has a real impact on the educational community. Indeed, it is quite likely that without the original article in *Ingenium* and its publication in the book [22] Picado's discovery might still be largely unknown to the world outside mathematicians, with a loss both for mathematics education and the general public (the urban myths about "the extra digit" still exist to this day!).

The Euro Banknotes, a Second Curious Episode

With the introduction of the euro in 2002, the European Central Bank (ECB) faced the obvious problem of counterfeiting of the banknotes in the biggest market in the world. The most advanced technology was used in the banknotes: special paper and silver bands, lasers and holograms; at least two of the anti-counterfeiting measures were kept secret by the ECB. One could expect the identification scheme used for euro banknotes would be equally sophisticated, truly 21st century, as it was also kept secret.

After the circulation of the euro banknotes on 1st January 2002, the first named author undertook a somewhat quixotic personal project: to discover the mathematical algorithm used to identify euro banknotes. He started collecting and annotating all the banknote numbers which he encountered, and recruited some friends as well. In the process of programming and entering the data, he noticed the structure of the numbers:

L-DDDDDDDDDD

where L stands for a letter of the alphabet from J to Z and D stands for a digit from 0 to 9, so the number is a alphanumeric string made up of one letter and 11 digits. However, strangely, *the last digit is never 0*. This alone indicated a mathematically special role for the last digit as a check digit.

At that point he published an article in *Ingenium* explaining the problem, giving preliminary results, and inviting readers to submit their recorded data. The response was enthusiastic: with this collective effort, he gathered thousands of numbers. In the process of programming and entering the data, the following was experimentally determined. If a fixed numerical value is attributed to the letter L at the start of the

Table 1 Correspondence between first letter of euro banknote codes, numerical value and country of emission

Letter	Value	Country
J	2	UK
K	3	Sweden
L	4	Finland
M	5	Portugal
N	6	Austria
O	7	–
P	8	Netherlands
Q	9	–
R	1	Luxembourg
S	2	Italy
T	3	Ireland
U	4	France
V	5	Spain
X	6	Denmark
W	7	Germany
Y	8	Greece
Z	9	Belgium

identification number, then the (last) control digit is simply determined by imposing that the sum of all the 12 numerical values of the digits is congruent with 0 (mod 9). Or, in plainer terms: the identification scheme for euro banknotes is simply the thousand-year old process of *casting out 9s*. This is all the more surprising since it is an extremely inefficient error-detecting algorithm, as we all know from elementary school (with a success rate of less than 90 %). So the banknotes with the most advanced physical systems against counterfeiting had also, embarrassingly, the most mathematically unsophisticated algorithm for error detection!

The reader can check this for him/herself by pulling out a euro banknote. All that is required is to know the correspondence between the leading letter L and its numerical value. This is given in Table 1. The numerical values of the letters are simply 1 to 9 in ascending order (again with no attribution of 0, since this is arithmetic mod 9). Moreover, a letter has no deeper meaning than simply identifying the country of origin of the banknotes. In fact, even though the UK, Denmark and Sweden are not part of the eurozone, they already have allotted places, should they wish to join!

This collective quest with the readership of *Ingenium* was a great success. It was possible to identify what may be called a second "mini-code" on the flip side of the notes. This consisted of L-DDD-LL in very small print. This was an even bigger mathematical disappointment: it is a mere serial number from the typography where the banknote is issued and has no mathematical content at all. All this information is gathered in book form [23] and, as far as we know, nowhere else.

This makes for wonderful material for communicating mathematics in popular lectures. Everybody (in the eurozone) handles banknotes, even small children. So it is possible to tell a story with meaningful mathematical content, by adapting the theoretical aspects of coding theory and identification systems to the audience, but always keeping in mind a very real example—the banknotes people handle. In this way a curious interface between mathematics education and engineering was created. The audience of mathematically educated engineers provided the data. A mathematician discovered the solution. This solution is ideally suited for the communication and popularization of mathematics. A very unlikely connection indeed!

Both case studies above illustrate this situation quite clearly. They deal, each in their own way, with the same kind of mathematical problem—identification systems and check digits, or in a more mathematical language, coding theory and cryptography. These are problems that we can be sure that everyone is acquainted with in one form or another: they arise in contexts from national ID cards (mandatory in Portugal) to credit cards, from Internet passwords to bar codes, from cell phones to DVDs, and increasingly so in the past 20 years. Moreover, they are both understandable and mathematically meaningful, and characterize problems that should interest the community. So coding theory and cryptography seem ideally suited as vehicles to convey the importance of useful mathematics to the public.

LPDJLQH D VHFUHW, a Film on Elliptic Curves and Cryptography

Cryptography refers to secure methods of transmitting and safeguarding secret and valuable information and it has been a popular subject in mathematics popularization. Computer advances in the last half century have made cryptography a fundamental part of contemporary financial life: it is used for credit cards, online bank transfers, email, etc. Since the late 1970s, the invention of the RSA public key system has been widely used. It is based on prime number theory and on the difficulty of factoring very large integers. During the World Mathematical Year 2000, the Isaac Newton Institute displayed every month a different math poster in the London Underground on the theme "Maths Predicts"; October's poster (Fig. 6) was exactly on cryptography [31].

In 1984, Hendrick Lenstra Jr introduced a new method for factoring large integers using elliptic curves and in the following year Neal Koblitz and Victor Miller independently proposed to use elliptic curves in cryptography [32]. Since then mathematical sophistication has been raised to a new level. The security of ECC (elliptic curve cryptography) algorithms is based on the discrete logarithm problem of elliptic curves, which seems to be a much harder problem in finite field arithmetic. Recent mathematical advances imply that a certain desired security level can be attained with significantly smaller keys, for instance, a 160-bit ECC key provides the same level of security as a 1024-bit RSA key [33]. Elliptic curves have deep and beautiful properties. They are plane curves of the type

$$y^2 = x^3 + ax + b,$$

Fig. 6 October's poster in the London Underground during the WMY2000 [31]

which have been studied since the 19th century. Nowadays they illustrate the beauty of the links between number theory, algebra and geometry and their theory provides a powerful mathematical tool to strengthen the security of e-commerce and secure communications.

During a workshop held at the Mathematisches Forschungsinstitute Oberwolfach, in February 2010, the second named author visited the IMAGINARY exhibition [34] at MiMa [35] and started to wonder how to use the computer programme "Surfer" to create beautiful animated surfaces, a challenge that he shared with Stephan Klaus. His idea became a project, under the initiative of the *Centro Internacional de Matemática*, when he met Armindo Moreira, an architect associated with the *Casa da Animação* from Porto. In that summer, the project became a film, 9:30 minutes long, of art and mathematics about elliptic curves and cryptography [36], with the help of Andreas Matt [34] and Bianca Violet, who produced the Surfer movies, and with the collaboration of Victor Fernandes, also from the *Casa da Animação* who, with Armindo Moreira, produced the digital film with original music and original mathematical animations (Fig. 7).

The film story [36], which can be read at the end of the film, starts with numbers appearing chaotically until they form sets of Pythagorean triples and then an animation of their correspondence, already well known to the ancient Babylonians around 1600 bc, to right triangles with integer sides and to the problem of splitting a given square number into two squares. Elliptic curves are introduced associated with Fermat's equation $a^N + b^N = c^N$ through the equation $y^2 = x(x - a^N)(x + b^N)$, the class of Frey curves that were introduced and studied in the 1980s. These curves were instrumental in the path that led to the complete proof, by A. Wiles and R. Taylor in 1994, of the famous observation written by Pierre de Fermat around 1637 in the margin of his copy of Diophantus' work: "No cube can be split into two cubes, nor any biquadrate into two biquadrates, nor generally any power beyond the second into two of the same kind."

The introduction of the z coordinate in the elliptic curve equation to render it homogeneous of degree three, $y^2z = x^3 + axz^2 + bz^3$, describes in space of a family

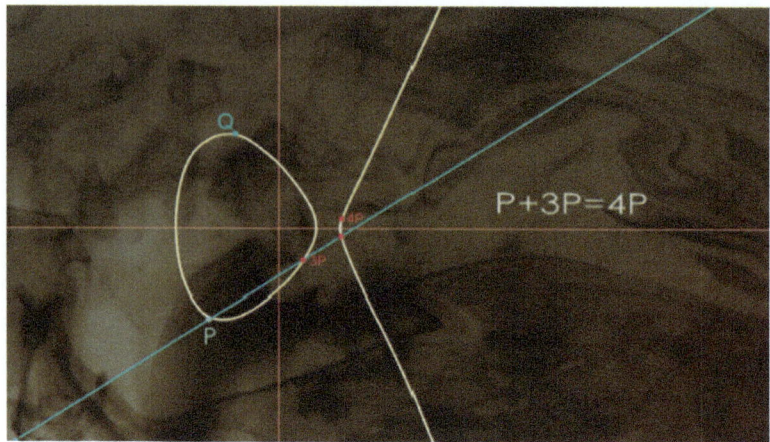

Fig. 7 A screenshot of the animated illustration of the group law on an elliptic curve

Fig. 8 The aesthetical value of the evolving surfaces produced with Surfer from the movie

of algebraic surfaces with two parameters a and b. The computational variation of these equations with the Surfer programme generated beautiful animations (Fig. 8) that were artistically integrated into the film in order to stimulate the public imagination and to evoke mathematical creativity. This central part of the film is purely artistic and has no direct mathematical meaning.

In the third part, two mathematical animations illustrate the group law on an elliptic curve in the rational plane (Fig. 7) and modular arithmetic is illustrated by a clock and an animated image of dots representing an elliptic curve in the finite field F_{23}. The final animation with 0s and 1s symbolize the data being scrambled mathematically by the computer to encipher and to decrypt a secret. Accompanying the whole film, without spoken words, the original music composed by Fernandes and

Moreira provides the right sound track to challenge the public with a final puzzle. Referring to the old and unreliable method of the Caesar cipher, which consists of a simple arithmetic operation to encipher a message in the usual Latin alphabet by means of the formula $d = c - 3(\mathrm{mod}\,26)$, the final paragraph of the text gives the key to decipher the title of this film: _ _ _ _ _ _ _ _ _ _ _ _ _ _ (words at the end of the film).

The first public presentation of the film *LPDJLQH D VHFUHHW* was on 26 September 2010 in Óbidos, during the opening of the workshop "Raising Public Awareness of Mathematics". The film text is available in four languages, English, German, Portuguese and Spanish, and can be freely downloaded in a high quality version from http://www.cim.pt/LPD-UHW and can also be seen directly on YouTube.

References

1. Mathematics and Industry, OECD report (2008). http://www.oecd.org/dataoecd/47/1/41019441.pdf
2. ESF Forward Look on "Mathematics and Industry", European Science Foundation, Strasbourg (2010). www.esf.org
3. Euler, L.: Lettres a une Princesse d'Allemagne sur divers sujets de Physique et de Philosophie. Petersburg, 1768, 1772, 3 vols. (a third edition of the English translation was published in Edinburgh in 1823 and is available at http://www.eulerarchive.org)
4. Thomson (Lord Kelvin), W.: Popular Lectures and Essays, vol. II, pp. 571–592. Macmillan and Co., London (1894)
5. Reis, F.E.: Enciclopedismo – conhecimento para um público diversificado. In: Reis, F.E. (ed.) Felicidade Utilidade e Instrução - A divulgação científica no jornal enciclopédico dedicado à rainha 1779, 1788–1793; 1806. Porto Editora, Porto (2005)
6. Caraça, B.J.: Conferências e outros escritos. Lisboa (1970)
7. Caraça, B.J.: Conceitos Fundamentais da Matemática, Edições Cosmos, Lisboa, vol. I (1941); vol. 2 (1942); Re-edition, Gradiva, Lisboa (1998)
8. Aristotle, Metaphysics, XIII 3.107b
9. Livio, M.: The Golden Ratio. Broadway Books, New York (2002)
10. Emmer, M. (ed.): The Visual Mind II. MIT Press, Cambridge (2005)
11. Sullivan, J.: Mathematical pictures: visualization, art and outreach. In: Behrends, E., Crato, N., Rodrigues, J.F. (eds.) Raising Public Mathematical Awareness. Springer, Berlin (2011)
12. Robinson, J.: Symbolic Sculpture, the Collected Works of John Robinson. http://www.bradshawfoundation.com/jr/genesis.php
13. Cipra, B.A.: The best of the 20th century: editors name top 10 algorithms. SIAM News 33(4) (2000)
14. Cannon, J.W.: Mathematics in marble and bronze: the sculpture of Helaman Rolfe Pratt Ferguson. Math. Intell. **13**(1), 30–39 (1991)
15. Para Além da Terceira Dimensão - Beyond the Third Dimension Catalogue of the exhibition by CMAF/Universidade de Lisboa (2000). http://alem3d.obidos.org
16. Banchoff, T., Cervone, D.: In and Outside the Torus (computer animated film). http://alem3d.obidos.org/i/torusio/tioro4.mpg or http://alem3d.obidos.org/i/torusio/tioro4.gif
17. Banchoff, T.F., Cervone, D.P.: A virtual reconstruction of a virtual exhibit. In: Borwein, J., Morales, M.H., Polthier, K., Rodrigues, J.F. (eds.) Multimedia Tools for Communicating Mathematics, pp. 29–38. Springer, Berlin (2002)
18. Polthier, K.: Visualizing mathematics—online. In: Bruter, C.P. (ed.) Mathematics and Art, pp. 29–42. Springer, Berlin (2002)

19. Polthier, K., et al. (eds.): MathFilm Festival 2008, a Collection of Mathematical Videos. Springer, Berlin (2008)
20. Rota, G.-C.: The phenomenology of mathematical beauty. In: Palombi, F. (ed.) Indiscrete Thoughts. Birkhauser, Basel (1997). Chap. 10, republished in pp. 3–14 of The Visual Mind II. MIT Press, Cambridge (2005)
21. von Neumann, J.: The mathematician. In: Newman, J.R. (ed.) The World of Mathematics, vol. 4, pp. 2053–2063. Simon and Schuster, New York (1956)
22. Buescu, J.: O Mistério do BI e Outras Histórias. Gradiva Publicações, Lisboa (2001)
23. Buescu, J.: Da Falsificação de Euros Aos Pequenos Mundos. Gradiva Publicações, Lisboa (2003)
24. Buescu, J.: O Fim do Mundo Está Próximo? Gradiva Publicações, Lisboa (2007)
25. Buescu, J.: Casamentos e Outros Desencontros. Gradiva Publicações, Lisboa (2011)
26. Ingenium, Boletim da Ordem dos Engenheiros (Bulletin of the Portuguese Engineers Association). Available online at http://www.ordemengenheiros.pt/Default.aspx?tabid=1234
27. Drezner, Z., Hamacher, H.W.: Facility Location: Applications and Theory. Springer, Berlin (2002)
28. Buescu, J., Casquilho, M.: A minimum distance: arithmetic and harmonic means in a geometric dispute. Int. J. Math. Educ. Sci. Technol. **142**(3), 399–405 (2011)
29. Gallian, J.A.: The mathematics of identification numbers. Coll. Math. J. **22**, 194–202 (1991)
30. Picado, J.: A álgebra dos sistemas de identificação. Bol. SPM **44**, 39–73 (2001)
31. Maths Posters in the London Underground. Isaac Newton Institute for Mathematical Sciences, University of Cambridge (2000). www.newton.ac.uk/wmy2kposters/
32. Koblitz, N.: The uneasy relationship between mathematics and cryptography. Not. Am. Math. Soc. **54**(8), 972–979 (2007)
33. Hankerson, D., Menezes, A., Vanstone, S.A.: Guide to Elliptic Curve Cryptography. Springer, Berlin (2004)
34. Matt, A.: IMAGINARY and the idea of an open source math exhibition platform. In: Behrends, E., Crato, N., Rodrigues, J.F. (eds.) Raising Public Awareness of Mathematics. Springer, Berlin (2012)
35. MiMa, museum for minerals and mathematics. http://www.mima.museum
36. Rodrigues, J.F.: "LPDJLQH D VHFUHHW", a film on elliptic curves and cryptography, CIM Bulletin # 29, January 2011, pp. 27–28

WWMD?

Barry Cipra

Abstract For many of us, the late Martin Gardner is like a god, whose powers of mathematical exposition are nonpareil, and whose second coming we all eagerly await. His columns for *Scientific American*, and the books they begat, framed the public awareness of mathematics for decades, and they will be enjoyed for centuries to come. I will preach here the humble sermon that there is always much to be learned about presenting mathematics to the laity by pondering carefully the words of Gardner, who strung them together so magisterially. Faced with the dilemma of explaining complex ideas to an often bewildered audience, mathematicians will do well to keep in mind the introspective question, What Would Martin Do?

> Escrever È como a droga que repugno e tomo, o vìcio que desprezo e em que vivo.
> —Fernando Pessoa, *Livro do desassossego*
> (Writing is like the drug I abhor and keep taking, the addiction I despise and depend on. *The Book of Disquiet*

Follow the Leader

"What Would Jesus Do?" is a catchphrase of American Christianity. Often reduced to the bumper sticker or wristband initialism, WWJD? (oftentimes without the question mark), the query is intended as an exhortation to rely on the example of Jesus when faced with life's conundra. Moral dilemmas (and their dicorollaries), the implication goes, can be solved with certainty by simply acting in accord with Christ. The slogan first arose (according to Wikipedia, our modern bible for all things factual) in Kansas (my home state, and one state north of Martin Gardner's Oklahoma) in the 1890s, but it has had a modern resurgence, at least in the United States.

The phrase cries out for parody, usually by substituting a less benign role model, such as What Would Genghis Do?, What Would Machiavelli Do?, What Would OJ Do?, and my own favorite, What Would Tony Soprano Do? This essay is a playful

B. Cipra (✉)
Northfield, MN, USA
e-mail: bcipra@rconnect.com

E. Behrends et al. (eds.), *Raising Public Awareness of Mathematics*,
DOI 10.1007/978-3-642-25710-0_24, © Springer-Verlag Berlin Heidelberg 2012

(analytic?) continuation of that tradition—not that I want to suggest that Martin Gardner is in the same category with Genghis Khan or Machiavelli or OJ Simpson or Tony Soprano. Well, maybe Machiavelli, who was also a good, clear writer of vast influence and undiminished popularity.

I'd better point something out up front. What Would Jesus Do? is often a thinly veiled appeal for the follower to attend not so much to what Jesus' example itself says, but rather to what the *exhorter* says that Jesus' example says. You can get damn near any theology you want out of the New Testament. I'm going to succumb to the same temptation: I really don't know what Martin would do, but I'll point to things he did that argue in favor of my own idiosyncratic exegesis. My best advice may simply be to ignore everything else I have to say here and just go read Gardner—you can't go wrong by reading Gardner.

Don't Ask, Don't Tell

The question we're faced with here is how to make the public aware of mathematics —presumably in a good way. (Early on in my career, mathematicians used to bemoan the fact that you just never saw a mathematician on the cover of, for example, *Time* or *Newsweek*. It always made me think of the old adage: "Be careful what you wish for because you just might get it." Then in 1996, it came to pass that a mathematician did make the cover of both magazines. His name was Ted Kaczynski, aka the Unabomber.) There is, of course, the underlying question, Why bother? Let's not dwell too long on this, because some of the answers are unflattering. There's pure vanity, for example: We like it when people appreciate the wonderful things we do. There's also jealousy: We hate it when the astronomers with their big telescopes or the particle physicists with their enormous accelerators get all those front-page headlines with four-color photographs. And there's the hunger for money and power: If only people properly understood the importance of mathematics, the theory goes, they'd shower us with funds to carry out our arcane, incomprehensible studies.

I might add here my own personal answer: It's a great way to make a living. As I see it, I let the really smart people do all the hard work, proving deep theorems and developing wonderful new applications of mathematics. Then I make them work even harder, explaining to me what they've done, sometimes over and over, until I misunderstand it in a way that makes sense to me. Then I make them work hard again, correcting my misunderstandings. I also make my editors work hard, fixing up my wording so that readers will have a chance of understanding what the smart people have done. The only work I do involves tapping at a keyboard. And a lot of the time all I do is stare at an empty computer screen, waiting for it to fill up with cogent sentences. Of course that's a little easier said than done. The American sportswriter Red Smith put it best: "There's nothing to writing. All you do is sit down at a typewriter and open a vein."

But even if we set aside the question Why, there's still the question, What exactly are we trying to accomplish? Only if we know what we're trying to do can we sensibly ask the question, How do we go about it? One problem, of course, is that are multiple objectives. We do want to recruit and retain the next generation of bright young minds to carry on the good work we've done. (I'm using the inclusive "we" here despite the confession of the previous paragraph ...) We also want to create a sustained diaspora of mathematically well-trained people in other walks of life—and by this I don't just mean scientists and engineers, but people in everything from corporate management to postmodern literary criticism. (Just think how much less twaddle there would be if our counterparts in the humanities were capable of writing correct, logical proofs. I had a student once in a first-year honors linear algebra course who told me years later that learning to writing proofs about vector spaces helped her not just in subsequent math classes, but also when it came to writing term papers in history and literature courses. She was, to be sure, a rather extraordinary student to begin with; toward the end of the semester, she wrote a proof about eigenvalues in rhyming couplets.)

These first two objectives comprise people who are actually going to do mathematics for a living, or a portion thereof. But there are at least two more objectives. One is to ensure that decisionmakers in business, industry, and government realize that mathematics can make key contributions toward solving many of the real-world problems they are charged with. These are people who, by and large, do not have the time, talent, or desire to do mathematics themselves. But it's by no means necessary to *be* a mathematician in order to appreciate what mathematics can do. After all, you don't have to be a lawyer or a plumber (to mention two highly paid professions of dubious repute) to recognize when you need one. (There is a wonderful line of dialog in the 1997 movie *Volcano*, in which Tommy Lee Jones, playing the head of a disaster-response agency, turns to an underling and shouts: "Get me a scientist!" One pines for a similar big-budget Hollywood extravaganza with the line "Get me a mathematician!")

Another objective—the last one I'll take note of here, and the one closest to Gardner's calling—is simply to nourish the enthusiasm for mathematics that a surprisingly broad spectrum of the public enjoys. I want to emphasize here the words "surprisingly broad." Mathematics has a reputation for being arcane and incomprehensible (see above), with a learning curve of such low slope that only years of trudging dedication enable one to appreciate its true beauty. (Note: People often speak of a *steep* learning curve as the barrier to mastery, but it's actually the opposite. A learning curve describes mastery as a function of time, so a steep curve is one where you learn a lot in a short amount of time. It's really the flat curves you want to avoid, not to mention the ones of negative slope!) And indeed it is a slog to become a professional mathematician. But much as you can play pick-up basketball without the endless hours of conditioning and practice, or dabble in watercolors without an MFA, or sing in the shower with no Carnegie Hall ambition, it's entirely possible to do some recreational mathematics without first excreting a doctoral dissertation. It's likewise as possible to enjoy a popular account of mathematical discoveries and ideas as it is to admire the artistry of LeBron James, Damien Hirst, or Philip Glass.

First Impressions

No one did a better job of nourishing mathematical enthusiasm than Martin Gardner. He had an outstanding pulpit from which to feed the multitudes. His column in *Scientific American* reached millions of readers each month. (In addition to hundreds of thousands of individual subscribers, *Scientific American* has long been a staple in college, high school, and public libraries. One never knows, of course, how many horses actually drink the water to which they're led, but it's probably safe to estimate Martin's devoted following in the many thousands. Jesus only needed twelve to get started, one of whom turned out to be a bad apple.)

It's well worth looking at how Gardner went about this mission. Consider the opening of his inaugural column, in the January, 1957, issue of *Scientific American*. The reader's eye is drawn to a grid of numbers:

19	8	11	25	7
12	1	4	18	0
16	5	8	22	4
21	10	13	27	9
14	3	6	20	2

The column begins:

> "Magic squares have intrigued mathematicians for more than 2,000 years. In the traditional form the square is constructed so that the numbers in each row, each column and each diagonal add up to the same total. Presented here is a magic square of a new type. It seems to have no system: the numbers appear to be distributed in the matrix at random. Nevertheless the square possesses a magical property as astonishing to most mathematicians as it is to laymen."

Gardner goes on to explain that if you pick five numbers from the grid so that you've got one in each row and one in each column, then add the five numbers you've selected, you always get the sum 57 (in honor of the year of publication). It might be a bit of an exaggeration to say this is astonishing to mathematicians—that claim, I would claim, is part of Gardner's strategy of inviting the reader to think of himself as a mathematician. What is astonishing is how complicated a mathematician's explanation of what's going on can be. Here, for example, is one attempt (taken from the comments section in Brian Hayes's bit-player.org posting on Martin Gardner, shortly after Gardner's death):

> "Each row can be obtained by adding a constant to the entries of the first row: for example, the second row is the first row minus 7. So if we label the entries in the first row as $x1, \ldots, x5$, and the constants as $c1, \ldots, c4$, any

combination of 5 values chosen from distinct rows and columns has to sum to $(x1 + \cdots + x5) + (c1 + \cdots + c4)$."

Compare this to Gardner's elegant explanation:

"Like most tricks, this one is absurdly simple when explained. The square is nothing more than an old-fashioned addition table, arranged in a tricky way. The table is generated by two sets of numbers: 12, 1, 4, 18, 0 and 7, 0, 4, 9, 2. The sum of these numbers is 57."

He goes on to explain, "You can construct a magic square of this kind as large as you like and with any combination of numbers you choose. It does not matter in the least how many cells the square contains or what numbers are used for generating it. They may be positive or negative, integers or fractions, rationals or irrationals."

Opening Lines

Let's wrap up with a look at just the first sentence or so of Gardner's next several columns (again all from 1957):

"Mathematical brain teasers are an ancient sport, and really good new ones do not come along very often." (February)

"Who has not as a child played ticktacktoe, that most ancient and universal struggle of wits . . . " (March)

"A paradox is a truth which cuts so strongly against the grain of common sense that it is hard to believe even when you are confronted with the proof." (April)

"To a mathematician few experiences are more exciting than the discovery that two seemingly unrelated mathematical structures are really closely linked." (May)

"As many readers of this magazine are aware, a Moebius band is a geometrical curiosity which has only one surface and one edge." (June)

"It is something of an occasion these days when someone invents a mathematical game that is both new and interesting. Such a game is Hex . . . " (July)

Each of these sentences draws the reader in with something familiar and inviting, but hints at something new the reader will learn about. Contrast this with the opening of a typical math paper:

"Let K be an algebra defined over Q, let π be the relative Galois group of K, let σ be an element of π and let C^σ be the collection of continuous functions on K invariant under σ."

Indeed, one wonders how Herman Melville might have approached the beginning of his classic novel, *Moby-Dick*, had he majored in mathematics instead of studying the classics. Here's Herman the great American novelist:

"Call me Ishmael."

And here's Melville the mathematician:

"Let me name be Ishmael, let the captain's name be Ahab, let the boat's name be Pequod, and let the whale's name be as given in the title."

(Note: I've lifted these jokes from an article I wrote for the *Mathematical Intelligencer*, "Andy Rooney, PhD," published in 1988. The "Let K be an algebra ..." line, I believe, came from an actual math paper, but I no longer remember which one.)

Gardner made it look easy. But it's harder than it looks. Let me try to give a sense of just how hard, with a little tour of the sausage factory: snippets of email correspondence I had with one of my editors a couple of years ago. I was reporting on some talks at the Joint Mathematics Meetings held in January, 2009. One of them was by Robert Niemeyer, a graduate student at the University of California, Riverside, who was working out some of the theory of billiards on fractals—a rather counterintuitive notion—with his advisor, Michel Lapidus. Here's the first line (in the jargon of journalism, the "lede") from my first draft:

" 'Clouds are not spheres, mountains are not cones, coastlines are not circles,' fractal pioneer Benoît Mandelbrot famously wrote. And, he might have added, pool tables are not polygons."

The editor wrote back a very welcome reply:

"Not much work needed on this one! Just have a look, and let me know if there's anything you'd like done before I send it to top edit."

("Top edit" is a process by which the editor's edits get edited.) I was pleased to hear my job was basically done. And indeed, my first draft had survived virtually intact, a rather rare phenomenon. The editor had made a single change—in the lede, which now read:

" 'Clouds are not spheres, mountains are not cones, coastlines are not circles,' fractal pioneer Benoît Mandelbrot famously wrote. And, he might have added, pool tables are not polygons—but mathematicians can ask 'what if?' "

The editorial alteration struck me as problematic, so I wrote back:

"The only thing I'd change is the end of the first paragraph to ...—so mathematicians can ask 'What then?' "

I had opened a can of worms. Mercifully, the editor finally took the bait. Here's the rest of our exchange.

Editor: But what is the "then" you're talking about? If they just ignore the fact that pool tables aren't polygons, it's more like saying "So what?"

Me: Is it necessary to add anything after the "pool tables aren't polygons"? How about"—which calls for a bit of mathematical hustle."?

Editor: How about, "... but mathematicians can ask, 'What if they were?'"

Me: Mathematicians are already studying billiards on polygons. What Neimeyer and Lapidus are aiming at (ultimately) is billiards on fractals ...

Editor: I see! Then it seems to me that the sentence is misleading in a different way ... There must be an elegant way to keep this lede on track.

Me: Pool tables, on the other hand, *are* polygons. But mathematicians can ask "What if they weren't?"

Editor: Eureka!

This all took about an hour. For one sentence. It reminds me of Pessoa, also from *The Book of Disquiet*: *Pasmo sempre quando acabo qualquer coisa* (I'm astounded whenever I finish something).

Rigour in Communicating Maths: A Mathematical Feature or an Unnecessary Pedantry?

Maria Dedò

Abstract The need for rigour in mathematical communication may raise non-trivial problems, when the target of this communication includes people with little mathematical background. In what follows we discuss some of these problems, and we argue how it may be possible to keep some kind of mathematical rigour also in an informal way of communicating maths.

Anyone involved with activities aiming to raise public awareness, sooner or later, will have to face the problem of mathematical rigour: how should we keep track of this rigour in mathematical communication for people who are not mathematicians, to what extent and with what concerns?

Mathematics is strictly linked with rigour: it is hard to imagine mathematics without rigour, and if someone doubts how it can be possible to communicate mathematics without being rigorous, these doubts are well justified. On the other hand, mathematical rigour, and in particular rigorous mathematical language, is a thorough enemy of mathematical communication: it is well known that non-mathematicians are scared if they even see a formula in a text; and, apart from these irrational fears that we may try to oppose, it is obvious that a phrase containing technical terms excludes people who do not know the meaning of these terms.

So these two requirements may seem to be deeply antagonistic, and partly they are: in [1] it is even proposed the existence of a constant K and a "theorem" asserting that the product of rigour and communicability will never be greater than K. This is just a metaphor, of course, as a "definition" of rigour (and of communicability) would surely be ... not rigorous; but we cannot ignore this conflict, which continuously poses "practical" questions in any initiative directed at raising public awareness. What we aim to do here—as it is obviously unreasonable to expect complete answers to these questions—is to use some examples in order to examine different facets of the problem, hoping to raise interest in a deeper analysis. We do think that such an analysis could be very useful, especially if it involves people fac-

M. Dedò (✉)
Università degli Studi di Milano, Milan, Italy
e-mail: maria.dedo@unimi.it

E. Behrends et al. (eds.), *Raising Public Awareness of Mathematics*,
DOI 10.1007/978-3-642-25710-0_25, © Springer-Verlag Berlin Heidelberg 2012

ing the problem from different points of view (mathematicians, both working or not working on RPA activities, journalists, lay users of some RPA initiative, ...).

When the *Centro matematita*[1] was born, six years ago, some colleagues objected to its proposed title (Research Centre for the Communication and the *Informal* Learning of Mathematics), arguing that it was a contradiction to use the adjective *informal* for a subject, like mathematics, which is based on formalism. However, at the birth of the Centre, its founders declared they were convinced that—in our present situation—there is a great demand and a great need (both in schools and in society) for the informal communication of mathematics. So, far from thinking that there is a contradiction, we think that it is necessary to learn how to communicate mathematics in an informal way, whilst trying to maintain, as much as possible, all the essential features of mathematics: and, among them, rigour, which has of course an important role.

What Do We Mean by Rigour?

This apparently harmless question is in fact very hard to answer, and much harder would have been to answer a question like "What is rigour?", which I wouldn't even dare to pose. Rather, what I want to do here is just to put some limits, as clearly as possible, to what we are going to discuss.

A first clarification is the distinction between rigour in mathematics and rigour in communicating maths (in RPA activities). Clearly we are speaking here (only) about the second; nevertheless, let me just make a quick remark regarding the first. I want to consider the debate in the Bulletin of the American Mathematical Society, which was raised, at the end of the last century, with the paper by Jaffe and Quinn on what they called *theoretical mathematics*, meaning a sort of heuristic aspect of mathematics, not necessarily depending on rigorous proofs. Their paper gave rise to a sequel including a paper by Thurston, a collection of different responses,[2] and an answer by Jaffe and Quinn: see [2–5].

We recall that discussion just to point out that we cannot take for granted that the problem of rigour arises only in mathematical communication to non-mathematicians, while, within mathematics itself, everything is clear, and we know exactly what rigour is and we use it correctly in communication among mathematicians. That debate shows that this is *not* the case, and that the problem of rigour in mathematics deserves discussion.

---

[1] All the examples discussed in the following originate from activities devised and carried out by the *Centro matematita*; we give a rough list of such activities in the Appendix.

[2] And it is worthwhile to explicitly give a list of names, as the number and the quality of the mathematicians involved in that debate were quite impressive; besides Jaffe, Quinn and Thurston, there were comments by: Atiyah, Borel, Chaitin, Friedan, Glimm, Gray, Hirsch, MacLane, Mandelbrot, Ruelle, Schwarz, Uhlenbeck, Thom, Witten and Zeeman.

Coming back to our theme, rigour in communication, we observe that communication in RPA activities very rarely requires proofs, and, when it goes into mathematical content, it mainly involves a description (of mathematical objects or phenomena), without the need for proofs, or, when justification of the results is required, it is not a formal proof, but, rather, aims at convincing, at making the result reasonable. One could then argue that this does not require rigour: however, we shall try to show, through various examples, that this sort of "plausible reasoning" also requires rigour, although of course it can't be the same kind of rigour we expect in a mathematical paper.

Another aspect which requires clarification concerns what I call the popular perception of rigour, that is, what the general lay person thinks is rigour (and mathematical rigour in particular). This often has very little to do (or nothing at all!) with rigorous mathematical thinking; however, people working in RPA activities need to be aware of these opinions (and try to change them . . . !).

I refer here to the fact that it is quite normal, other than for mathematicians, to think that something is "rigorous" if (and only if!) it is written in an incomprehensible way, using many symbols, and formulas and difficult words. This is sometimes even used on purpose, for example, some foolish commercial advertisement stating that there is a "73 % increase in gloss" for a toothpaste, suggesting (alas, with some reason!) that the simple use of numbers or percentages makes people imagine "this is mathematics, so it is rigorous thinking, so the product must be good" instead of thinking, for example, that, in order for the phrase to make sense, we should know (at least) what it is meant by gloss and how it is measured.

Unfortunately, this misunderstanding, of associating rigour (just) with the usage of technical mathematical language, can be quite common and extends also to mathematics classes in schools. It is not uncommon to find in schools an obsessive insistence for correct mathematical terminology. It is of course an important and positive matter that students master the proper usage of (a few!) technical terms, but there is a borderline, beyond which this insistence is too formal, and becomes pedantry. Sometimes, it can be even worse: it can become a mask which just hides the lack of real mathematical content.

The borderline we are referring to is related to the capacity for transmitting (mathematical) ideas, which give meaning to the concepts introduced. As long as the meaning is clear, students are not at risk of repeating empty phrases, while, when the meaning is not mastered, the insistence on the usage of technical terms is not only useless, but risks also producing negative effects.

Everyone who has experienced school, at different levels, has their own collection of foolish phrases made by students and repeated by their teachers as jokes. I don't want to offer my own collection; I just want to point out that, in the great majority of these situations (and also in very different ones), the crucial element they have in common is the fact that some words happen to have no meaning at all.

Correctly being able to repeat that "the square root of a number is that number, which, when squared, gives back the original number" does not necessarily imply being able to answer the question "what is the result of $(\sqrt{5})^2$?". The two things are connected only if we have been able to give a meaning to what we are doing;

otherwise, although it may seem paradoxical, they may live in separate, and non-communicating, rooms.

Another element we should be aware of, regarding the public perception of (mathematical) rigour, is what I call the fear of rigour, which is extremely common in people who have not mastered (elementary) mathematics. With the words of Peano [6]: "*chi non conosce bene i fondamenti d'una parte qualunque della matematica rimane sempre titubante, e con una esagerata paura del rigore*", that is "someone who does not master the basic elements of any part of mathematics always remains hesitant and with an exaggerated fear of rigour".

This aspect is particularly evident (and dangerous) in formal teaching, and it may be one of the reasons of the pedantry which we sometimes meet in school classes; in [7], I report on some examples, which I found meaningful, of this kind of fear and its consequences.

An Example

Let us examine a game which we use[3] in different situations (workshops for students of different ages, but also for their teachers). Two small groups (2–3 people each) sit in front of each other, at the two sides of a table, separated by a divider, which should prevent the first group seeing the objects held by the second group (and vice versa). We give a "mathematical object" to the first group and we ask them to describe it to the second group, so that they can build a copy of it. The same request is applied the other way round and, at the end, the divider is taken away so that the two groups can see if what they built has any similarities with what they were meant to build.

The "mathematical object" can be, for example, a polyhedron, which should be not too complicated, but also not too standard (for example we wouldn't use a cube, or a prism, or a pyramid). Typically, uniform polyhedra (like we see in Fig. 1) are good examples to use. Of course, the two groups should have at their disposal some material that allows them to build the objects described to them; often, it is useful that each group first builds the polyhedron they were given, before giving instructions to the other group about how to build it.

Other "mathematical objects" that can be successfully used for this game are pictures, for example a picture drawn on squared paper, or built on a geoboard; or a picture (like Fig. 2) with a relevant symmetry that we are obliged to notice and to describe in order to give a simple way of reconstructing it; or even—if we want the game to be more demanding—a picture of a knot, or a physical knot itself (like image 2556).[4]

---

[3] Here and throughout, the pronoun "we" refers to the *Centro matematita*: see the Appendix.

[4] The image can be found at the address http://www.matematita.it/materiale/index. php?p=cat&im=2556. In the following, all images marked with a number can be found in the collection *Images for Mathematics*, by the *Centro matematita*, at a similar address, obtained by replacing the last number.

Fig. 1 Uniform polyhedra
(image 11108)

Fig. 2 A rosette (image
13494). Drawing by
Francisco Martín Casalderrey

In all these cases (and other similar ones we can propose) the process of "describing" the object to the other group requires a non-trivial usage of (normal) language: the two groups have first of all to distinguish the relevant features of the object that have to be transmitted to the other group; then, they often discover (with some surprise) that a minimal variation in what they say may produce a significant difference in what the other group builds. So they understand that "rigour" in language is necessary in order to obtain a unambiguous communication. Also, the (possible) insistence on using a particular word instead of another, which seemed just an obsession of mathematics teachers, may begin to make sense after this game.

Of course, this kind of analysis is more useful if the two groups are forced to record the messages they send to the other group, so that—if there is a mistake—they can go back and find the word or phrase that caused the misunderstanding, and the different choices that could lead to a correct solution.

This is an example of how to build rigorous mathematical thinking: the participants realize—and this happens also if their phrases do not contain all the precise terms—what is necessary for the other group to understand the shape of the given object; and by having a definite purpose (and being involved in the game for this purpose) keeps the meaning of the words in mind: there is no risk of repeating empty "definitions"! So, the mathematical instrument (a precise definition, an unambiguous term) arrives exactly an the moment when it is needed, and when it can be appreciated; not before.

The Mathematical Language

The mathematical language is an enormous achievement of human mankind: if one reads for example a description of an elementary standard problem, like a second degree equation, before the use of a letter for an unknown became common, one can realize the amazing power of good notation. But ...

But things get simpler if (and only if!) the new level of abstraction is taught when the person feels the need for that abstraction: otherwise, it is just an empty formalism and ... we may find students saying they studied equations with "x" only, so they are not able to solve an equation like $3y^2 + 2y - 2 = 0$.

Another thing to note is that mathematical language, which is precious when used by people who have mastered it, may become a dangerous instrument when used by people who have not mastered it. This is particularly risky in a situation (like an exhibition or an activity for general public) where we are speaking to "everybody" (in contrast to a formal learning situation, where we are talking to a group more or less homogeneous in their knowledge and level of abstraction, although of course everybody has his/her own ways of thinking, and his/her own times of learning).

When we are choosing the text to accompany a poster in an exhibition (for example), or for a page on a website, we should remember this (rather obvious) fact: sometimes we are "tempted" to use mathematical language, we may think that by saying something in a formal way we salve our consciences, which otherwise would be a bit disturbed when realizing that we can't, in that text, say all that we feel we "should" say, in the right order and in the right place. However, we should always keep in mind that a communication is something with two ends, with the message going from one to the other. And, in order to judge the correctness of a message, it is not enough to judge how correct the message is when it starts its journey, but it is necessary to judge how the message arrives at the other end and how it is interpreted, and understood, by the person who receives it.

It is certainly difficult to consciously abandon mathematical language; and it gets even more difficult if we don't want this to also mean abandoning mathematically rigorous thinking. When we use normal language, we lose much of the precision: things are so much easier with the language of mathematics (when it is mastered) and especially, it is so much easier to check if a phrase is correct or not.

As a test: try to write the definition of a group (for example; or anything else) in "normal" language, without using any symbol or any word that is learnt at university level; then ask ten different mathematicians if it is correct or not. It is easy to imagine that there will be different opinions (while it is unlikely of course to find different opinions about the definition of a group written in normal mathematical terms). And, moreover, in this respect, mathematicians' opinions should not be the only ones sought: we do need to test our definition with people who do not know what a group is, in order to bring out any possible misunderstanding.

However, when we abandon formal language and we try to communicate the ideas of mathematics informally, there are some benefits, and we should learn how to profit from the best of these.

The main negative aspect of "normal" language compared to mathematical language is its potential ambiguity: but the same ambiguity can also lead to a richness in

Fig. 3 Poster from the exhibition (image 10663). Images by Gian Marco Todesco, poster by Daniela Gaggero

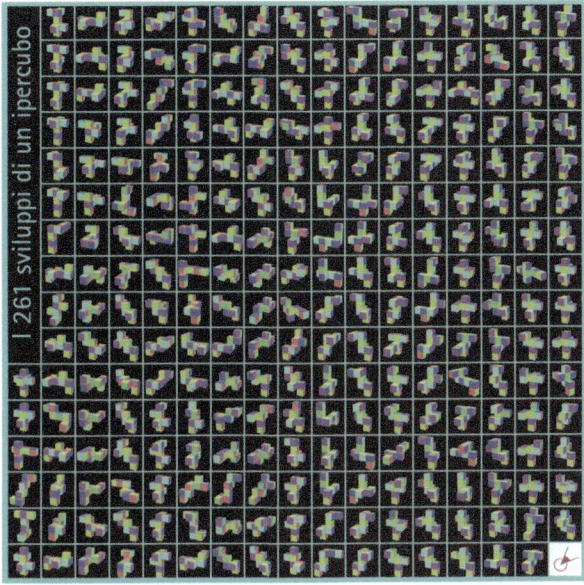

I 261 sviluppi di un ipercubo

many respects. First of all, because it allows us to explore regions that would surely be inaccessible for non-mathematicians if we used formal language. We successfully created an exhibition about the 4th dimension,[5] and of course this was largely constructed with an analogy: we explained the nets of a hypercube (see Fig. 3 or image 10187) starting from the nets of a cube (see image 9926); or we can build a 120-cell from 120 dodecahedra (as in Fig. 4) underlining the parallel with the building of a dodecahedron from 12 pentagons; or we can show a star polytope (like Fig. 5), just because it is beautiful, and use a video (recalling the story of Flatland) to transmit the idea of a four-dimensional object passing through our three-dimensional world, which would show a changing three-dimensional section of it.

We used in that exhibition another possible advantage of an ambiguous message, because ambiguity may enhance mental associations, and mental associations are a precious instrument in transmitting ideas and building up one's own mental images and building abstract concepts. Of course, if we decide to accept ambiguity, and sometimes we deliberately play with it (in a particular type of mathematical communication) and if at the same time we don't want to completely give up rigour, we should be conscious that we are accepting a risk, and that we are walking a very narrow path: so we should also increase the controls we use in order to check how the message can be interpreted.

An obvious consequence of this is that we need to listen, as much as possible, to the reactions of the public to the different parts of an informal communication: this is a fundamental point in judging whether a mathematical concept arrives at the other end of the communication in a substantially correct way.

---

[5] See the Appendix.

Fig. 4 A 120-cell (image 10835). Image by Gian Marco Todesco

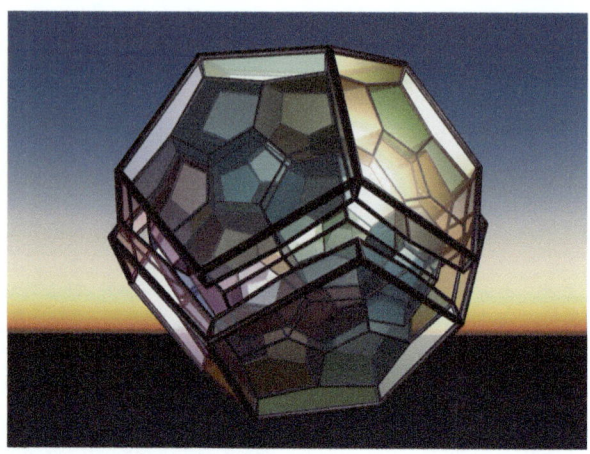

Fig. 5 A regular stellated polytope (image 3494). Image by Gian Marco Todesco

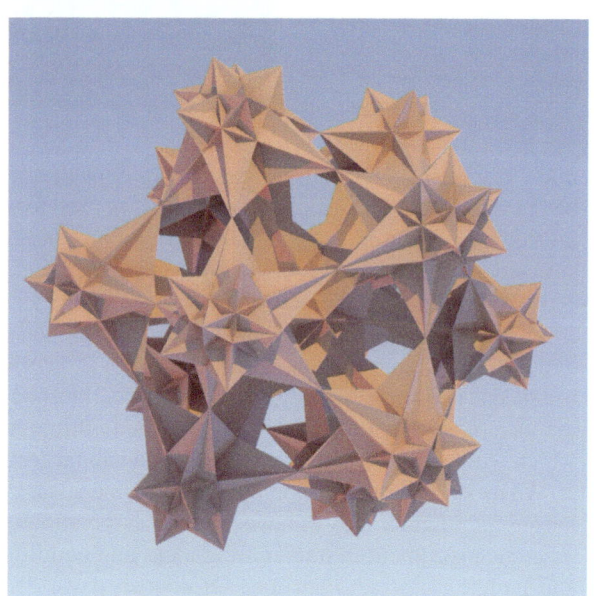

Another Example

Let us take a detour with an example from the exhibition "Symmetry, playing with mirrors".[6] The main objects of this exhibition are three mirror boxes (as in image 1508, where one can "see" the three possible different groups of plane isometries generated by the reflections from the sides of a plane triangle: as in Fig. 6, and images 13900 and 13880) and three three-dimensional kaleidoscopes (as in Fig. 7,

---

[6]See the Appendix.

Fig. 6 Inside a mirror box (image 1520). Photo by Sabrina Provenzi

Fig. 7 Kaleidoscopes (image 1544)

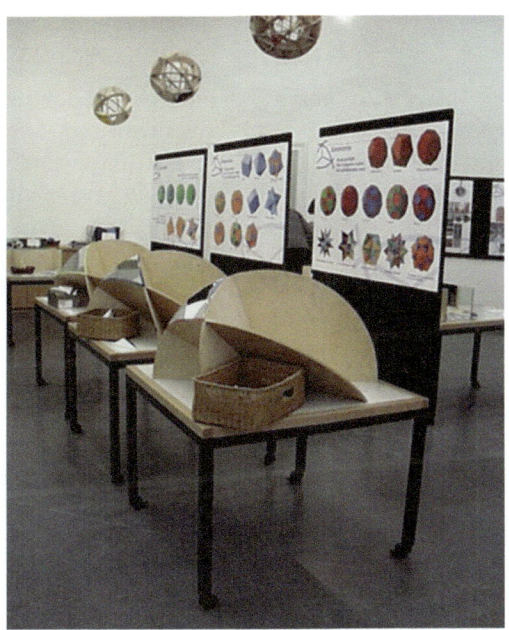

where one can "see" the three possible[7] different groups of space isometries generated by the reflections from the sides of a spherical triangle: as in Fig. 8, and images 1576 and 13216). The public's reactions in front of the mirror boxes are usually dominated by the surprise of "seeing infinity". If we further analyse this impression, we can see that, in fact, the "infinite" that we "see" in a mirror box is much nearer to the abstract concept of infinity than the use of the word "infinite" in normal colloquial language (where it usually just means "very many", which is not infinity at all!).

---

[7]Excluding the prismatic ones, corresponding to spherical triangles with two right angles.

Fig. 8 A stellated
dodecahedron (image 1595)

If we try to "count" the number of repetitions we see of a given image in a mirror box (as in Fig. 6), we discover with some surprise that there are not very many at all, we don't even easily arrive at 10 (depending on how clean the mirrors are ... !): in spite of this, without a doubt we have the impression of infinity, which evidently is not due to the number of repetitions. Why do we get this impression and what causes it?

I remember a dialogue between a small boy and his father: the father asked the boy if he liked it, and the boy answered: "Yes but... I feel a bit lost" (see Fig. 9);[8] and he then explained further: "because you can go there, and then go beyond ... and still beyond ... and still beyond ... ".

What the boy was expressing very clearly is that this impression of infinity is because of (not the number of repetitions, but) the fact that we can always go farther: it is just symmetry, which transmits the idea of (potential) infinity; the symmetry of the pattern we see in the mirrors allows us to be sure that it does not end there, because we know how to go on ...

This comment helped us to realize how the concept of infinity seen (informally, of course!) in the mirrors was in fact quite correct (can we say rigorous?) and this was an assumption, which allowed us to consider the idea of using mirrors in a quite different way to explore a comparison between infinities using a bijection.[9] And ... we know historically this has been quite a great idea!

---

[8]The text in Italian that you can see in the image is a famous verse of the poem *L'infinito* by the Italian poet Giacomo Leopardi: "e il naufragar m'è dolce in questo mare".

[9]We can start with a finite situation, a three-dimensional kaleidoscope, such as Fig. 10, to ask: "How many balls do you see?" or, better: "Are there more red balls or blue balls or yellow balls?" This is quite nice, because it brings to mind the symmetry of a cube: the rings of 8 red (or 6 blue, or 4 yellow) balls correspond to the faces (or to the vertices, or to the edges) of a cube, so there are six (or eight, or twelve) of them; and $6 \times 8 = 8 \times 6 = 12 \times 4 = 48$. But there is a more elegant way to prove that the three numbers are equal, if we realize that there is a bijection between balls of different colors. Of course this is not the way that an average person will say it; he/she will

Fig. 9 A boy looking at "infinity" in a mirror box (image 13081)

Fig. 10 How many balls? (image 1566)

probably express this fact in a more confused way, but in any case it is quite evident that, as long as we put one (and just one) ball of each color between the mirrors, then the number of virtual images will be the same for each color. So we can now pose another question, and ask whether in a mirror box (as in Fig. 11) we can see more red shapes, or blue, or green. Here we cannot count any more, and we have to compare infinities, but the observation we just made for the finite case can give us an idea about how to make sense of comparing infinities.

Fig. 11 Inside a mirror box
with angles 30°, 60° and 90°
(image 1517). Photo by
Sabrina Provenzi

Non-verbal Languages

Coming back to our main subject, a point which needs further comment is the role
of non-verbal languages, that is the possibility of using images (or virtual animations, or three-dimensional models that can be touched and manipulated), in order
to convey information about mathematical concepts.

This is a very delicate theme, especially as regards rigour, as it is obvious that
any image or any three-dimensional model is, necessarily, "a bit false". However,
we know very well that images have an enormous strength in suggesting ideas, so
that it would really be a pity not to use them, especially in informal communication:
instead, we should study how to use them, what kind of precautions we can adopt in
order to minimize the risks of misunderstanding, what kind of verbal messages can
be useful to go with them, etc.

A first obvious observation is that there should be many images:[10] the reason is
that a mathematical image deals with an abstract concept, so it is not representative,
but it is rather evocative, it helps the imagination, it suggests something. And, often,
the "something" which has to be evoked can arise naturally through a comparison
with different images: we could not say that in a flower like Fig. 12 we see a group,
but we can well suggest the idea of the dyhedral group with 10 elements by showing
many different images, all sharing the same symmetry group: a pear like image
3564, a wheel like image 4826, another flower like image 578, a sea urchin like
image 5649, a rosette like image 13475, etc.

But … can we really say that we "see" a dihedral group with 10 elements in
a flower like Fig. 12 (for example)? Of course a formal answer must be negative,
and for more than one reason: first of all, any example of symmetry coming from
nature would be just an "approximate symmetry", which of course does not exist
if we define symmetry in a rigorous way. Obviously it would be possible to define
(rigorously) a notion of approximate symmetry, but this makes things much more
complicated, so it is probably not what we want to do in a popularization activity.

---

[10]And this is the motivation behind the big collection of *Images for mathematics* (http://www.
matematita.it/materiale/index.php) by the *Centro matematita*: see the Appendix.

Fig. 12 Apple blossom
(image 4440). Photo by
Franco Valoti

Fig. 13 A rosette (image
1792)

Another reason for answering negatively is that it is risky to confuse the three-dimensional object (the flower)—where the (eventual) symmetry operations are rotations of the space around a line and reflections with respect to a plane—and the two-dimensional image seen on the paper or on the screen—where the symmetry operations are rotations of the plane around a point and reflections with respect to a line; the three-dimensional symmetry of the object is like that of the two-dimensional image only if the photo is taken from a very particular point of view.

So, there are many good reasons to avoid the use of such images if we want to keep some rigour in our message. Nevertheless, we all know very well not only that images can be powerful in transmitting an idea, but, moreover, that the photo of a flower can be (much more beautiful and) much more powerful in this respect in comparison with an image like Fig. 13, which would be more "correct", because the problems we noticed for the flower do not apply to the drawing.

So, what should we do? Our answer (of course!) depends on the context. Just to give an example of what we mean, in the kit[11] about symmetry which the *Centro matematita* lends to schools for workshop activities on the theme of symmetry, we enclose a CD with a collection of "symmetric" images. The images are divided into two different folders: one titled "observing symmetry" and the other one titled "classifying symmetry". In the first, we felt free to insert images like the photo of the flower, while the second contains only images with "perfect" symmetry (no photos

---

[11] See the Appendix.

of real objects, but only drawings or symbols like image 12042). In the comments for teachers, we explain this and we recommend using only images from the second folder in activities where pupils are asked to detect the symmetry type of something. When teachers are aware of the underlying problems, they can choose to use images from the first folder with some kind of ambiguity, thus provoking useful discussions among students, about these ambiguities.[12]

This distinction allowed us to use all the suggestions, with the fascination because of the mix with "real life", meanwhile keeping a distinction when one refers to a strictly technical aspect (classifying images). Often there exist solutions like the one described in this example, which allow a reasonable compromise between the desire to use the power of real life examples and the need to be sufficiently rigorous.

As a tribute to the wonderful Portuguese town, which hosted the workshop originating this book, and in particular to the museum where we held our meeting, I will mention another example, that is the problem of "what is a point?" Of course the mathematical concept of a point is not the same thing as a "point" as used in normal language and an image (for example a photo like Fig. 14) can be an occasion to dot the i's about this concept.

A Final Example

Let's examine another example, mathematically less trivial: can we say that the sculpture by Max Bill (Fig. 15) or the one by Josep Canals (Fig. 16), or also other ones like images 7943 or 68, represent Möbius bands?

Here the problem can be quite delicate. The average reaction of people with a good mathematical background (e.g. mathematics university students or teachers) goes through (at least) three phases.

Phase 1: "Of course, yes, they are Möbius bands!" (maybe with doubts with respect to the sculpture of Canals). In some sense, it is because we already know what is a Möbius band which is misleading here; probably, our brain does not even "see" the thickness, because it already has a (conceptual) image of a Möbius band, as something with no thickness at all; and our mental image "wins" with respect to the "real" image that we see.

Phase 2: "Of course not! These objects have thickness and Möbius bands are surfaces, and have no thickness." Once the thickness (maybe through the sculpture of Canals, or through somebody's questions) has been noticed by their brain, a person with a mathematical background (much more so than a layman!) can't accept compromises, so the answer will be negative for *all* the examples, with no relevant difference among them. It is independent of the (relative) measure of this thickness; even if the object is very thin, it has a thickness, so it can't be a "real" Möbius band.

---

[12]In http://www.matematita.it/materiale/index.php?p=pathc.sub8, you can see a journey through images made by 14-year-old pupils with a limited mathematical background, with this sort of discussion; quoting one of them: "Look at the differences! How can we expect the fruit to be symmetric as our math teacher would like?".

Fig. 14 What is a point?
(image 13864)

Possible phase 2.5 (which is very risky!): "*As* there is a thickness, which is a small interval, these objects are (not Möbius bands, but) the product of a Möbius band and a small interval." This seems very natural, but is definitely *false*.[13]

We noticed these reactions (for example) in students collaborating with the *Centro matematita* during public exhibitions: the positive aspect is that, once one goes beyond this mistake and finally (phase 3!) realizes that topologically all these three-dimensional objects are nothing else than solid tori, this deeply affects the permanent understanding of the subject, and in particular of the relation between orientability (which is an intrinsic property, related to the Möbius band) and two-sidedness (which is an extrinsic property, so can be applied not to an object in itself, but only to a couple of objects, as a Möbius band in a given ambient space; and the answer may be different depending on the ambient space, just as a circle is two-sided in a cylinder and one-sided in a Möbius band).

But the point of interest for us here is another one: conscious of the problems, can we use these objects to represent Möbius bands? Or is it better to leave out this possibility? Here again there is not a universal answer, but the answer depends on the context.

---

[13]If there existed (an "object", that is) a subspace of our ordinary space homeomorphic to the product of a Möbius band and an interval, this would imply that the Möbius band is two-sided in ordinary space (as it is obviously two-sided in the 3-manifold represented by the product of a Möbius band and an interval); but we know that Möbius bands are one-sided in ordinary 3-space, so this 3-manifold can't be a subspace of ordinary space.

Fig. 15 A sculpture by Max
Bill (image 7947). Photo by
Simone Piuri

Fig. 16 A sculpture by Josep
Canals (image 8888). Photo
by Simone Piuri

Fig. 17 Cylinder or Möbius
band? (image 11368). Image
by Gian Marco Todesco

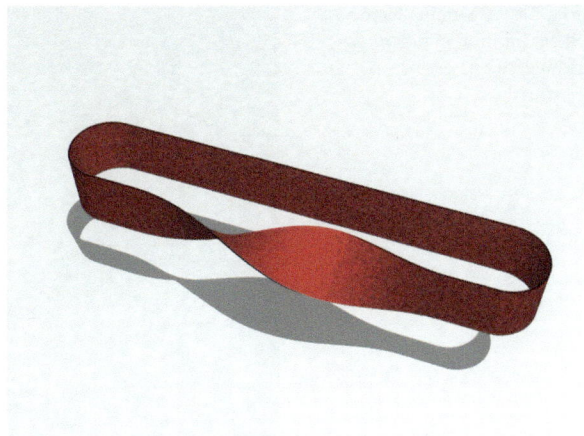

What I would like to point out here is that mathematicians' reactions can be different from laypeople's reactions (and we should be fully aware of this!); in particular, for anyone with some mathematical background, once the thickness is noticed (phase 2) it exists for *any* example, independently of how thin it is (and, paradoxically, this works also when the same person did not even notice—in phase 1—this thickness); this is definitely not true for laypeople, who may notice the thickness (and notice it much more than a mathematician in phase 1, because they don't have a conceptual image superposing the "real" one), depending really on how thick it is.

So, in the end, in order to convey the idea of a Möbius band, the problem of "how thick it is", although mathematically irrelevant, is, instead, crucial from the point of view of correct communication and possible misunderstandings, both in building a model or in making a picture (as Fig. 17). Of course, here as with other examples, we should be conscious of the problem and use all the possible corresponding cautions and antidotes .... For example, if we build a (model of a) "Möbius band" as Fig. 18 for an exhibition, the person who is going to decorate it should be aware of the phenomenon in order to paint exactly the same scene on the "two sides"!

Conclusions

Let's try to draw some (provisional) conclusions for this analysis.

The examples we examined illustrate that, although it is certainly true that in an informal communication, aiming to transmit mathematical ideas and mathematical meaning, we should abandon mathematical language, we should try also, in contrast, to maintain rigorous mathematical thinking, at least as much as possible. We should also make clear to our target audience, again as much as possible, what "rigorous mathematical thinking" is, especially when this conflicts with the average perception of rigour (and mathematics): making clear to lay people what is *not* rigour can very usefully be one of the messages (or even the most important one) in a given RPA activity.

When we say that rigour should be maintained "*as much as possible*", we are
also implicitly asserting that our method of understanding rigour is dynamic; we
cannot just say that rigour is there or is not there, but the problem may be more
complicated than that and we have to build rigour, through a continuous process, by
going through different levels of approximation.

This opens a big, not trivial problem, as regards "measuring" whether a given
communication is rigorous enough; the examples we quoted were meant to illustrate the fact that rigour in an informal communication should be "measured" in a
different way to a formal one, with special attention paid to the context and to the
way the message may be interpreted by the given recipient.

But this measure will always be a matter of opinion and we must be prepared
for the fact that different mathematicians will have different opinions in judging
if a page of mathematics written in "normal" language is acceptable or not. This
leaves an open problem to discuss, and I think that mathematicians who are working on RPA activities should be the first ones to propose methods—as objective as
possible—for deciding which "right or wrong" criteria to use, in domains where
unambiguous language is not possible.

I conclude with a tentative partial answer, which can be "deduced" from this
quote by Giuseppe Peano: "Il rigore matematico è molto semplice. Esso sta
nell'affermare tutte cose vere, e nel non affermare cose che sappiamo non vere.
Non sta nell'affermare tutte le verità possibili." that is "Mathematical rigour is very
simple. It consists in asserting true statements and in not asserting what we know is
not true. It does *not* consist in asserting every possible truth."

Coherently with Peano's phrase, we could judge if a phrase is "rigorous enough"
by imagining a change of recipient[14] and passing, for example, from the general
public to maths university students: if we adjust the message by simply adding new

---

[14]I first heard this said by Manuel Arala Chaves, regarding his own experiences and not relating to
Peano's quotation.

information, and we have no need to change anything, then this means that the original phrase was "rigorous enough". Otherwise, if we have to change something, ... maybe there is something we need to change also for the previous recipient.

Appendix

We list in this appendix some of the outputs of the *Centro matematita*, in order to give an idea of the kind of experiences that gave rise to the comments discussed in this paper.

matematita (http://www.matematita.it) is an interuniversity research centre for the communication and informal learning of mathematics, which became active in 2005, building on the experiences of four different universities (Milan, Milan-Bicocca, Pisa and Trento).

One of the main aims of *matematita* is to design, develop and promote exhibitions, books, journals and multimedia materials, and to see how these work both in formal teaching situations (schools) and with the public in general.

I include here a rough sketch of its activities.

Activities for the great public include:

- the conception and realization of (temporary and permanent) exhibitions: http://www.matematita.it/realizzazioni/mostre.php?NL=en gives a list and a rough description of the main ones;
- the organization on request of exhibitions or other activities in science festivals or other such events.

Activities for schools include:

- courses for teachers;
- workshops, in the university, for pre-university students (11–18 years old);
- self-contained kits for workshop activities on different subjects (for students 6–18 years old); the kits can be lent to schools; see: http://www.matematita.it/realizzazioni/materiale_didattico.php#libri_i;
- MATh.en.JEANS: from the French experience of more than twenty years (see http://mathenjeans.free.fr/amej/accueil.htm), a similar scheme in Italy (http://www.mathenjeans.it/).

The editorial activities of the Centre include:

- books and CDs relating to the exhibitions (see: http://www.matematita.it/realizzazioni/pubblicazioni.php);
- a series of books (*Quaderni a quadretti*) which aims to be useful for teachers at primary and secondary schools (see: http://www.quadernoaquadretti.it/quaderno/pubblicati.php);
- *XlaTangente*, a new maths journal for young people, 16–20.

The online activities of the Centre include:

- http://www.xlatangente.it/xlatangente/index.do the web site of the journal *Xla-Tangente*;
- http://www.quadernoaquadretti.it/ a web site for the book series *Quaderni a quadretti*; for nearly ten years the web site has hosted online games for school pupils (6–14 years old);
- http://www.matematita.it/materiale/index.php the web site *Images for mathematics*, with a collection of images and animations.

References

1. Greco, P.: L'idea Pericolosa di Galileo, storia della comunicazione della scienza nel Seicento UTET (2009)
2. Jaffe, A., Quinn, F.: Theoretical mathematics: toward a cultural synthesis of mathematics and theoretical physics. Bull. Am. Math. Soc. **29**(1), 208–211 (1993)
3. Thurston, W.P.: On proof and progress in mathematics. Bull. Am. Math. Soc. **30**(2), 161–177 (1994)
4. Atiyah, M., et al.: Responses to "Theoretical mathematics: toward a cultural synthesis of mathematics and theoretical physics" by A. Jaffe and F. Quinn. Bull. Am. Math. Soc. **30**(2), 178–207 (1994)
5. Jaffe, A., Quinn, F.: Response to comments on "Theoretical mathematics". Bull. Am. Math. Soc. **30**(2), 208–211 (1994)
6. Peano, G.: Opere scelte, vol. III a cura di U.M.I (in "Sui fondamenti dell'analisi", p. 273)
7. Dedò, M.: Più matematica per chi insegna matematica. Boll. Unione Mat. Ital. **8**(4-A), 247–275 (2001)

Mathematics Between Research, Application, and Communication

Gert-Martin Greuel

Abstract Possibly more than any other science, mathematics of today finds itself between the conflicting demands of research, application, and communication.

A great part of modern mathematics regards itself as searching for inner mathematical structures just for their own sake, only committed to its own axioms and logical conclusions. To do so, neither assumptions nor experience nor applications are needed or desired.

On the other hand, mathematics has become one of the driving forces in scientific progress and moreover, has even become a cornerstone for industrial and economic innovation. However, public opinion stands in strange contrast to this, often displaying a large amount of mathematical ignorance.

In this article, which is addressed to a readership without any special mathematical education, I shall look at these tensions and try to reveal some of the causes that lie underneath. I will also try to explain a current scientific research topic, but mainly focus on the question whether it is possible or necessary to transmit an understanding of mathematics to the general public.

I am aware that my ideas on mathematics, which are presented here as theses in a rather dense form, certainly do need further elaboration. Nevertheless I hope that they are interesting enough to stimulate further discussions.

Wise Words

Let me introduce my conception regarding research, application, and communication by first quoting some celebrated personalities, and developing my own point of view afterwards.

Extended version of talks held in Hannover, October 2009, Kiev, November 2009, and Óbidos, September 2010.

G.-M. Greuel (✉)
University of Kaiserslautern, Kaiserslautern, Germany
e-mail: greuel@mathematik.uni-kl.de

E. Behrends et al. (eds.), *Raising Public Awareness of Mathematics*,
DOI 10.1007/978-3-642-25710-0_26, © Springer-Verlag Berlin Heidelberg 2012

Research

In this context, I mean by research pure scientific work carried out at universities and research institutes. Trying to explain concrete mathematical research to a non-mathematician is one of the hardest tasks, if at all possible. But it is possible to explain the motivation of a mathematician to do research.

Therefore, my first thesis refers to this motivation and is a quote from Albert Einstein (physicist 1879–1955) from 1932:

> "The scientist finds his reward in what Henri Poincaré calls the joy of comprehension, and not in the possibilities of application to which any discovery of his may lead." [1]

Indeed the "joy of comprehension" is both motivation and reward at the same time and from my own experience I know that most scientists would fully agree with this statement.

Application

One could hold numerous lectures on the application of mathematics, probably forever. The involvement of mathematics in other sciences, economics, and society is so dynamic that after having demonstrated one application one could immediately continue to lecture on the resulting new applications.

The thesis concerning the application of mathematics consists of three quotes, by Galileo Galilei (mathematician, physicist, astronomer; 1564–1642), Alexander von Humboldt (natural scientist, explorer; 1769–1859), and Werner von Siemens (inventor, industrialist; 1816–1892), in chronological order:

> "Mathematics is the language in which the universe is written." [2]
> "Mathematical studies are the soul of all industrial progress." [3]
> "Without mathematics one is left in the dark." [4]

These statements seem to indicate that there is a direct relation between mathematical research and applications. In fact, recent research directions in geometry are motivated by new developments in theoretical physics, while research in numerical analysis and stochastics is often directed by challenges from various fields of application. On the other hand, the development of an axiomatic foundation of mathematics is guided by trying to formalize mathematical structures in a coherent way and not by the motivation to understand nature or to be useful in the sense of applications. Partly due to this development, it appears that the relationship between research-orientated or pure mathematics on the one hand and application-orientated or applied mathematics on the other hand is not without its strains. Some provocative statements in this article will illustrate this.

Communication

Each of us, whether a mathematician or not, is aware how difficult it is to communicate mathematics. Hans Magnus Enzensberger (German poet and essayist, born 1929) discussed the problem of communicating mathematics on a literary basis in 1999. He writes:

> "Surely it is an audacious undertaking to attempt to interprete mathematics to a culture distinguished by such profound mathematical ignorance." [5]

The exhibition *IMAGINARY—Through the Eyes of Mathematics* is one attempt to interpret and communicate mathematics to a broad audience and there exist many other examples of successful communication. Nevertheless, the problem remains and will be discussed later when I shall give some reasons why it is so difficult.

Let me now start with a more detailed analysis of the above topics.

On the Birth of Mathematical Ideas

Mathematical research has several aspects indeed, but here I am going to have a closer look at only two of them: the research topics themselves and the way in which mathematical research is performed. The latter aspect concerns the way in which mathematical ideas arise. This is an extremely creative process, which happens quite frequently in interaction with other mathematicians.

I am the Director of the Mathematisches Forschungsinstitut Oberwolfach (MFO), an internationally renowned research institute situated in the heart of the Black Forest. Those who are not familiar with mathematical research might get an impression of the process by the following description of a research center which is specially designed to foster interactions among mathematicians and to inspire creative thinking. Mathematicians simply refer to the institute after the village where it is located: Oberwolfach (Fig. 1).

The MFO has become world famous as a birthplace for mathematical ideas; some people call it a "paradise of mathematics". The following anecdotes will illustrate the degree of awareness and esteem of the institute. One time at lunch, when I asked a young American mathematician whether she had known Oberwolfach before, she answered: "To be honest, Oberwolfach is the only German word I know". And a well-known senior mathematician said with a twinkle in his eyes that the only invitations he accepts without his wife's permission, are those to an Oberwolfach workshop.

The Oberwolfach model has become so successful that many institutes all over the world have followed its example. So let me explain the main aspects of this model.

Fig. 1 General view of the Oberwolfach Institute [6]

How Do Mathematical Ideas Arise in Oberwolfach?

Mathematical research mainly studies the structure and inner relationships of mathematical objects and tries to develop more comprehensive theories about them. Many mathematical questions derive from the effort to describe nature in mathematical terms, but it often happens that the mathematical frame was created before the applications. The process of research, when successful, leads to mathematical theorems, whose proofs are typically complicated.

Historically, coincidence also plays a big role. Improving the chances for progress by coincidence is one of the main purposes of the meetings at the Oberwolfach Institute. When getting to know the background of an important result during a talk, one can suddenly have a flash of insight, perhaps leading to considerable progress in one's own research. Small group discussions, inviting the fresh thoughts and comments of colleagues, can lead to a sharing of these insights and to finding the right direction for further work. It happens quite often that two or three colleagues, during such discussions, become aware that they, though coming from different backgrounds and with different motivations, are interested in similar problems and are able to unify their ideas in order to establish a common research project.

This happens nearly daily in these workshops, so that a great number of important papers have been initiated at Oberwolfach in this manner. In contrast to the typically large conferences all over the world, the small workshops at Oberwolfach focus on active research where open questions abound [7].

The final write-up of a proof is usually best done at the home institute, but the development of a mathematical theory and, within such a theory, the promising idea for a proof, is an extremely creative process depending very much on intuition and experience and certainly benefiting from an exchange of ideas. This is why personal contact between the researchers is a crucial point at Oberwolfach. The famous

"Oberwolfach atmosphere" is completely free from distractions and enables discussions among specialists, but also between young mathematicians at the beginning of their career and famous experts.

History of Oberwolfach

In order to be able to understand the myth of Oberwolfach we have to look back to its beginnings after World War II. As early as 1946 the first small meetings were held at the old hunting lodge "Lorenzenhof". Among the participants were mathematicians like the Frenchman Henri Cartan, whose home country had been an "arch-enemy" for centuries. His family had suffered tremendously under the regime of the National Socialists so that his participation was not at all a matter of course. The first famous guests visiting the Lorenzenhof were Heinz Hopf (a world-famous topologist from Zurich, a German of Jewish descent who had moved from Germany to Switzerland in 1931) and Henri Cartan (the "grand maître" of complex analysis from Paris). It was said that "Without Hopf and Cartan Oberwolfach would have remained a summer resort for mathematicians, where in a leisurely atmosphere dignified gentlemen would polish classic theories" [8].

In August 1949 a group of young "wild" Frenchmen met in Oberwolfach who had taken up the cause of totally rewriting mathematics as a whole, based on the axiomatic method and aiming at a new unification. It was a truly bold venture that only young people would dare to take up. Some of their names have become famous, including Henri Cartan, Jean Dieudonné, Jean Pierre Serre, Georges Reeb, and René Thom. A photo from that time was only discovered a few years ago (Fig. 2). It shows part of the group in the autumn of 1949. Cartan himself could not come, due to the consequences of a car accident.

On the far left you can see René Thom, the later Fields Medalist and founder of "Catastrophe Theory", and in the middle Jean-Pierre Serre, also later Fields Medalist and winner of the first Abel Prize.

The Gospel According to St Nicolas and the Freedom of Research

In the first guest book they wrote down the *Evangile selon Saint Nicolas*, Fig. 3, a humorous homage on the Lorenzenhof and its famous Oberwolfach atmosphere, endorsed with mathematical hints. The name *Evangile selon Saint Nicolas* is an allusion to the works of Nicolas Bourbaki, an alias for that group of French mathematicians who wanted to rewrite mathematics entirely from scratch. During that time, almost no one in Germany had heard of Bourbaki. During the Nazi period, the so-called "Deutsche Mathematik" simply missed some important developments in mathematics, for instance in topology, the theory of distributions, and in complex and algebraic geometry. It is one of the most extraordinary achievements of the

Fig. 2 From left to right: René Thom, Jean Arbault, Jean-Pierre Serre and his wife Josiane, Jean Braconnier and Georges Reeb [9]

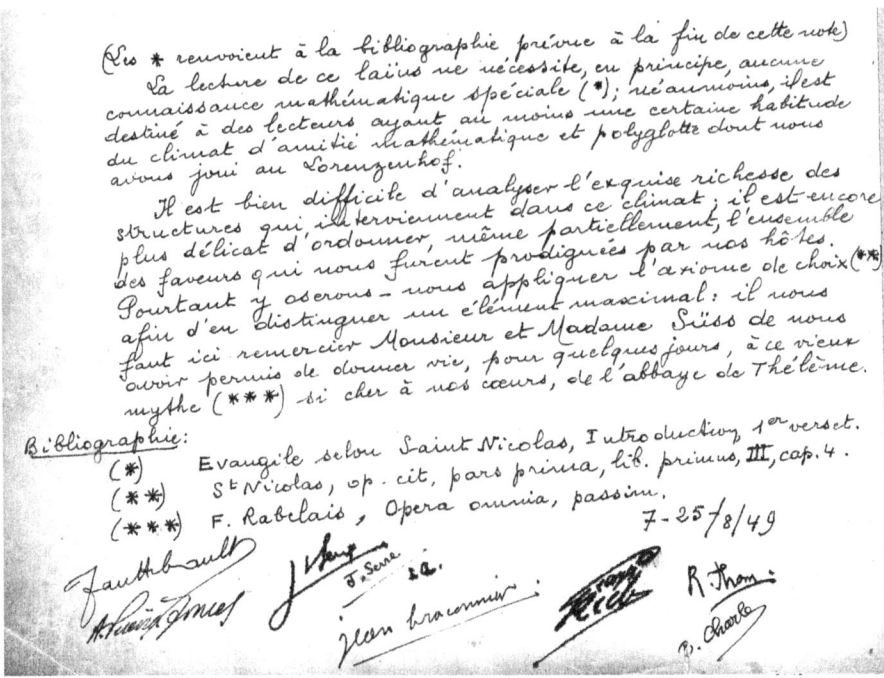

Fig. 3 Evangile selon Saint Nicolas from the first guest book of Oberwolfach [11]

small Oberwolfach workshops that those mathematicians who stayed in Germany and were not expelled by the Nazis were able to join the world's elite again [10].

The myth of the Abbaye de Thélème mentioned by the authors of the *Evangile selon Saint Nicolas* refers to the motto of the Abbey of Thélème, a utopic and idealized "anti-monastery" from Rabelais' *Gargantua*: the motto was "Fay ce que voudras" (Do as you please). Even today the Oberwolfach Institute is sometimes compared to an isolated monastery where mathematicians live and work together, only devoted to their science. Bourbaki has become a synonym for the modern development of mathematics being interested only in the development of its internal structures based on a few basic axioms. This restriction of the scientific objective implies a great freedom from external forces but, implicitly, also from responsibility for the consequences of its research. I think it is not a coincidence that the young Bourbaki group refers to the myth of the Abbaye de Thélème after the end of World War II.

Mathematical Research—Popularization Versus Communication

Having described some of the process of mathematical research, let me now consider the challenge of communicating mathematics and its research results.

The Popularization of Mathematics is Impossible

I would like to start with a provocative quotation by Reinhold Remmert (mathematician; born 1930) from 2007:

> We all know that it is not possible to popularize mathematics. To this day, mathematics does not have the status in the public life of our country it deserves, in view of the significance of mathematical science. Lectures exposing its audience to a Babylonian confusion and that are crowded with formulas making its audience deaf and blind, do in no way serve to the promotion of mathematics. Much less do well-intentioned speeches degrading mathematics to enumeration or even pop art. In Gauss' words mathematics is "regina et ancilla", queen and maidservant in one. The "usefulness of useless thinking" might be propagated with good publicity. Insights into real mathematical research can, in my opinion, not be given. [12] (See Fig. 4.)

Remmert's statement about the popularization of mathematics has a point. However, we have to distinguish between popularization and communication. While his statement may apply to popularization, it does, in my opinion, not apply to the communication of mathematics. Before I explain why, let me start with the difficulties that we face when trying to communicate mathematics.

Fig. 4 "Usefulness of useless thinking" [14]

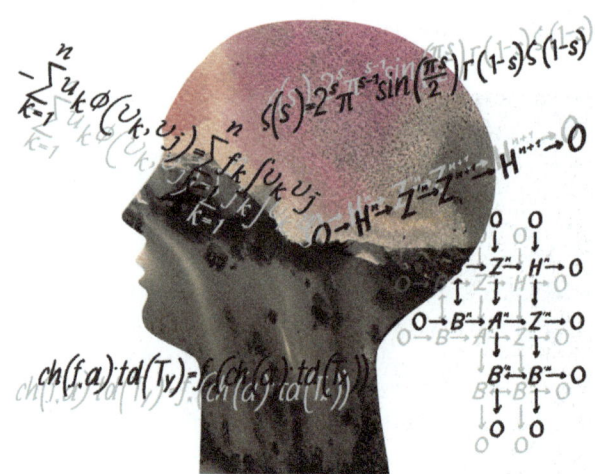

Structural Difficulties

First of all we may ask why it is not be possible to communicate mathematics. What is different in mathematics compared to other sciences? One could argue that for any research, regardless of whether it's in physics, chemistry, or biology, very specialized knowledge is required so that the popularization of research on the one hand and profoundness and correctness on the other do not go well together. Nevertheless, due to your own experience you will all have the feeling that mathematics might fall into a special category. In my opinion there are two significant structural reasons why it is so difficult to communicate or even popularize mathematics.

The first reason is that objects in modern mathematics are abstract creations of human thought. I do not wish to enter into a discussion of whether we only discover mathematical objects, which exist independently of our thoughts, or whether these objects are abstractions of human experience. Except for very simple ideas, like natural numbers or elementary geometrical figures, mathematical objects are not perceived, even if one can argue that they are not independent of perception. Objects like e.g. groups, vector spaces, or curved spaces in arbitrary dimension cannot be experienced with our five senses. They need a formal definition, which does not rely on our senses. Gaining an understanding of mathematical objects and relations is only possible after a long time of serious theoretical consideration.

Another reason is that mathematics has developed its own language, more than any other science. This is necessarily a result from the previous point that mathematics cannot be experienced directly. Therefore, each mathematical term needs a precise formal definition. This definition includes further terms that must be defined, and so on, so that finally a cascade of terms and definitions is set up that make a simple explanation impossible. But even in ancient times the abstraction from objects of our perception has always been a decisive part of mathematics, which made it difficult to comprehend. In Euclid's words: "There is no royal road to geometry" [13].

The language of mathematics requires an extremely compact presentation, a symbolism that allows replacing pages of written text by a single symbol. The peak of mathematical precision and compact information is a mathematical formula. But mathematical formulas frighten and deter. Stephen Hawking (physicist; born 1942) wrote in 1988: "... each equation I included in the book would halve the sales" [15].

The importance of "closed" mathematical formulas or equations might change in the future, being at least partly replaced by computer programs. However, this will not make communication easier.

These structural reasons support the thesis of Remmert that the nature of mathematical research cannot be popularized. And all mathematicians engaged in research would agree, that it is nearly impossible to feasibly illustrate to a mathematically untrained person the project one is currently working on. Actually, this experience applies not only to mathematically untrained people but even to mathematicians working in a different field.

The Need to Communicate Mathematics

Nonetheless, the statement that insights into the nature of mathematical research are not possible for a non-mathematician is for me hard to accept. Because this also implies quite a lot of resignation. As much as this statement might be true when limited to genuine mathematical research, it is not true when you take into consideration the fact that mathematical research has become a cultural asset of mankind during its development over 5,000 years.

Furthermore, it is my impression that everyone has a feeling for mathematics even if it is developed to different degrees. Each of you who has been around small children would know that already from an early age they take great pleasure in counting and natural numbers and have a quantitative grasp of their surroundings. They often love to solve little calculations. Regrettably, this interest often gets lost during schooling. I would even go so far as to introduce the following:

Thesis *In an overall sense, mathematical thinking is, after speech, the most important human faculty. It was this skill especially that helped the human species in the struggle for survival and improved the competitive abilities of societies. I believe that mathematical thinking has a special place in evolution.*

By mathematical thinking I mean analytic and logical thinking in a very broad sense, which is certainly not independent of the ability to speak. Of course, the development of mathematics as a science is a cultural achievement but, in contrast to languages, it developed in a similar way in different societies. We can face the fact that the importance of mathematics for mankind has grown continuously over the centuries, regardless of the cultural and social systems. No modern science is possible without mathematics and societies with highly developed sciences are in general more competitive than others. Attaching this value to mathematics, one must conclude the following:

Thesis *Society has the fundamental right to demand an appropriate explanation of mathematics. And it is the duty of mathematicians to face this responsibility.*

However, if mathematicians want to make their science easier to understand it will be at the expense of correctness. And that's a problem for mathematicians. All their professional training is necessarily based on being exact and complete. Mathematicians simply abhor to be inexact or vague. But in order to be understood by society, they will have to be just that [16]. I admit that this remains a continual conflict for every mathematician.

How Can We Raise Public Awareness in Mathematics?

In my experience there are two approaches for raising public interest in mathematics and demonstrating its significance: First, by examples that show the applicability of mathematics, and second, by examples that demonstrate the beauty and elegance of mathematics.

The first approach is certainly the favored one and it is often the only one accepted by politicians. However, we should not underestimate the second approach: it is often much more appealing and even crucial if we wish to get children interested in mathematics.

The elegance of a mathematical proof can really be intellectually fulfilling, e.g. the proof that the square root of 2 is an irrational number, or that there are infinitely many prime numbers. Both proofs can be given in advanced school classes. More accessible and therefore even more suitable for a larger audience is the beauty of geometrical objects. An example of this kind is the mathematical exhibition IMAGINARY with its beautiful pictures. For a more detailed description of IMAGINARY and the experiences of this travelling exhibition see the chapter in this book by Andreas Matt.

Mathematical Research—Intuition and Rigor

In the following I would like to try to explain a recent problem in my own research field of algebraic geometry and singularity theory. Although the explanation will mainly be by showing pictures, I am unable to avoid formulas. Nevertheless, I will be vague and I will simplify.

Algebraic Geometry

Algebraic geometry is, generally speaking, about the description of the set of solutions of a system of polynomial equations. Here I will focus on one equation only.

Fig. 5
$129/8x^4y - 85/8x^2y^3 +$
$57/32y^5 - 20x^4 - 21/4x^2y^2 +$
$33/8y^4 - 12x^2y + 73/8y^3 +$
$32x^2 = 0$

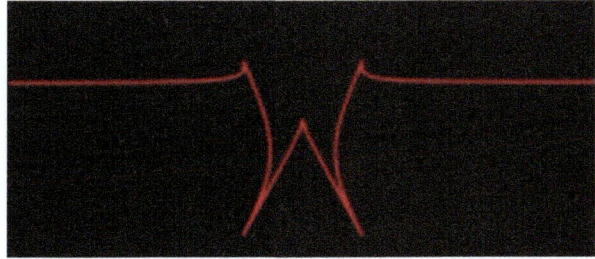

You might still remember from school the equation $y - x^2 = 0$ or $y = x^2$ for a parabola and $y^2 + x^2 - 1 = 0$ for a circle. Whereas a circle and a parabola are smooth curves, the next curve has singularities.

The equation of the curve in Fig. 5 is of degree 5, called a quintic, where the degree is the maximum sum of exponents in any of its terms. The curve has 5 peaks, called cusps, the maximum possible number for a quintic. The equation is rather complicated and we do neither realize its shape nor anything else from the equation, although it contains all the information.

Singularities

Before I get to a concrete recent research problem, let us have a look at some pictures demonstrating that singularities occur in our daily life. A parabolic mirror has an exact focus; the reflected rays meet at exactly one point, a "singularity" (usually "singular" refers to non-regular behavior, demonstrating an exception).

If the mirror is not a parabola, a focal curve, called a caustic in geometrical optics, develops and replaces the ideal focal point. This focal curve has its own singularity, a peak. Looking at the pictures, we can see the significance of a singularity: in the caustic curve it is the point where light energy has its highest intensity, the point of highest temperature.

One of the best-known caustics can be observed on sunny days in your cup of coffee, see Fig. 6. There even exist solar cookers, Fig. 7, which benefit practically from this singularity.

Curves with Many Singularities

A mathematical research problem in connection with singular curves is the following:

How many singularities can a plane curve of degree d have at most?

That an upper bound for the number of singularities should exist may be seen from the simplest example, a curve of degree $d = 1$. It is given by a linear equation,

Fig. 6 Coffee cup with caustic [17]

Fig. 7 Solar cooker [18]

which means that the curve is a line and hence has no singularities. With a bit more effort we can see that a curve of degree $d = 2$ (or $d = 3$) can have at most 1 (or 3) singularities, and these curves are realized by the union of 2 (resp. 3) lines which intersect in 1 (resp. 3) crossing points. In fact, the simplest singularities on a curve are crossing points, called nodes. The next simple ones are peaks, called cusps.

Now let C denote a plane curve of degree d with n nodes and k cusps. It was classically known and proved around 1920 by the Italian geometer Francesco Severi (1879–1961) that such a curve must satisfy

$$k + 2n \leqq 1/2d^2 + 3/2d$$

for very large d [19]. That is, the number of nodes plus two times the number of cusps can grow at most quadratically with the degree d, when d goes to infinity. However, it remained open for quite a while, whether curves with so many crossings and cusps really do exist. It was only known that $k + 2n$ can grow linearly in d, but it

Table 1 Upper and lower bounds for the number of nodes on a surface of degree d	d	1	2	3	4	5	6	7	8	d
	$n \leq$	0	1	4	16	31	65	104	174	$4/9d^3$
	$n \geq$	0	1	4	16	31	65	99	168	$5/12d^3$

was unknown whether there are curves of arbitrary high degree with $k + 2n$ growing quadratically with d.

This problem was finally solved affirmatively in 1989. In fact it was shown that for any k and n such that $k + 2n \leq 1/2d^2 - 2d + 3$ there exist curves with n nodes and k cusps [20]. The proof of this result required profound theorems of modern algebraic geometry such as vanishing theorems of cohomology and the resolution of singularities but also computer computations of concrete examples (using the computer algebra system SINGULAR [21]). It is worth noting that for any given degree the strongest results are obtained by combining theoretical results with computer computations or, in other words, by combining theory and practice.

World Record Surfaces

A similar question to the one for curves can be posed for surfaces or for algebraic varieties of any dimension, namely what is the maximum number of singularities on a variety of given degree? For such varieties of higher dimension the problem is more difficult than for curves. At the moment we do not know whether for surfaces there is an asymptotic behavior for the maximum number of singularities similar to that for curves.

Instead of looking for the asymptotic behavior of the number of singularities when the degree goes to infinity, we may look for "world record surfaces", that is, for surfaces of a small degree with the maximum possible number of singular points. This is a current research topic. A complete answer is known only up to degree $d = 6$ where the theoretical upper and the known lower bounds for the number of nodes coincide. For higher d there are upper and lower bounds but the exact maximum number of nodes is unknown for $d > 6$. The Table 1 presents the known results, where the second row shows the theoretical upper bound and the third row the maximum known lower bound.

Surfaces with singularities look amazingly attractive. For example, the Barth sextic [22] is a beautiful surface of degree 6 with the symmetry of an icosahedron (and with a terrible complicated equation) Fig. 8. It holds the world record with 65 singularities and this record can never be improved for $d = 6$. From the table you can see that for degree 7 the maximum possible number is 104, but the actual known number of singularities is only 99. To fill in this gap for $d \geq 7$ is a topical research problem in algebraic geometry.

Fig. 8 Barth sextic [23]

Geometry Versus Algebra

The reader may wonder whether producing a nice picture like Fig. 8 is the essence of algebraic geometry. One may ask: what significance do such pictures have for research? It might be surprising, but it is a fact that images are nearly irrelevant in research in modern algebraic geometry, at least as far as proofs are concerned.

However, for many mathematicians like me, pictures are an important source of intuition. Geometry and algebra stimulate different parts of your brain. By means of images you will get ideas, which you then try to prove rigorously by means of algebra. Pictures are also a means of communication. I am tempted to give a digression on algebra versus geometry (or topology), an apparent conflict passing through the history of mathematics from its beginning until now. Consider the well-known quotation by Hermann Weyl (mathematician; 1885–1955) from 1939:

> "In these days the angel of topology and the devil of abstract algebra fight for the soul of every individual mathematical domain." [24]

More than sixty years later Sir Michael Francis Atiyah (mathematician; born 1929) wrote in a similar vein:

> "Algebra is the offer made by the devil to the mathematician. The devil says: 'I will give you this powerful machine, it will answer any question you like. All you need to do is give me your soul: give up geometry and you will have this marvelous machine.' [...] when you pass over into algebraic calculation, essentially you stop thinking; you stop thinking geometrically, you stop thinking about the meaning." [25]

I must suppress a further digression but instead I would like to introduce my own thesis into this conflict, including the role of computers:

Thesis *Geometry gives intuition, algebra provides rigor, and the computer forces merciless precision. Step by step, from geometry to the computer, we are gaining certainty but we are losing some of the liberty in our thinking. Rigor and precision are prerequisites for correctness, but they are of limited value if they are applied without intuition.*

Research and Application—Theory and Practice

It cannot be denied and is a simple and easily verifiable fact that mathematics is applied in our everyday life. But the reason behind this fact remains hidden, as emphasized by Bourbaki:

> "That there is an intimate connection between experimental phenomena and mathematical structures, seems to be fully confirmed in the most unexpected manner by the recent discoveries of contemporary physics. But we are completely ignorant as to the underlying reasons for this fact (supposing that one could indeed attribute a meaning to these words) and we shall perhaps always remain ignorant of them." [26]

I am afraid we do not know much more about this connection than Bourbaki in 1950. On the other hand, Bourbaki clearly believed that the formal axiomatic method is a better preparation for new interpretations of nature, at least in physics, than any method that tries to derive mathematics from experimental truths. Of course, this view creates tensions. In the following I am going to touch on the historical and current tensions between pure research and applications, between theory and practice.

Everyday Applications of Mathematics

When we talk about the application of mathematics, we have to face the fact that mathematics is essential for new and innovative developments in other sciences as well as in the economy and for industry. I do not claim that only mathematics can provide innovation, but it is no exaggeration to claim that mathematics has become a key technology behind almost all common and everyday applications, which includes the design of a car, its electronic components and all security issues, safe data transfer, error correction codes in digital music players and mobile phones, and includes the design and optimization of logistics in any large enterprise. We may say that

Mathematics is the technology of technologies.

However, since the mathematical kernel of an innovation is in most cases not visible, the relevance of mathematics is either not acknowledged by the general public or simply attributed to the advances of computers. In 2008, the German Year of Mathematics, the book *"Mathematik—Motor der Wirtschaft"* [27], (*Mathematics—Motor of the Economy*) was published, giving 19 international enterprises and the German Federal Employment Agency a platform to describe how essential mathematics has become for their success. The main point of this publication was not to demonstrate new mathematics, but to show that, in contrast to a great proportion of the general public, the representatives of economy and industry are well aware of the important role of mathematics.

Hilbert's Vision

The application of mathematics in industry and the economy is certainly a part of our utilization of nature but, according to David Hilbert (mathematician; 1862–1943), mathematics, and *only* mathematics, is the foundation of nature and our culture in a fundamental sense:

> "The tool implementing the mediation between theory and practice, between thought and observation is mathematics. Mathematics builds the connecting bridges and is constantly enhancing their capabilities. Therefore it happens that our entire contemporary culture, in so far as it rests on intellectual penetration and utilization of nature, finds its foundation in mathematics."
> [28]

Based on his belief, Hilbert tried to lay the foundation of mathematics on pure axiomatic grounds, and he was convinced that it was possible to prove that they were without contradictions. The inscription on his gravestone in Göttingen expresses this vision with the words: *"We must know—we will know"*. Today it is no longer possible to fully adhere to Hilbert's optimism, due to the work of Gödel on mathematical logic showing that the truth of some mathematical theories is not decidable within mathematics. But Hilbert's statement about the *utilization* of nature is truer than ever.

On the other hand, this is no reason to glorify mathematics or to consider it superior to other sciences. First of all, the utilization of nature is not possible with mathematics alone. Many other sciences contribute, though differently, in the same substantial way. Secondly, the utilization of nature cannot be considered as an absolute value, as we know today. We are all a part of nature and utilization, as necessary as it is, can destroy nature and therefore part of our life.

Pure Versus Applied Mathematics

I use the terms "pure" and "applied" mathematics although it might be better to say "science-driven" and "application-driven" mathematics. In any case, here is a very provocative and certainly arrogant quotation of Godfrey Harold Hardy (mathematician; 1877–1947) from his much quoted essay *A Mathematician's Apology*:

> "It is undeniable that a good deal of elementary mathematics [...] has considerable practical utility. These parts of mathematics are, on the whole, rather dull; they are just the parts which have least aesthetic value. The 'real' mathematicians, the mathematics of Fermat and Euler and Gauss and Abel and Riemann, are almost wholly 'useless'." [29]

Hardy distinguishes between "elementary" and "real" (in the sense of interesting and deep) mathematics. The essence of his statement has two aspects: elementary mathematics, which can be applied, is unaesthetic and dull, while "real" mathematics is useless.

I think that Hardy is wrong in both aspects. Of course there exist interesting and dull mathematics. Mathematics is interesting when new ideas and methods prove to be fruitful in either solving difficult problems or in creating new structures for a deeper understanding. Routine development of known methods almost always turns out to be rather dull, and it is true that many applications of mathematics to, say, engineering problems are routine. However, this is not the whole story. Before applying mathematics, one has to find a good mathematical model for a real world problem, and this is often not at all elementary or trivial but a very creative process. This point is completely missing in Hardy's essay. Maybe, because pure mathematicians do not consider this as mathematics at all.

His other claim must also be refuted. Very deep and interesting results of "real" mathematics have become applicable, as we now know. That is, the border between interesting and dull mathematics is not between pure and applicable mathematics, but goes through any sub-discipline of mathematics, independent of whether it is applied or pure. In fact, many great mathematicians worked in pure as well as in applied mathematics. Let me quote Felix Klein (mathematician; 1848–1925) writing about Gauss, one of the greatest mathematicians ever:

> *"The work of Gauss in the field of applied mathematics I would like to call the high point of his life's work. The true core and basis of his achievements is founded in pure mathematics, a field he dedicated himself to youth."* [30]

Klein is seen by many as one of the last great mathematicians who combined both applied and pure mathematics in his work. Like Gauß, he started in pure mathematics and then later turned to applied mathematics. In my opinion, studying first pure and then applied mathematics has its advantages.

Applications Cannot Be Predicted—The Lost Innocence

Nowadays we know better than in Hardy's time that his statement about the uselessness of pure mathematics is wrong. The following quote by Hardy concerns his own research field, number theory:

"I have never done anything 'useful'. No discovery of mine has made, or is likely to make, directly or indirectly, for good or ill, the least difference to the amenity of the world." [31]

Shortly after Hardy's death the methods and results from number theory became the most important elements for public-key cryptography, which today is used millions of times daily for electronic data transfer in mobile phones and electronic banking. For his claim that deep and interesting mathematics is useless, Hardy calls Gauß and Riemann and also Einstein his witnesses:

"The great modern achievements of applied mathematics have been in relativity and quantum mechanics, and these subjects are, at present at any rate, almost as 'useless' as the theory of numbers. It is the dull and elementary parts of applied mathematics, as it is the dull and elementary parts of pure mathematics, that work for good or ill. Time may change all this. No one foresaw the applications of matrices and groups and other purely mathematical theories to modern physics, and it may be that some of the 'highbrow' applied mathematics will become 'useful' in as unexpected a way." [32]

The last sentence indicates that Hardy himself was skeptical about his own statements, although he did not really believe in the possibility of "real" mathematics becoming useful. However, the development of GPS, relying on the deep work of Gauss and Riemann on curved spaces and on Einstein's work on relativity, proves the applicability of their "useless" work.

This is not about blaming Hardy because he did not foresee GPS or the use of number theory in cryptography. Nevertheless, his strong statements are somewhat surprising as he was of course aware that, for example, Kepler used the theory of conic sections, a development of Greek mathematics without intended purpose, in order to describe the planetary orbits. So, what was the reason that Hardy insisted on the uselessness of "real" mathematics?

In my opinion we can understand Hardy's strong statements, made in 1940 at the beginning of World War II, only if we know that he was a passionate pacifist. It would have been unbearable for him to see that his own mathematics could be useful for the purpose of war. He was bitterly mistaken.

We all know that nowadays the most sophisticated mathematics, pure and applied, is a decisive factor in the development of modern weapon systems. Without GPS, and hence without the mathematics of Gauss and Riemann, this would have been impossible. Even before that, the atomic bomb marked the first big disillusionment for many scientists regarding the innocence of their work. If there was ever a paradise of innocence with no possibility for mathematics to do 'good or ill' to mankind, it was lost then.

Applicability Versus Quality of Research

History shows, and the statements of Hardy prove this, that it is impossible to predict which theoretical developments in mathematics will become "useful" and will have an impact on important applications in the future. The distinction between pure and applied mathematics is more a distinction between fields than between applicability. Quite often we notice that ideas from pure research, only aiming to explore the structure of mathematical objects and their relations, become the basis for innovative ideas creating whole new branches of economic and industrial applications. Besides number theory for cryptography, I would like to mention logic for formal verification in chip design, algebraic geometry for coding theory, computer algebra for robotics, and combinatorics for optimization applied to logistics, to name just a few.

Although the list of applications of pure mathematics could be easily enlarged, it is also clear that some parts of mathematics are closer to applications than others. These are politically preferred and we can see that more and more national and international programs support only research with a strong focus on applications or even on collaboration with industry.

I think the above examples show that it would be a big mistake, if applicability were to become the main or even the only criterion for judging and supporting mathematics. In this connection I like to formulate the following:

Thesis *The value of a fundamental science like mathematics cannot be measured by its applicability but only by its quality.*

History has shown that in the long run, quality is the only criterion that matters and that only high-quality research survives. It is worthwhile to emphasize again that any kind of mathematics, either science driven or application driven, can be of high or low quality.

In view of the above and many more examples, one could argue that we would miss unexpected but important applications by restricting mathematical research to a priori applicable mathematics. This is certainly true, it is, however, not the main reason why I consider it a mistake to judge mathematics by its applicability. My main reason is that it would reduce the mathematical sciences to a useful tool, without a right to understand and to further develop the many thousands of years of cultural achievements of the utmost importance. This leads us to reflect on freedom of research.

Freedom of Research and Responsibility

Freedom of research has many facets, for example in the sense of "Fay ce que voudras" of the Abbaye de Thélème, mentioned earlier, or just emphasizing unconditional research. It implies in any case that the scientist himself defines the direction of research.

In mathematics there is an even more fundamental aspect. Today's mathematics is often searching for inner mathematical structures, only committed to its own axioms and logical conclusions and thus keeping it free from any external restriction.

This was clearly intended by the creators of modern axiomatic mathematics. Georg Cantor (mathematician; 1845–1918), the originator of set theory, proclaims: "The nature of mathematics is its freedom" [33] and David Hilbert considers this freedom as a paradise: "Nobody shall expel us from the paradise created for us by Cantor" [34].

This kind of freedom was certainly felt by the group of young Bourbakists meeting in Oberwolfach in 1949 when they referred to the myth of the Abbaye de Thélème mentioned above, and many mathematicians of today feel the same way.

On the other hand, there are reasons to question this freedom as an absolute value into question, because it does also imply freedom from responsibility. However, we must emphasize that this does not excuse the individual scientist as a human being from his responsibilities. The physicist Max Born wrote in 1963:

> "Although I never participated in the application of scientific knowledge to any destructive purpose, like the construction of the A-bomb or H-bomb, I feel responsible." [35]

It may be argued that the self-referential character of the science is, at least partially, responsible for the lack of responsibility. This is emphasized by Egbert Brieskorn (mathematician; born 1937) who not only deplores this character but even goes a step further in claiming that this attribute implies the possibility of assuming and misusing power:

> "The restriction on pure perception of nature by combining experiment and theoretical description by means of mathematical structures is the subjective condition to evolve this science as power. The development of mathematics as self-referential science enforces the possibility to seize power for science as a whole. [...] It belongs to the nature of the human being to prepare and to take possession of the reality. We should not feel sorrow about that, however, we should be concerned that the temptations of power is threatening to destroy our humanity." [36]

The self-referential character appears clearly in Hilbert's and Bourbaki's concept of mathematical structures based on the axiomatic method. This concept was of great influence in the development of mathematics in the twentieth century. It was, however, never without objections and nowadays it is certainly not the driving force anymore. In theoretical mathematics the most influential new ideas arise from a deep interaction with physics, in particular with quantum field theory. Atiyah even calls this the "era of quantum mathematics". [37] Applied mathematics such as numerical analysis or statistics, on the other hand, has always been too heterogeneous to be adequately covered by Bourbaki's approach. It is often driven by challenging problems from real world applications. But I do not see that this fact makes it less vulnerable to the temptations of power, maybe even to the contrary.

Not denying this threat for any kind of mathematics, I like to point out that freedom of research is a precious gift, related to freedom of thought in an even broader sense. Mathematicians for example are educated to use their own brains, to doubt any unsubstantiated claim, and not to believe in authority. A mathematical theorem is true not because any person of high standing or of noble birth claims it, but because we can prove it ourselves. In this sense I like to claim:

Thesis *Mathematical education can contribute to freedom of thought in a broad sense.*

On the other hand, being aware of the "lost innocence" and the fact that mathematics can be "for good or ill" to mankind, freedom must be accompanied by responsibility. The responsibility for the impact of their work, though not a part of science itself and not easy to recognize, remains the task for each individual mathematician.

Thesis *Freedom of research must be guaranteed in mathematics and in other sciences. It has to be defended by scientists, but it must be accompanied by responsibility.*

IMAGINARY—Mathematical Creations and Experiences

Let me return to the communication of mathematics. As explained above, this is by no means an easy task, but mathematicians themselves have to make the effort to communicate their science. In fact, many mathematicians do so with remarkable success. The present book is a proof of these efforts.

IMAGINARY is mathematics as art: geometry presented as an attractive visual world. It started as a travelling exhibition, created by the Mathematisches Forschungsinstitut Oberwolfach, in the German Year of Mathematics 2008. Its aim is to interest people in mathematics by showing them the beauty of mathematical objects and to fill them with inspiration and stimulate their imaginations. The exhibition has been shown in more than 35 cities in different countries with several hundred thousand visitors, and its success has been overwhelming. It is interesting that only journalists have asked for applications while the other visitors have experienced the unexpected beauty and the "joy of comprehension". I would like to show two pictures from the IMAGINARY art gallery and otherwise refer the reader to the chapter in this book by Andreas Matt [38], and the web page www.imaginary-exhibition.com (see Figs. 9 and 10).

The pictures were created for an online competition in cooperation with a German science magazine, using the free IMAGINARY software *surfer* and won the first and third prizes. Details, including the equations, can be found at Spektrum der Wissenschaften, Mathematik-Kunst-Wettbewerb [38].

It seems that the success of IMAGINARY will continue. Besides the interest in the ongoing exhibitions there is a surprisingly high demand for online programs which allow individual mathematical experimentation, and there is also a demand for further background information. So far we have registered about 250,000 downloads of the IMAGINARY software and about 70,000 downloads of mathematical

Fig. 9 Tropenwunder
(tropical wonder)

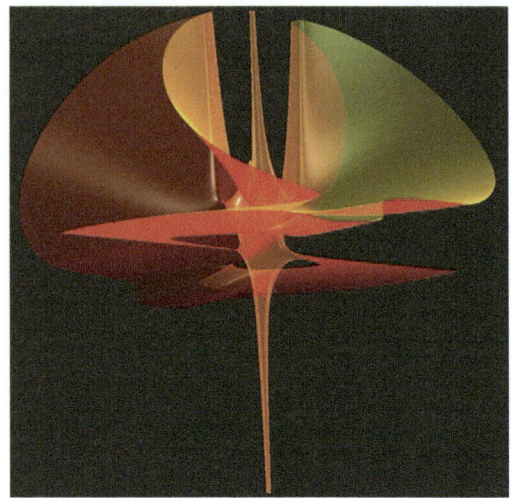

Fig. 10 Ikosidodekaeder
(icosidodecahedron)

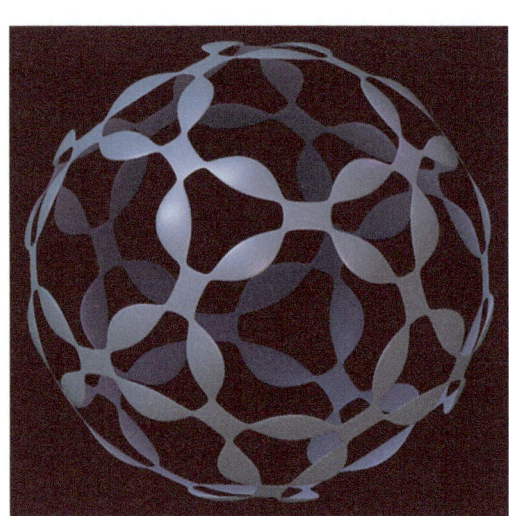

material [39]. In particular the figures demonstrate that a substantial proportion of the general public is interested in mathematics, if it is presented in an appealing manner.

Acknowledgements   I would like to thank Stephan Klaus, Andreas Matt, Andrew Ranicki, Reinhold Remmert, and Jose-Franzisco Rodrigues for their helpful comments.

Notes and References

Several of the quotations in this article are common knowledge but, when cited, their origin is often not documented. I have made special effort to give the original source or, when this turns out to be impossible, to provide a reliable "second-hand" source. Moreover, whenever I have access to the original source of a quotation and when I find it helpful, I cite also some of the surrounding text. In order to allow a smooth reading I give in the main text an English translation when the original quotation is not in English, while the References contain the original version.

1. "What the scientist aims at is to secure a logically consistent transcript of nature. Logic is for him what the laws of proportion and perspective are to the painter, and I believe with Henri Poincaré that science is worth pursuing because it reveals the beauty of nature. And here I will say that the scientist finds his reward in what Henri Poincaré calls the joy of comprehension, and not in the possibilities of application to which any discovery of his may lead."

   Albert Einstein, in: *Epilogue, A Socratic Dialogue*, p. 211, Interlocutors: Max Planck, Albert Einstein, James Murphy. In: Max Planck, "Where is Science Going?" Norton, New York, 226 pages (1932)

2. "Philosophy is written in that great book which ever lies before our eyes—I mean the universe—but we cannot understand it if we do not first learn the language and grasp the symbols, in which it is written. This book is written in the mathematical language, and the symbols are triangles, circles and other geometrical figures, without whose help it is impossible to comprehend a single word of it; without which one wanders in vain through a dark labyrinth."

   Galileo Galilei, in: *Opere Complete di Galileo Galilei*, Firenze, 1842, ff, vol. IV, p. 171, as quoted by Edwin Arthur Burtt: *The Metaphysical Foundations of Modern Science*, p. 75, Dover Reprint, New York, 352 pages (2003)

3. "Ich komme aus dem Lande, das Land der Mathematiker geblieben ist, [...] auch der mathematischen Studien, welche die Seele aller industriellen Fortschritte sind."

   Attributed to Alexander von Humboldt, in: Roland Z. Bulirsch, *Weltfahrt als Dichtung*, p. 14, in: "Dokumentation zur Verleihung des Literaturpreises der Konrad-Adenauer-Stiftung e.V. an Daniel Kehlmann", 52 pages (2006)

4. " 'Ohne Mathematik tappt man doch immer im Dunkeln', das schrieb vor mehr als 150 Jahren Werner von Siemens an seinen Bruder Wilhelm."

   Quote taken from: Peter Löscher, *Siemens AG*, p. 99, in: "MATHEMATIK – Motor der Wirtschaft", Eds. G.-M. Greuel, R. Remmert, G. Rupprecht; Springer Verlag, 125 pages (2008)

5. "Es gehört eine gewisse Kühnheit dazu, in einer Kultur, die sich durch ein profundes mathematisches Nichtwissen auszeichnet, derartige Übersetzungsversuche zu unternehmen." Hans Magnus Enzensberger, *Zugbrücke außer Betrieb - Drawbridge Up*, p. 44, A K Peters LTD, Natick, MA, 48 pages (1999)

6. Archives of the Mathematisches Forschungsinstitut Oberwolfach

7. The text of this section is, slightly modified, taken from: *Editorial of the Ober-wolfach Reports*, published by the Mathematisches Forschungsinstitut Oberwolfach in cooperation with the European Mathematical Society

8. "Der erste große Name im Gästebuch ist Heinz Hopf (Zürich), ein Topologe von Weltruf. Im November 1946 war Henri Cartan, dessen Familie während der deutschen Besatzung großes Leid erdulden mußte, zu Besuch. Ohne Hopf und Cartan wäre Oberwolfach damals vielleicht eine "Sommerfrische für Mathematiker" geblieben, wo würdige Herren in beschaulicher Ruhe klassische Theorien polierten. Gott sei Dank kam es anders."

   Reinhold Remmert, *Mathematik in Oberwolfach – Erinnerungen an die ersten Jahre*, p. 1. Grußwort zur Einweihung der Bibliothekserweiterung am 5. Mai 2007, published by the Mathematisches Forschungsinstitut Oberwolfach, 27 pages (2008). See also Annual Report 2007 of the Mathematisches Forschungsinstitut Oberwolfach, http://www.mfo.de/scientific-programme/publications/annual-publications

9. Archives of the Mathematisches Forschungsinstitut Oberwolfach

10. For a historical appreciation of the "Deutsch-Französische Arbeitsgemeinschaft", a meeting of young German and French mathematicians in Oberwolfach, see: Maria Remenyi, *Oberwolfach im August 1949: Deutsch-Französosische Sommerfrische*, Math. Semesterber., Mathematische Bildergalerie, Springer (2011)

11. "Das Lesen dieser Zeilen erfordert im Prinzip keinerlei spezielle (∗) mathematische Kenntnisse, dennoch sind sie für Leser bestimmt, die zumindest ein gewisses Gefühl entwickelt haben für die mathematisch und vielsprachig freundschaftliche Atmosphäre, an der wir uns auf dem Lorenzenhof erfreut haben. Es ist äußerst schwierig, die auserlesene Vielfalt der Strukturen, die diese Atmosphäre zustande bringt, zu analysieren; es ist zudem noch viel schwieriger, die Gunstbeweise, die uns durch unsere Gastgeber zu Teil wurden, in ihrer Gesamtheit auch nur zum Teil einzuordnen. Dennoch wagen wir es hier, das Auswahlaxiom (∗∗) anzuwenden, um ein maximales Element auszuzeichnen: unseren Dank an Herrn und Frau Süss, die es uns ermöglichten, für einige Tage diesem alten Mythos (∗ ∗ ∗) der Abbaye de Thélème Leben zu verleihen, der uns so sehr am Herzen liegt.

   Literaturangaben:

   (∗) Sankt Nikolaus Evangelium, Einleitung, 1. Vers
   (∗∗) Sankt Nikolaus, op. cit. pars prima, lib. primus, III, Kapitel 4
   (∗ ∗ ∗) F. Rabelais, Opera omnia, passim.

   Jean Arbault, Jean-Pierre Serre, René Thom, A. Pereira Gomez, Josiane Serre, Georges Reeb, Bernard Charles, Jean Braconnier."

   Guestbook of the Mathematisches Forschungsinstitut Oberwolfach, No 1, p. 2. German translation see [8], p. 7 and [10]. Online at http://oda.mfo.de/view/viewer.jsf

12. "Wir alle wissen, daß Mathematik nicht popularisierbar ist. Sie hat bis heute im öffentlichen Leben unseres Landes nicht die Stellung, die ihr Kraft der

Tragweite ihrer Inhalte zukommt. Vorträge, wo die Hörer vom babylonischen Sprachgewirr und Formelgestrüpp taub und blind werden, eignen sich nicht für Werbung. Noch weniger helfen gut gemeinte Reden, wo Mathematik zu einer Rechenkunst oder gar Pop-Kultur erniedrigt wird. Mathematik ist nach Gauß 'regina et ancilla', Königin und Magd in einem. Die 'Nützlichkeit nutzlosen Denkens' kann man vielleicht öffentlichkeitswirksam propagieren, Einblicke in das Wesen mathematischer Forschung lassen sich nach meiner Erfahrung nicht geben." [8], loc. cit., p. 20

13. According to Proclus, a neo platonist (412–485 A.D.), Euclid replied to King Ptolemy, who asked whether he could not learn geometry more easily than by studying the *Elements*: "There is no royal road to geometry."

    Quote taken from http://www.1902encyclopedia.com/E/EUC/euclid-mathematician.html

14. Poster of the Mathematisches Forschungsinstitut Oberwolfach. Design by Boy Müller

15. "Someone told me that each equation I included in the book would halve the sales. I therefore resolved not to have any equations at all. In the end, however, I did put in one equation, Einstein's famous equation, $E = mc^2$. I hope that this will not scare off half of my potential readers."

    Stephen W. Hawking, *A Brief History of Time*, "Acknowledgments", Bantam Dell Publishing Group, 224 pages (1988)

16. Compare this with the dialogue from the preface of Ian Stewart's *The Problems of Mathematics* (Oxford Univ. Press, 1987) where a mathematician is chatting with a fictional layman "Seamus Android":

    – *Mathematician*: It's one of the most important discoveries of the last decade!
    – *Android*: Can you *explain* it in words ordinary mortals can understand?
    – *Mathematician*: Look, buster, if ordinary mortals could understand it, you wouldn't need mathematicians to do the job for you, right? You can't get a feeling for what's going on without understanding the technical details. How can I talk about manifolds without mentioning that the theorems only work if the manifolds are finite-dimensional paracompact Hausdorff with empty boundary?
    – *Android*: Lie a bit.
    – *Mathematician*: Oh, but I couldn't do that!
    – *Android*: Why not? Everybody *else* does.
    – *Mathematician* (tempted, but struggling against a lifetime's conditioning): But I *must* tell the truth.
    – *Android*: Sure. But you might be prepared to bend it a little, if it helps people understand what you're doing.
    – *Mathematician* (sceptical, but excited at his own daring): Well, I suppose I could give it a *try*.

    Quote taken from: [5], loc. cit., p. 45–47

17. Picture by Christian Ucke, http://users.physik.tu-muenchen.de/cucke

18. Picture from http://www.atlascuisinesolaire.com

19. Francesco Severi, *Vorlesungen über algebraische Geometrie*, Teubner, 408 pages (1921)
20. Gert-Martin Greuel, Christoph Lossen, Eugenii Shustin, *Plane curves of minimal degree with prescribed singularities*, Invent. Math. 133, 539–580 (1998)
21. Gert-Martin Greuel, Gerhard Pfister, Hans Schoenemann, *SINGULAR—A Computer Algebra System for Polynomial Computations*, free software, http://www.singular.uni-kl.de (1990–to date)
22. Wolf Barth, *Two Projective Surfaces with Many Nodes Admitting the Symmetries of the Icosahedron*, J. Alg. Geom. 5, 173–186 (1996)
23. Picture produced with the ray tracer *surfer*, free software, http://www.imaginary-exhibition.com/surfer.php
24. "In this purely algebraic way based on the adjunction argument we master the orthogonal and the symplectic invariants. This procedure has even stood the test in certain special cases where the statement of full reducibility breaks down.

    In these days the angel of topology and the devil of abstract algebra fight for the soul of each individual mathematical domain. [. . .].

    I feel bound to add a personal confession. In my youth I was almost exclusively active in the field of analysis; the differential equations and expansions of mathematical physics were the mathematical things with which I was on the most intimate footing. I have never succeeded in completely assimilating the abstract algebraic way of reasoning, and constantly feel the necessity of translating each step into a more concrete analytic form. But for that reason I am perhaps fitter to act as intermediary between old and new than the younger generation which is swayed by the abstract axiomatic approach, both in topology and algebra."

    Hermann Weyl, *Invariants*, pp. 500–501, Duke Mathematical Journal 5, 489–502 (1939)
25. "One way to put the dichotomy in a more philosophical or literary framework is to say that algebra is to the geometer what you might call the 'Faustian offer'. As you know, Faust in Goethe's story was offered whatever he wanted (in his case the love of a beautiful woman), by the devil, in return for selling his soul. Algebra is the offer made by the devil to the mathematician. The devil says: 'I will give you this powerful machine, it will answer any question you like. All you need to do is give me your soul: give up geometry and you will have this marvellous machine.' (Nowadays you can think of it as a computer!) Of course we like to have things both ways; we would probably cheat on the devil, pretend we are selling our soul, and not give it away. Nevertheless, the danger to our soul is there, because when you pass over into algebraic calculation, essentially you stop thinking; you stop thinking geometrically, you stop thinking about the meaning."

    Sir Michal Atiyah, *Special Article—Mathematics in the 20th Century*, p. 7, Bull. London Math. Soc. 34, 1–15 (2002)
26. Nicholas Bourbaki, *The Architecture of Mathematics*, p. 231, Amer. Math. Monthly 57, No. 4, 221–232 (1950)
27. Gert-Martin Greuel, Reinhold Remmert, Gerhard Rupprecht, Eds., *MATHEMATIK – Motor der Wirtschaft*, see [4]

28. "Das Instrument, welches die Vermittlung bewirkt zwischen Theorie und Praxis, zwischen Denken und Beobachten, ist die Mathematik; sie baut die verbindende Brücke und gestaltet sie immer tragfähiger. Daher kommt es, dass unsere ganze gegenwärtige Kultur, soweit sie auf der geistigen Durchdringung und Dienstbarmachung der Natur beruht, ihre Grundlage in der Mathematik findet."

    David Hilbert, *Naturerkennen und Logik*, Versammlung Deutscher Naturforscher und Ärzte in Königsberg, 1930. Quote taken from: http:// quantumfuture.net/gn/zeichen/hilbert.html, linking to an mp3 version of the original speech by Hilbert. For the English translation see: http://math.ucsd.edu/ ~williams/motiv/hilbert.html

29. Godfrey H. Hardy, *A Mathematician's Apology*, pp. 32–33, Cambridge University Press, 52 pages (1940)

30. "Die besprochenen Arbeiten von Gauß auf dem Gebiet der angewandten Mathematik möchte ich als Krönung seines Lebenswerkes bezeichnen. Der eigentliche Kern und das Fundament seiner Leistungen aber liegt auf dem Gebiet der *reinen Mathematik*, der er sich in seinen Jugendjahren widmete."

    Felix Klein, *Vorlesungen über die Entwicklung der Mathematik im 19. Jahrhundert*, p. 24, Reprint, Springer Verlag, 208 pages (1970)

31. See [29], loc. cit. p. 49

32. See [29], loc. cit. p. 39

33. "Dagegen scheint mir aber jede überflüssige Einengung des mathematischen Forschungstriebes eine viel größere Gefahr mit sich zu bringen und eine um so größere, als dafür aus dem Wesen der Wissenschaft keinerlei Rechtfertigung gezogen werden kann; denn das *Wesen* der *Mathematik* liegt gerade in Ihrer *Freiheit*."

    Georg Cantor, *Gesammelte Abhandlungen*, p. 182, Ed. Ernst Zermelo, Springer, 486 pages (1932)

34. "Aus dem Paradies, das Cantor uns geschaffen hat, soll uns niemand vertreiben können." David Hilbert, *Über das Unendliche*, p. 170, Math. Ann. 95, 161–190 (1926)

35. "Obwohl ich an der Anwendung naturwissenschaftlicher Kenntnis für zerstörerische Zwecke, wie die Herstellung der A-Bombe oder der H-Bombe, nicht teilgenommen habe, fühle ich mich verantwortlich."

    Max Born, *Erinnerungen und Gedanken eines Physikers*, in: Max und Hedwig Born, Der Luxus des Gewissens: Erlebnisse und Einsichten im Atomzeitalter, p. 73, Nymphenburger Verlagshandlung, 200 pages (1969)

36. "Die Beschränkung auf reine Naturerkenntnis durch die Verbindung von Experiment und theoretischer Beschreibung mit Hilfe mathematischer Strukturen ist die subjektive Bedingung der Möglichkeit der Entfaltung dieser Wissenschaft als Macht. Die Entwicklung der Mathematik als selbstreferentielle Wissenschaft verstärkt die Machtförmigkeit der Wissenschaft insgesamt. [...] Daß der Mesch sich der Wirklichkeit bemächtigt, daß er sie sich zurechtmacht, gehört zu seinem Wesen. Darüber soll man nicht traurig sei, wohl aber darüber, daß die Verführung der Macht unsere Menschlichkeit zu zerstören droht."

Egbert Brieskorn, *Gibt es eine Wiedergeburt der Qualität in der Mathematik?*, p. 257–258, in: Wissenschaft zwischen Qualitas und Quantitas, Ed. Erwin Neuenschwander, Birkhäuser Verlag, 444 pages (2003)

37. "I have said the 21st century might be the era of quantum mathematics or, if you like, of infinite-dimensional mathematics. What could this mean? Quantum mathematics could mean, if we get that far, 'understanding properly the analysis, geometry, topology, algebra of various non-linear function spaces', and by 'understanding properly' I mean understanding it in such a way as to get quite rigorous proofs of all the beautiful things the physicists have been speculating about."

Sir Michal Atiyah, *Special Article—Mathematics in the 20th Century*, p. 14, Bull. London Math. Soc. 34, 1–15 (2002)

38. The picture Tropenwunder was created by Hiltrud Heinrich and Ikosidodekaeder was created by Martin Heider. See http://www.spektrum.de/blatt/d_sdwv_extra_artikel&id=947549&_z=798888&_z=798888

39. See the IMAGINARY website: www.imaginary-exhibition.com

Keeping Mathematical Awareness Alive

Vagn Lundsgaard Hansen

Abstract Caring for the development of abstract mathematics is not at first sight useful and in modern society the applications of mathematics are more sophisticated and more difficult to grasp than in earlier times. This may easily lead to a decline in the emphasis on mathematical knowledge and thought in society, in particular in the educational system. The increasing pressure from governments for more focus on immediate applications of mathematics in education and in mathematical research at universities makes it a huge task to inform the public about the usefulness and necessity of understanding the abstract mathematical thinking required to develop fundamental new applications of mathematics. How can we handle it?

Introduction

With the possible exception of the ancient Romans, most civilizations throughout history have been aware of the usefulness and importance of mathematics in connection with the welfare, the development and survival of their civilization. The Romans were practical people and did not care much about the development of mathematical ideas and theory; see e.g. [1]. Prior to the Romans, scholars in the ancient Greek civilization, represented by such giants as Euclid and Archimedes, developed both mathematical ideas and solved practical problems. After the Romans, scholars in the Arab civilization again paid attention to both abstract (theoretical) and concrete (practical) aspects of mathematics.

Caring for the development of abstract mathematics (mathematical ideas) is not at first sight useful to people and in modern society the applications of mathematics are more sophisticated and more difficult to grasp than in earlier times. This makes it increasingly difficult to communicate even elements of contemporary mathematics to the public and hence stimulates the attitude that new developments and discoveries in mathematics are things you do not have to bother about. This may easily lead

Article presented at the RPAM conference in Óbidos, Portugal, 26–29 September 2010.

V.L. Hansen (✉)
Technical University of Denmark, Copenhagen, Denmark
e-mail: v.l.hansen@mat.dtu.dk

E. Behrends et al. (eds.), *Raising Public Awareness of Mathematics*,
DOI 10.1007/978-3-642-25710-0_27, © Springer-Verlag Berlin Heidelberg 2012

to a general decline in the emphasis on mathematical knowledge and thought in society, in particular in the educational system. Some will say that we have already begun the decline with the increasing pressure from governments for more focus on immediate applications of mathematics in education and in mathematical research at universities. It is a demanding task to inform the public about the usefulness and necessity of understanding the abstract mathematical thinking required to develop fundamental new applications of mathematics that may be of great importance for the future welfare of modern society.

Raising the Public Awareness of Mathematics in the Past

In his *Elements* from around 300 BC, Euclid was the first mathematician in history to produce a substantial account of abstract mathematics paying no attention to the reasons for producing the work and with no indication of the scope of applications of the material developed. Supposedly it was unnecessary to show the motivation at a time where only very few occupied themselves with intellectual work. As long as the mathematical ideas were accessible to the scholars who knew the context of the problems under discussion, there was little need to know the motivation of new abstract mathematical ideas. The relation to concrete phenomena was intimate and provided the motivation.

In the natural sciences, it is fair to say that the abstraction process was first fully exploited by the great natural philosophers René Descartes (1596–1650), Galileo Galilei (1564–1642) and Isaac Newton (1642–1727), in connection with the formulation of the basic laws of mechanics. Maybe one can even go further and assert that the application of abstract reasoning in the quantitative study of concrete phenomena has its origin here, since qualitative abstract thinking prevailed and totally dominated quantitatively based explanations of concrete natural phenomena in the natural philosophy of Aristotle (384–322 BC) and his followers. Starting with Galileo the experiment became an integral part of the scientific process in the natural sciences; see e.g. [2, 3]. This does not imply that experiment became a substitute for abstract reasoning—on the contrary. Galileo's conclusion that heavy and light bodies fall to the ground with the same speed was not only based on actual experiments, but also on abstract reasoning that if a body is divided into smaller parts, then each of the parts will fall with the same speed as the body itself.

From the beginning of the nineteenth century, mathematics experienced a dramatic change in the importance of abstract ideas mainly from algebra and new, very abstract mathematical disciplines emerged. By the end of the twentieth century, the mathematics used in society had reached a level of abstraction only accessible to very few but of fundamental importance for society. This has called for more concerted actions for making mathematics visible to the great public since taxpayers after all provide most of the money for basic research in mathematics.

Fig. 1 Poster for the World
Mathematical Year 2000

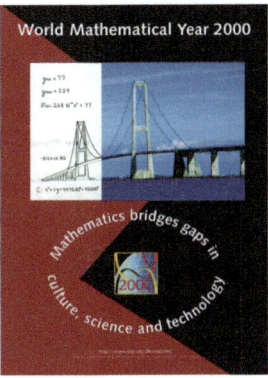

Mathematics in Culture, Science and Technology

For the World Mathematical Year 2000, I designed a poster with the text "Mathematics bridges gaps in culture, science and technology" (Fig. 1). The image on the poster was provided by my good friend Marit Hvalsøe Schou. It shows the bridge connecting the Danish islands Fyn and Sjælland ("Storebæltsbroen"), such that half of the bridge is shown as a photo and the other half as a "technical" drawing. The image shows to the public in an intuitive way that abstract mathematical ideas and scientific laws lie behind the impressive technological constructions, which enrich our culture and are of great importance for modern society.

I suppose that most people will connect the image on the poster to engineering and not to mathematics. Pictures alone will, most likely, never make the layman think about mathematics. A text on a poster is therefore important if you want to convey a message about mathematics, and the text needs to be formulated in terms that can be remembered by a layman if you intend to stimulate further thoughts.

The images on a poster will most surely have to come from applications of mathematics and will at first sight therefore only stimulate the application aspects of mathematics. This should be remembered by all kinds of mathematicians since the investigation of abstract pure structures in mathematics in isolation from the concrete origins of the structures will not in itself provide sufficient evidence for the importance of mathematics in society. On the other hand the level of complexity of mathematical models used in contemporary society is strongly dependent on abstract mathematical ideas from very recent pure mathematical research. Luckily enough, most mathematicians nowadays will agree that mathematics has the dual nature of combining abstract thinking with the description of concrete phenomena and that the strength of mathematics is to be found in the symbiosis between the abstract structures and the concrete applications.

Abstract mathematics has become a driving force in many sciences and lies behind many deep and important developments in society. In addition to applications of vintage mathematics, probabilistic and stochastic ways of thinking, and even very recent mathematical results, are now an integral part of high-tech constructions and decision-making of all kinds.

Mathematical Conversations

Most people find mathematics impossible as a topic for conversation and some will even resent it strongly. On the other hand, there are also many people who are prepared to engage in a mathematical conversation if you can make it short and interesting. In fact, I believe that a substantial part of the population in most countries will be interested under the right circumstances.

Many kinds of toys, games and puzzles touch the same mental processes as mathematics, such as imagination, intuition, systemization of experiences, combination of information, etc. And rational decisions in daily life and society at large also indirectly use mental processes relating to mathematics, like analyzing an argument, making a hypothesis, testing the validity, reaching conclusions and finally making forecasts. If you can convince people that the abilities which they do not think of as related specifically to mathematics are among the most basic ideas pursued in mathematics, they will not immediately be frightened away from having a mathematical conversation.

One of the difficult issues in a discussion with an educated layman is always to explain what mathematics is about. In [4], I offer the following explanation:

"Mathematics is the science of structures and patterns. Mathematicians look for structures in nature, patterns in daily life, structures essential for the functioning of society, etc., seeking rational explanations of the phenomena, and building abstract models by which to predict and control them."

In a speech [5] to the Balzan Prize Foundation in 1999, Mikhael Gromov formulated his views of mathematics slightly differently but in the same spirit:

"The task of mathematics and mathematicians is to articulate the visible regularities in the physical and mental worlds, and to find new structural patterns unperceivable by direct intuition and common sense."

There are many surprising and counter-intuitive phenomena that you can discuss. From geometry, I can mention among others *curves of constant width* and the *Dirac's string problem* (see e.g. [6]). Briefly, Dirac's String Problem is to explain mathematically why you can untwist a double twisting of loose (or elastic) strings with fixed ends without cutting it up, and why, on the other hand, you cannot do it with strings having only a single twist (Fig. 2). In [7], Ehrhard Behrends presents several good examples from geometry as well as from other subjects.

Mathematical Storytelling

When it comes to mathematical storytelling, the masterpiece *Alice's Adventures in Wonderland* must be the greatest book of all. It was written by the Oxford mathematician Charles Dodgson under the pen-name Lewis Carroll in 1865. You can get a vivid impression of the circumstances behind this book by reading Robin Wilson [8].

Alice's Adventures in Wonderland contains scenes inspired by genuine mathematical results even fairly advanced results. My favorite is the one in which Alice

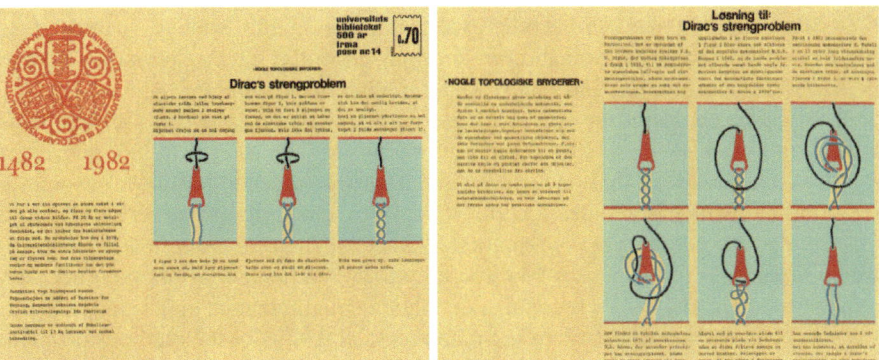

a Front of bag   **b** Back of bag

Fig. 2 Dirac's string problem on a Danish shopping bag produced in 1982

Fig. 3 Alice and the caterpillar

meets a caterpillar (Fig. 3), who helps her to regain her normal height after she has been reduced to a height of three inches in an earlier scene. The advice to Alice is to eat from a magic mushroom about which the caterpillar says: "One side will make you grow taller, and the other side will make you grow shorter". Alice hopefully breaks off a piece from each of the two sides of the mushroom, which is difficult since the mushroom is entirely round and has no sides. First she takes a bit of one of the pieces and gets a shock when her chin slams down onto her foot. In a hurry she eats a bit of the other piece and shoots up to become taller than the trees. Alice now discovers that she can become exactly the height she wants by carefully eating from one piece of the mushroom and then from the other, alternating between getting taller and shorter, and finally regaining her normal height.

The scene is inspired by mathematical results discovered by the German mathematician Dirichlet in 1837 on what is nowadays known as *conditional convergent infinite series*, namely, that one can assign arbitrary values to infinite sums with alternating signs of magnitudes tending to zero, by changing the order in which the magnitudes are added. This is remarkable since *Alice's Adventures in Wonderland* was written in 1865 when this was contemporary mathematics. On the other hand, non mathematicians hardly recognize that there is mathematics in the story and only very few can see in it the underlying mathematical results or the methods of mathematics. So what we really strive for in raising the public awareness of mathematics is something more.

I think there is a demand for short mathematical stories to enrich teaching in the educational system at all levels. Such short stories are not easy to conceive if they are to contain serious mathematical ideas and results. I certainly know that my short story [9] was a stroke of inspiration when walking down a main street in Copenhagen on a sunny day having a particular lecture in mind that I was on my way to deliver. This story has a complete proof of the isoperimetric problem for quadrilaterals, arranged as a competition among quadrilaterals with a prescribed perimeter who compete to enclose the largest area.

The proof starts with an arbitrary quadrilateral with a given perimeter and then shows that you can change it stepwise into quadrilaterals with the same perimeter but with increasing area. After a few steps you end up with a square, which is the quadrilateral enclosing the largest area. This proof of the isoperimetric problem for quadrilaterals satisfies my ideal of a good proof, namely, that it gives an *explanation*, which is more than just a correct formal proof. As mentioned the proof can be told as a short story and it can also easily be arranged as a play with students playing the roles of various quadrilaterals. I know of at least one French school where this has actually been done, based on a French version of the story. I have myself tested the story on primary school children in the upper grades as well as on adults with a positive attitude to mathematics but without specific mathematical knowledge. It seems to work.

Over the years, I have witnessed several episodes in daily life, and in political decision-making, where people have overlooked mathematics, or mathematical reasoning, in situations where even a simple mathematical calculation, or an elementary logical argument, could have prevented impossible undertakings, or meaningless decisions. I imagine that several stories with a clear mathematical angle could be written for such daily life situations.

Classical school problems from ancient Egypt and Babylonia involving calculations of areas and volumes are actually met in practice if you e.g. happen to have the fortune of being the chair of your local community. It is stunning how unwilling people are to use mathematics in arguments and therefore they let other people get away with obnoxious assertions. Amusing stories with a clear mathematical angle might be useful in showing to the public the benefits of rational reasoning, including the use of abstraction and mathematical model building.

Fig. 4 Ajax and Achilles playing a board game

The Importance of Mathematics Teaching

The main concern in teaching a subject is to raise the level of knowledge and to develop the skills of individuals. On the other hand, raising public awareness has more to do with raising enthusiasm. Teaching a subject is therefore not the same as raising public awareness of the subject but the two tasks are mutually beneficial.

There can be little doubt that the best way of keeping mathematical thoughts and knowledge alive is to make sure that mathematics teaching functions well from the very first day of school for every child. Mathematics teachers must be comfortable with a substantial catalogue of fundamental mathematical ideas and results in order to raise the awareness of the importance of mathematics and to create enthusiasm in the classroom.

The most important issue for raising the public awareness of mathematics may very well be to consolidate and develop the mathematical knowledge of teachers of mathematics at all levels. To enhance this process, mathematical conversations and mathematical storytelling are useful since good mathematical stories can be retold over and over again and they will disseminate to wider and wider circles. Mathematics in art is a good topic for mathematical conversations. In the amphora shown in Fig. 4, a planar reflection symmetry is realized by spatial rotation about the axis of symmetry. Discussing mathematics is an important stage on the way to grasping mathematics and, eventually, thinking mathematically.

Raising the Awareness of Mathematics by Research Mathematicians

Many university mathematicians believe that raising the public awareness of mathematics is other people's responsibility. Their research is so important, maybe in their own minds even to mankind, that they cannot waste a moment of their precious time for this kind of activity. The few colleagues who do work at raising the public awareness of mathematics can easily feel frustrated and may sometimes wonder whether this kind of activity is valued at all. There can be little doubt that popularization activities alone will not qualify someone for a research position in mathematics. But there are good reasons that such activities should count in connection with appointments to university positions.

The foremost reason is that without the understanding and support of the public, mathematics will be at risk. Mathematicians must prove their worth in society. The mathematicians who pursue their own research in isolation from the rest of society depend on other mathematicians speaking their case to the public, and hence they still benefit from popularization activities. University administrators should be aware of this.

Another good reason for valuing popularization activities is that they enrich both the research and the ordinary teaching for those who engage in them. Popularization activities are sufficiently demanding that they will almost surely broaden your view of mathematics and make you a more knowledgeable and inspired university teacher. On the other hand, you should never forget that you are a research mathematician, since this is what makes your view of mathematics interesting.

I think that it is necessary to make university teachers and administrators aware of the importance of raising the public awareness of mathematics. I often feel that research mathematicians are reluctant to do awareness-of-mathematics activities not because they think that they are unimportant but because they find them difficult. Many research mathematicians are so narrow in their general knowledge of mathematics that they are unable to communicate the essence of mathematics to others. Advanced books with popularization material could help to broaden the general knowledge of mathematics among research mathematicians, and may even in the longer perspective help to improve the way we communicate research mathematics. The beautiful and comprehensive books [10] and [11] are valuable in this respect.

Mathematics in the Media

Recently I have begun wondering whether radio is a better medium for conveying mathematical ideas and thoughts than television, even including geometrical topics. Words help to guide the attention of an audience in the intended direction and stimulate the imagination of an engaged listener. Creating pictures in your own mind is necessary in order to absorb and appreciate abstract material and cannot be substituted with pictures delivered by a television program with many other distractions.

Even so, high-quality television programmes about mathematics have been shown in many countries. It is, however, a self-perpetuating fact that the public basically expects to be entertained by television and hence serious programmes about mathematics will not be broadcast during prime time. On the radio there seems to be a little more room for serious programmes that enrich and stimulate the mind without there being a central focus on entertainment. Since more and more programmes on radio and television are podcasted, problems with broadcast times may soon disappear.

Concluding Remarks

Keeping the awareness of mathematics alive is a great challenge. The themes within mathematics of immediate appeal to the great public tend to be quickly written about by professional science writers without a proper mathematical background. Very often you feel there is a lack of authenticity in such writing, which seldom catches the essentials of the mathematics and in most cases fails to show that mathematics is one of the most creative areas of human thinking.

Fortunately there are an increasing number of prominent mathematicians who care about disseminating the pearls of both classical and contemporary mathematics to a wider audience. Their success is of the utmost importance for convincing new generations of young people that mathematics is a subject worthy of their attention during their education and even to persuade some young people to engage seriously with mathematics throughout their life.

It is undoubtedly true that seeing mathematics in action helps to keep it alive in the minds of people. However, the result of an application of mathematics is usually easier to describe than the mathematics behind the application, though, an application of basic school mathematics is usually overlooked and is not counted. Consequently, the mathematics behind an application is typically not presented in any degree of detail at all, giving the public the feeling that mathematics is mysterious and inaccessible to everybody other than a select few.

Stochastics is a good example of a young branch of mathematics with many spectacular applications in other sciences, which is built on advanced and difficult abstract mathematics. There are several good examples of brilliant expositions of some of the mathematical ideas behind stochastics, which has convinced me that it is possible to present essential aspects from all branches of mathematics. It is important and rewarding to do so.

References

1. Kline, M.: Mathematical Thought from Ancient to Modern Times. Oxford University Press, London (1972)
2. Hansen, V.L.: Geometry in Nature. AK Peters, Wellesley (1993)
3. Kline, M.: Mathematics and the Search for Knowledge. Oxford University Press, London (1985)

4. Hansen, V.L.: Mathematics Alive and in Action, Theme on "Mathematics: History, Concepts, and Foundations" (edited by H. Araki), in Encyclopedia of Life Support Systems (EOLSS). http://www.eolss.net. Publication developed under the auspices of the UNESCO by Eolss Publishers, Oxford, UK (2005)
5. Gromov, M.: Speech to the Balzan Prize Foundation (1999), recorded at the webpage http://www.balzan.org/en/prizewinners/mikhael-gromov/berne-16-11-1999_454_459.html
6. Hansen, V.L.: Popularizing mathematics: from eight to infinity. In: Tatsien, L.I. (ed.) Proceedings of the International Congress of Mathematicians, Beijing 2002, vol. III, pp. 885–895. Higher Education Press, China (2002)
7. Behrends, E.: Fünf Minuten Mathematik. Vieweg & Sohn Verlag, Braunschweig (2006). Translations are available in many languages
8. Wilson, R.: Lewis Carroll in Numberland. Allen Lane, London (2008)
9. Hansen, V.L.: I am the greatest. Mathematics in School **25**(4), 10–11 (1996). See also the webpage http://www.mat.dtu.dk/people/V.L.Hansen/square.html
10. Engquist, B., Schmid, W. (eds.): Mathematics Unlimited—2001 and Beyond. Springer, Berlin (2001)
11. Gowers, T., Barrow-Green, J., Leader, I. (eds.): The Princeton Companion to Mathematics. Princeton University Press, Princeton (2008)

On the Importance of Useless Mathematics

António Machiavelo

Abstract It is very difficult to convey the purpose of large and important pieces of mathematics to the general public. Even college students of mathematics frequently ask, in an exasperating way: "What is this useful for?" In this short essay it is argued that in order to appropriately answer this question, one must first dispel some common-sense notions that just happen to be wrong. Fundamentally, one has to confront, head on, the question of what mathematics really is and exactly what things it deals with, which can only be satisfactorily understood within an evolutionary perspective.

Introduction

To explain the importance and beauty of mathematics to a layperson is quite a challenge. There are, of course, some good reasons for this. One of them is the fact that mathematics has a very long and rich history, which is cumulative unlike any other subject, so that to fully appreciate its profound beauty requires a non-trivial background. Another one, is that it requires the listener to pay attention very carefully for more than a couple of minutes, and think a little bit before understanding and appreciation can begin. However, there are also some all too common attitudes of mathematicians that do not help at all. One of them is to justify the importance of mathematics simply by quoting its technological by-products, mentioning that it takes sometimes centuries for a mathematical idea to prove useful. While of course true, it is clearly not very satisfying as a justification. And, of course, most of the time the listener (and, sometimes, the talker too) is unable to grasp the piece of mathematics involved, and so no idea is given of how exactly mathematics is used.

Another problematic attitude is the almost universal disregard mathematicians have for philosophical questions about what they do. And this entails that the mathematical community fails to come to grips with the rather crucial question of what mathematics is about exactly. It is then hard to sustain its importance in human affairs. It is, however, vital to clarify this, not only in order to justify funding for the

A. Machiavelo (✉)
Centro de Matemática da Universidade do Porto, Porto, Portugal
e-mail: ajmachia@fc.up.pt

E. Behrends et al. (eds.), *Raising Public Awareness of Mathematics*,
DOI 10.1007/978-3-642-25710-0_28, © Springer-Verlag Berlin Heidelberg 2012

Fig. 1 What exactly is this?

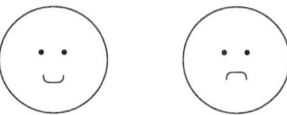

continuation of basic research: it is also the moral duty of the mathematical community. As Carl Sagan wrote regarding the burning of the Library of Alexandria ([1], p. 335):

> Science and learning in general were the preserve of a privileged few. The vast population of the city had not the vaguest notion of the great discoveries taking place within the Library. New findings were not explained or popularized. ... Science never captured the imagination of the multitude. There was no counterbalance to stagnation, to pessimism, to the most abject surrenders to mysticism. When, at long last, the mob came to burn the Library down, there was nobody to stop them.

We have to make sure this does not happen ever again. In particular, that it does not happen to this magnificent and beautiful edifice that is mathematics, the pillars of which go back to the very beginnings of human civilization. We must, therefore, popularize mathematics and tackle the thorny ontological questions about what it is about, and why it is important.

Some Misconceptions

Everyone seems to think that abstract things are necessarily difficult, and hence only accessible to some few *illuminati*. However, it is surprisingly easy to show this to be completely false. Consider the smileys in Fig. 1. Although any relation between these rather crude drawings and a real human face are extremely abstract (just please stop a moment to ponder this), we all see not only human faces, but even emotions on them! Now, that is extraordinarily abstract! We all can do this automatically and unconsciously because abstraction and pattern recognition are *Homo sapiens* specialties. They are our main evolutionary tools to deal with the world around us.

To offer just another example, people seldom notice how natural numbers themselves are incredible abstractions. To see this, let us first point out the distinction between a number, e.g. 6, and its representation. In fact, "6" is not the number six (Fig. 2). So, what is the object represented by "6"? What does it refer to? More precisely, what exactly is the number six? Well, it is a certain "quantity", which is an extremely abstract property of a collection of objects. Nevertheless, even very young children have no difficulty grasping this abstraction. In fact, most people are surprised, even puzzled, when I make the observation that "6" is not the number six, but just one representation for it. And this surprise stems from the fact that we humans abstract so easily that we are not even aware that we are abstracting!

For a human to say, as one hears all too often, that he or she does not like abstract things but only "concrete" ones, makes as much sense as a fish saying it does not

Fig. 2 Several
representations of the
number 6

like to swim, or a bird saying it does not like to fly! So what do they mean? As I see it, they usually mean one of two things. Either it is a complaint that something has been presented out of context, or exactly what one would think of a fish that did not like to swim, or of a bird that did not like to fly: that they are just plain lazy!

In my experience in talking about mathematics to several kinds of audiences, from seventh graders to senior citizens, I have found that, contrary to common belief, people do appreciate abstract proofs and formulas. This happens as long as they are presented in an appropriate context, and it seems that one of the best ways to do so is to slightly follow the historical development of the particular subject one is talking about. This is, however, quite hard to do, since history is quite complex, with immense ramifications and interconnections, and not linear as most like to make it. Consequently, there is a all too common tendency to oversimplify history, in part out of laziness, and thus to grotesquely deform the evolution of some mathematical ideas.

In conclusion, it seems to me that it is not the fact that mathematics is abstract that poses an obstacle to its appreciation, as is often believed. Besides the need for an historical approach, which emphasizes the genesis of the underlying ideas, what I often feel needs to be addressed is the question of what mathematics is really all about, what are its objects of study. In short: what does it all mean?

A Philosophical Conundrum

But the problem is that, as Marvin Greenberg wrote more than three decades ago ([2], p. 245), and which is still true today:[1]

> there is at present no intelligible account of what the statements of pure mathematics are about. The philosophy of mathematics is in a mess!

The main nagging questions are, of course, very well known. Are mathematical objects real or imaginary, discovered or created by humans? If they are mere products of our minds, then how can mathematics have so many and quite impressive applications to the "real" world? But if they are discovered, in what form do they exist? Can they be located in space and time? Where exactly are they?

Lots of philosophical "theories" have been conceived to try to answer these puzzling questions. The names of some of them are given in Fig. 3. Although all of them do make pertinent and interesting remarks on these matters, none of them seems to quite yield satisfactory answers. And one of the main reasons for this, it seems to me, rests on the fact that we humans have a more than natural tendency to attribute

---

[1]See [3–9].

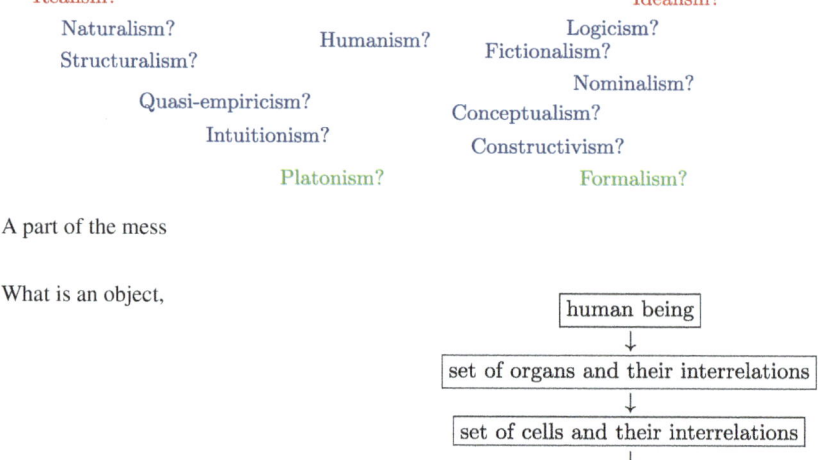

Fig. 3 A part of the mess

Fig. 4 What is an object, really?

reality, or existence, overwhelmingly to material objects that our senses can directly detect. Now, even if one tries to restrict the notion of "real objects" to things that are "physical" in some sense, one immediately runs into some difficulties, as for example: are electromagnetic waves "real" objects? What about gravity? These do seem to have a form of existence that is quite different from, say, a rock.

But even a "physical" object like a person, for example,[2] has layers of complexity that, although well known, are seldom thought of. To see this, let us consider some of the levels at which one can describe a human being (see Fig. 4). To a doctor, he is but a set of organs and its interrelations—let me draw here the reader's attention to the importance of these interrelations: rearranging the organs has absolutely dramatic consequences! Now, to a biologist, he is a set of cells and their (vital!) interrelations. To a physicist, he is but a set of atoms and their (crucial!) interrelations. But, and I find this extremely curious, according to the 1932 Nobel laureate in physics, Werner Heisenberg (1901–1976), elementary particles are "mathematical forms" ([10], p. 36) and, in general, (p. 51):

> The 'thing-in-itself' is for the atomic physicist, if he uses this concept at all, finally a mathematical structure.

To complicate things even further, let us observe that a human being is a set of atoms, together with their very special interrelationships, that varies with time! In an address to the National Academy of Sciences of the USA, in 1955, titled *The*

---

[2] I do not believe in things for which there is no real evidence for their actual existence, so I disregard beliefs in the existence of supernatural (whatever that means!) components of humans or other animals.

Value of Science (included in [11], pp. 240–248), Richard Feynman[3] (1918–1988) noted:

> [the] phosphorus that is in the brain of a rat—and also in mine, and yours—is not the same phosphorus as it was two weeks ago. ... the atoms that are in the brain are being replaced: the ones that were there before have gone away.
>
> So what is this mind of ours: what are these atoms with consciousness? Last week's potatoes! They now can *remember* what was going on in my mind a year ago—a mind which has long ago been replaced.
>
> ... the thing which I call my individuality is only a pattern or dance ... The atoms come into my brain, dance a dance, then go out—there are always new atoms, but always doing the same dance, remembering what the dance was yesterday.

That is, humans and animals in general are much more like rivers that like rocks: they are patterns, rather than "fixed" physical objects.

From all of this, what I want here to emphasize is that relations between "physical" objects are as real and important as the objects themselves. There are laws, patterns, interrelations, which are as real as anything else, and mathematics captures some of these inner relations that are just not visible to the naked eye. This has, of course, been noted many times, by many authors. For example, Rudy Rucker, in p. 4 of [12], writes:

> Mathematics is the study of pure pattern, and everything in the cosmos is a kind of pattern.

And everyone knows the passage from Galileo's 1623 book *Il Saggiatore*:

> Philosophy is written in this grand book—I mean the universe—which stands continually open to our gaze, but it cannot be understood unless one first learns to comprehend the language and interpret the characters in which it is written. It is written in the language of mathematics, and its characters are triangles, circles, and other geometric figures, without which it is humanly impossible to understand a single word of it.

But not many seem to grasp the full meaning of these words. We, who popularize mathematics, must explain them better.

An Evolutionary Perspective

What I tried to convey in the last section, may be rephrased as follows. The "forms" of Plato do exist in this world, not in a mysterious and intangible ideal world. They are the laws governing the interconnections of matter and energy, and the laws governing the interconnections between those "first-order" laws, maybe even some

---

[3] Nobel laureate in physics, 1965.

"higher-order" laws. They are part of a sort of inner structure of our cosmos. Mathematical objects (not their representations!) are elements of that structure. After one realizes this, then it is not hard today to answer what was a difficult question in Plato's time: how do we have access to the "mathematical world"?

The answer, given more than two millennia after Plato, is perfectly summarized in the so called "notebook M" of Charles Darwin (1809–1882),[4] in which one can find (p. 128, in an entry dated 4 September 1838):

> Plato says in Phaedo that our "necessary ideas" arise from the preexistence of the soul, are not derivable from experience—read monkeys for preexistence.

The mechanism of "natural selection and descent with slow modification", the seminal discovery made by Darwin,[5] or the "theory of evolution", as it is commonly known,[6] explained so many things that were previously completely baffling, and made intelligible an huge array of data and observations about living organisms previously scattered and mystifying. It stimulated, and continues to stimulate, fruitful research in several areas of biology.[7] However, after more than 150 years, is it still not properly understood by many people, and there are too many misconceptions[8] and completely wrong ideas about it.

Among the main erroneous ideas that interfere with an understanding of the theory of evolution, let us mention the following: (a) life evolves purely randomly, which arises from not realizing that there is a sharp distinction between the randomness of mutations and the mechanism of natural selection, which is anything but random; (b) to be "fit", meaning well adapted to a particular environment, implies to be ruthless and strong; (c) evolution implies a continuous progress from "inferior" animals to "superior" ones; (d) it justifies mean, cruel and immoral behavior; (e) it justifies the "law of the jungle".

These wrong ideas, together with the fact that evolution removes humans from a central pedestal above all other living creatures (which hurts our natural anthropocentric feelings), lead to an emotional denial, be it conscious or unconscious, of the "transcendently democratic"[9] and profound consequences of the insights of Darwin. A perfect example of this, and quite relevant for the subject of this essay, is the following passage from [17][10] (p. 19), where Roger Penrose clearly states why he prefers Plato's intangible, ideal world:

---

[4] Available online at http://darwin-online.org.uk.

[5] [13]. See also [14], Chaps. 3 and 4.

[6] Although Darwin never used the term "evolution", for good reasons, since it gives the wrong idea of a "progress".

[7] See [15], Chaps. 7–10.

[8] See [16] and http://evolution.berkeley.edu/evolibrary/misconceptions_faq.php.

[9] See [14], p. 67.

[10] Which is, nevertheless, an amazing book, a true tour de force!

How do I really feel about the possibility that all my actions, and those of my friends, are ultimately governed by mathematical principles of this kind? I can live with that. I would, indeed, prefer to have these actions controlled by something residing in some such aspect of Plato's fabulous mathematical world than to have them be subject to the kind of simplistic base motives, such as pleasure-seeking, personal greed, or aggressive violence, that many would argue to be the implications of a strictly scientific standpoint.

This shows that the author felt into the trap of some of the above mentioned misconceptions.[11] It comes then as no surprise that he writes a little later:[12]

it remains a deep puzzle why mathematical laws should apply to the world with such phenomenal precision.

In a Darwinian perspective this mystery starts to fade away, since, as Carl Sagan explains in [21], pp. 232–233:

we can imagine a universe in which the laws of nature are immensely more complex. But we do not live in such a universe. Why not? I think it may be because all those organisms who perceived their universe as very complex are dead. Those of our arboreal ancestors who had difficulty computing their trajectories as they brachiated from tree to tree did not leave many offspring.[13] Natural selection has served as a kind of intellectual sieve, producing brains and intelligences increasingly competent to deal with the laws of nature. This resonance, extracted by natural selection, between our brains and the universe may help explain a quandary set by Einstein: The most incomprehensible property of the universe, he said, is that it is so comprehensible.

I have always found it rather curious that everyone is so amazed with the extraordinary fine-tuning between some characteristics of some animals and their environment, and do not notice that the same applies to the human animal. They seem to assume, explicitly or, most of the time, implicitly, that there is a fundamental separation between our mental capabilities and Nature. The mind is the product of a natural selection that operated over a vast period of time, and is an integral part of Nature as anything else. It contains remarkable adaptations of humans to their environment, including the capabilities of pattern detection, abstraction, and the organization of information. As Rudy Rucker so well puts it, in [12], p. 16:

---

[11] And also, it seems to me, some lack of knowledge about the theory of evolution and its consequences. In page 22, Penrose also writes something that shows that he is unaware of the evolutionary basis of moral values, namely: "Morality has a profound connection with the mental world, since it is so intimately related to the values assigned by conscious beings and, more importantly, to the very presence of consciousness itself. It is hard to see what morality might mean in the absence of sentient beings". See, on this matter, [18, 19] and Darwin himself, in [20], Chaps. 3–5.

[12] [17], pp. 20–21.

[13] This is obviously simply a caricatural example.

Mathematics has evolved from certain simple and universal properties of the world and the human brain. That our mathematics is effective for manipulating concepts is perhaps no more surprising than that our legs are good for walking.

The Universe, Mathematics and Us

To summarize, the universe clearly seems to have a sort of inner mathematical "texture", and we humans are the product of a natural selection process that produced our brains, which can, through pattern recognition and abstraction, access at least part of that texture. This has allowed our species to uncover some parts of our universe that totally escape detection by our senses.

A striking example of this is the discovery by James Clerk Maxwell (1831–1879), on paper and with the paramount help of mathematics, around 1864, of electromagnetic waves. He realized that light is such a wave, and that there are many more kinds of these waves, whose existence was only experimentally confirmed more than two decades later, by Heinrich Hertz (1857–1894),[14] who wrote:[15]

> It is impossible to study this wonderful theory without feeling as if the mathematical equations had an independent life and an intelligence of their own, as if they were wiser than ourselves, indeed wiser than their discoverer, as if they gave forth more than he had put into them . . .

Other examples would be: the discovery, in 1900, of the quantum nature of the atomic world, by Max Planck (1858–1947), who was literally forced by mathematics to accept physical interpretations he did not like in the least;[16] Riemannian geometry, developed by Bernhard Riemann (1826–1866), around 1854, inspired by the work of Gauss (1777–1855), and later elaborated by Beltrami (1835–1900), Christoffel (1829–1900), Lipschitz (1832–1903), Ricci (1853–1925) and Levi-Civita (1873–1941), which played, more than half a century later, a crucial role in the general theory of relativity of Albert Einstein (1879–1955);[17] the prediction in 1928 of anti-matter made by Paul Dirac (1902–1984),[18] which was experimentally confirmed four years later. All of these are instances of what is illustrated by Fig. 5: that through the amazingly rich evolutive heritage that allows our brains to capture the mathematical structure of the universe, we have enlarged our horizons, by uncovering parts of that universe previously unknown to us, such as radio waves, for example.

---

[14] See [22], Chap. XX, and [23], Chap. 6.

[15] Quoted in [24], p. 101.

[16] [10], p. 4.

[17] See [25], Chap. 37, and §4 of Chap. 48.

[18] See [26], p. 392.

Fig. 5 The universe, human beings and mathematics

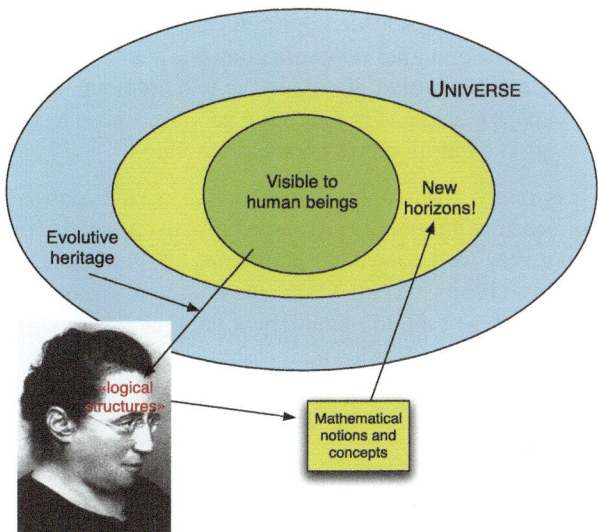

But what about the so-called "pure" mathematics? Let us focus on an example. No one really knows why (some of) the ancient Greeks became interested in "perfect numbers". A perfect number is a number that equals the sum of its divisors, itself excluded. The first two perfect numbers are 6 and 28. Euclid (c. 300 BCE), in the *Elements*, includes the following result: if n is a number such that $2^n - 1$ is a prime, then the number $2^{n-1}(2^n - 1)$ is perfect. This result raises the question of finding out for which numbers n the number $2^n - 1$ is prime. Almost certainly, it was while thinking about this that Pierre de Fermat (1601?–1665) discovered the now called Fermat's "little" theorem (see [27], Sect. 1.8). This result was put to use in 1978 (see [28]) to create digital signatures that are today essential for the security of almost all internet communications and are essential for internet business.

However, it is not the "practical" applications that should justify the effort to unravel the secrets of perfect numbers. To see why, consider the oldest open problem in mathematics: whether or not there are odd perfect numbers. No one has, so far, been able to find one, or prove that no such number exists. Now, what does this matter? I am almost sure that the answer will have no business use or direct utilitarian application. But if anyone finds the answer, I am certain that the methods would be novel and extremely interesting. Why? Because more than two millennia of the history of mathematics shows that every time someone solves a problem, especially one that has resisted several previous attempts, new and valuable ideas are brought forward.

And there is another more subtle "application" of the odd perfect number problem. To work on such a problem helps to sharpen some of our main evolutionary tools, because it serves to test our intellectual limits. It poses a very important question: are we able to solve this problem, or is it beyond our reach? This brings to mind what Carl Gustav Jacobi (1804–1851) wrote in a letter dated 2 July 1830, addressed to Adrien-Marie Legendre (1752–1833):

> M. Fourier avait l'opinion que le but principal des mathématiques était l'utilité publique et l'explication des phénomènes naturels; mais un philosophe comme lui aurait dû savoir que le but unique de la science, c'est l'honneur de l'esprit humain, et que sous ce titre, une question de nombres vaut autant qu'une question du système du monde.[19]

It is really our honour, the honour of our species, which is at stake when dealing with an open problem: are we intelligent enough? Moreover, work on such problems may very well contribute in creating the tools that can be used to overcome some of those limits! That has happened in the past, and it is invaluable. The evolutionary advantages of all this should now be obvious.

Of course, any information about perfect numbers will be a tiny and very humble piece of knowledge about the intimate structure of our universe, but it is still worthwhile. Leonhard Euler (1707–1783) said it best [29]:

> knowledge of every truth is a worthy matter in itself, even of those which seem unrelated to popular use; we have seen that all truths, at least those which we are able to understand, are so greatly connected with one another, that we cannot consider any one of them altogether useless without some rashness.
>
> And so, even if a certain proposition seems to be this way, so that regardless of whether it turns out to be true or false, it would be of no benefit to us anyway, still the method itself, by which we would established its truth or falsity, nevertheless may be useful in opening up the way for us to discover other, more useful truths. For that reason, I firmly believe that I have not uselessly expended work and effort in investigating the proofs of certain propositions. Hence, this theory of divisors does not lack any use, but rather may at some time show a utility in analysis that cannot be scorned.

Euler wrote this at the beginning of a paper where he offers a proof of Fermat's "little" theorem. He could not be more right, in that time did indeed show the utility of his analysis! In this regard, one should also never forget the old wisdom of Laozi, who pointed out that "a journey of a thousand miles begins with a single step".[20]

Closing Remarks

Behind the recurrent question concerning mathematical matters "What is it useful for?" is the question "What is the point of it all?" I believe that an answer to these questions requires an explanation of what mathematics is ultimately about, and what is the nature of its objects.

---

[19]"Fourier had the opinion that the main aim of mathematics was its public utility and the explanation of natural phenomena, but a philosopher like him should have known that the sole purpose of science is the honour of human mind, and in this regard, a question on numbers is as worthy as a question on the system of the world."

[20]From Chap. 64 of the *Tao Te Ching*.

What I have tried to argue above is that, in order to do that, one must first realize that, besides physical objects (whatever they really are), and as importantly, the world contains some sort of intrinsic logical inner structure. And our brains have been selected to apprehend it, to some extent. Working on a mathematical problem, as abstract as it may be, is to uncover a tiny piece of that inner structure. In that regard, mathematics is one of the greatest adventures of humankind, and it is imperative that we share some of its allure and charms with as many members of our species as possible.

Acknowledgements This work was partially supported by Fundação para a Ciência e Tecnologia (FCT) through Centro de Matemática da Universidade do Porto. Substantial parts of this paper were adapted from an article by the author, [30], published in Portuguese.

References

1. Sagan, C.: Cosmos. Random House, New York (1980)
2. Greenberg, M.: Euclidean and Non-Euclidean Geometry: Development and History, 2nd edn. Freeman, New York (1980)
3. Brian Davies, E.: Let platonism die. Newsletter of the European Mathematical Society, 24–25 (2007)
4. Hersh, R.: On platonism. Newsletter of the European Mathematical Society, 17–18 (2008)
5. Mazur, B.: Mathematical platonism and its opposites. Newsletter of the European Mathematical Society, 19–21 (2008)
6. Mumford, D.: Why I am a platonist. Newsletter of the European Mathematical Society, 27–30 (2008)
7. Davis, P.J.: Why I am a (moderate) social constructivist. Newsletter of the European Mathematical Society, 30–31 (2008)
8. Gardner, M.: Is Reuben Hersh 'Out there'? Newsletter of the European Mathematical Society, 23–24 (2009)
9. Brian Davies, E.: Some recent articles about platonism. Newsletter of the European Mathematical Society, 24–27 (2009)
10. Heisenberg, W.: Physics and Philosophy: the Revolution in Modern Science. Penguin, Baltimore (1989) (original from 1958)
11. Feynman, R.: What Do You Care What Other People Think? Bantam Books, New York (1989)
12. Rucker, R.: Mind Tools: the Five Levels of Mathematical Reality. Houghton Mifflin, Boston (1987)
13. Darwin, C.: On the Origin of Species by Means of Natural Selection, or the Preservation of Favoured Races in the Struggle for Life. Murray, London (1859). Available in The Complete Work of Charles Darwin Online at http://darwin-online.org.uk
14. Sagan, C., Druyan, A.: Shadows of Forgotten Ancestors. Ballantine Books, New York (1993)
15. Moore, J.A.: Science as a Way of Knowing: the Foundations of Modern Biology. Harvard University Press, Cambridge (1993)
16. Gregory, T.R.: Understanding natural selection: essential concepts and common misconceptions. Evol. Education Outreach **2**, 156–175 (2009)
17. Penrose, R.: The Road to Reality: a Complete Guide to the Laws of the Universe. Vintage, New York (2005)
18. Allen, C., Bekoff, M.: Animal play and the evolution of morality: an ethological approach. Topoi **24**, 125–135 (2005)
19. Frank, R.H.: Passions Within Reason: the Strategic Role of Emotions. Norton, New York (1988)

20. Darwin, C.: The Descent of Man, and Selection in Relation to Sex, 2nd edn. Murray, London (1882). Available in The Complete Work of Charles Darwin Online at http://darwin-online.org.uk
21. Sagan, C.: The Dragons of Eden: Speculations on the Evolution of Human Intelligence. Hodder & Stoughton, London (1977)
22. Kline, M.: Mathematics in Western Culture. Oxford University Press, London (1964) (original from 1953)
23. Osserman, R.: Poetry of the Universe: A Mathematical Exploration of the Cosmos. Anchor (1996)
24. Hon, G., Goldstein, B.: Hertz's methodology and its influence on Einstein. In: Wolfschmidt, G. (ed.) Heinrich Hertz (1857–1894) and the Development of Communication: Proceedings of the Symposium for History of Science, Hamburg, October 8–12, 2007, pp. 95–105. Books on Demand (2008)
25. Kline, M.: Mathematical Thought from Ancient to Modern Times. Oxford University Press, London (1990)
26. Heisenberg, W.: Development of concepts in the history of quantum theory. Am. J. Phys. **43**, 389–394 (1975)
27. Edwards, H.M.: Fermat's Last Theorem: a Genetic Introduction to Algebraic Number Theory. Springer, Berlin (1977)
28. Rivest, R.L., Shamir, A., Adleman, L.: A method for obtaining digital signatures and public-key cryptosystems. Commun. ACM **21**, 120–126 (1978)
29. Euler, L.: Theoremata circa divisores numerorum (E134). Novi Commentarii academiae scientiarum Petropolitanae 1, 1750, pp. 20–48. Reprinted in *Opera Omnia*: Series 1, Volume 2, pp. 62–85. Original article available online, along with an English translation by David Zhao, at www.eulerarchive.org
30. Machiavelo, A.: A natureza dos objectos matemáticos. Gaz. Mat. **161**, 7–16 (2010)

GPSR Compliance

*The European Union's (EU) General Product Safety Regulation (GPSR)
is a set of rules that requires consumer products to be safe and our
obligations to ensure this.*

*If you have any concerns about our products, you can contact us on
ProductSafety@springernature.com*

In case Publisher is established outside the EU, the EU authorized
representative is:

Springer Nature Customer Service Center GmbH
Europaplatz 3
69115 Heidelberg, Germany

Batch number: 09491150

Printed by Printforce, the Netherlands